Agriculture and
Economic Development

THE JOHNS HOPKINS STUDIES IN DEVELOPMENT
Vernon W. Ruttan, Consulting Editor

Asian Village Economy at the Crossroads: An Economic Approach to Institutional Change, by Yujiro Hayami and Masao Kikuchi

The Agrarian Question and Reformism in Latin America, by Alain de Janvry

Redesigning Rural Development: A Strategic Perspective, by Bruce F. Johnston and William C. Clark

Energy Planning for Developing Countries: A Study of Bangladesh, by Russell J. deLucia, Henry D. Jacoby, et alia

Women and Poverty in the Third World, edited by Mayra Buvinić, Margaret A. Lycette, and William Paul McGreevey

The Land Is Shrinking: Population Planning in Asia, by Gayl D. Ness and Hirofumi Ando

Agricultural Development in the Third World, edited by Carl K. Eicher and John M. Staatz

Agriculture and Economic Development, by Subrata Ghatak and Ken Ingersent

Agriculture and Economic Development

Subrata Ghatak
Senior Lecturer in Economics
University of Leicester

Ken Ingersent
Senior Lecturer in Economics
University of Nottingham

The Johns Hopkins University Press
Baltimore, Maryland

© 1984 by Subrata Ghatak and Ken Ingersent
All rights reserved

First published in the United States of America in 1984 by
The Johns Hopkins University Press
Baltimore, Maryland 21218

First published in Great Britain in 1984 by
Wheatsheaf Books Ltd
A Member of the Harvester Press Group

Library of Congress Cataloging in Publication Data

Ghatak, Subrata, 1939–
  Agriculture and economic development.

  (The Johns Hopkins studies in development)
  Includes bibliographies and index.
  1. Agriculture – Economic aspects – Developing countries.
2. Economic development.   I. Ingersent, K. A.
II. Title.   III. Series.
HD1417.G46   1984   338.1′09172′4   83–49200
ISBN 0–8018–3225–X
ISBN 0–8018–3226–8 (pbk.)

Photoset in 10 pt Times by Thomson Press (India) Limited, New Delhi
Printed in Great Britain by Biddles Ltd, Guildford and King's Lynn

To
Anita and Elizabeth

Contents

Preface xi

Chapter 1 Introduction 1

Chapter 2 Structure and Characteristics of Agriculture in LDCs 4
    1. Traditional agriculture 4
    2. Access to non-labour resources 18
    3. The farming environment: natural hazards and economic uncertainties 21
    4. Summary 23
    References 24
    Notes 25

Chapter 3 Role of Agriculture in Economic Development 26
    1. A framework of analysis 26
    2. Product contribution 27
    3. Market contribution 39
    4. Factor contribution 43
    5. Foreign exchange contribution 66
    6. Conclusions 69
    Appendix: Estimation of labour force growth rates 71
    References 71
    Notes 73

Chapter 4 Theory of Rent and the Concept of 'Surplus' 75
    1. Introduction 75
    2. Economic rent 75
    3. The theory of rent 76
    4. Rent and quasi-rent 78

	5. The Ricardian 'corn rent'	78
	6. The rental market	79
	7. Agricultural surplus	82
	8. Characteristics of landownership in underdeveloped agriculture	86
	9. The theory of share tenancy	87
	10. Some extensions of the share tenancy model	94
	References	95
Chapter 5	Agriculture in Dualistic Development Models	97
	1. Introduction	97
	2. The Lewis model	98
	3. The Fei–Ranis (FR) model	105
	4. The Jorgenson model	112
	5. Kelley, Williamson, Cheetham model	114
	6. Some concluding remarks on the dual economy models	119
	References	121
Chapter 6	Resource Use Efficiency and Technical Change in Peasant Agriculture	123
	1. Efficiency of resource utilisation	123
	2. Technological change in agriculture	142
	3. Generation of new agricultural technology	147
	4. Factor-biased technological change and its distributional consequences	155
	5. Agricultural technical change and agricultural employment: empirical evidence	160
	6. Agricultural resources and technical change in LDCs: policy conclusions	164
	References	169
	Notes	171
Chapter 7	Supply Response	172
	1. Introduction	172
	2. The Cobweb model: an illustration	174
	3. 'Perverse supply response' in backward agriculture	179
	4. A simple supply response model	181
	5. Supply response in the underdeveloped agricultural labour market	185

	6. The concept of 'marketed surplus': some methods of estimation	189
	7. Some criticisms of Krishna's method and the alternative approach of Behrman	192
	8. Perennial crops	199
	9. Conclusion	214
	References	214
Chapter 8	Institutional Constraints on Agricultural Development and Remedial Policies	217
	1. Inequitable landownership and land reform	217
	2. Capital and finance in underdeveloped agriculture	227
	3. Marketing imperfections and marketing policy	237
	References	250
Chapter 9	Population and Food Supplies	252
	1. The classical model	253
	2. Contra-Malthusian model	256
	3. Ecological disequilibrium	262
	4. Synthesis of population and food supply theories	263
	5. Malnutrition in developing countries	265
	Appendix: Formalisation of Boserup's model	275
	References	276
	Notes	278
Chapter 10	Agriculture and International Trade	279
	1. Introduction	279
	2. Main features of trade in agricultural goods	280
	3. Trade policies in developed countries and their impact on agricultural trade	282
	4. Welfare gains from price stabilisation	286
	5. Export instability and economic growth: some macro issues	289
	6. A survey of the literature	290
	7. Some measurement problems	293
	8. Prebisch's hypothesis	294
	9. The agricultural self-sufficiency argument	296

	10. Cartels in commodity trade and welfare gains and losses	299
	11. Integrated commodity agreement (ICA) schemes	301
	12. The compensating financing schemes	306
	References	307
	Notes	308
Chapter 11	Planning Agricultural Development	309
	1. Introduction	309
	2. Meaning and objectives of economic planning in LDCs	309
	3. Macro-planning: comparison of Soviet and Chinese models	311
	4. Choice of planning strategy for agriculture	317
	5. Planning techniques	322
	6. Agricultural project planning	329
	References	362
	Notes	363
Chapter 12	Agricultural Development: an Overview	364
Index		372

Preface

Many textbooks are available in the general field of 'development economics' and several on 'agricultural economics' with a developed country orientation. But very few texts have been written centring on the economic problems of agriculture in *developing* countries and virtually none in recent years. This book is our attempt to fill this gap in the literature.

The principal objectives of the book are to analyse agriculture's role in the development of Third World countries, to identify barriers to agricultural development and to examine critically remedial agricultural policies to foster more rapid development.

The book is primarily written for students of economics and agricultural economics who are already reasonably well versed in the principles of economics as taught to first and second year undergraduates in British universities. We expect that many of our readers will already have some knowledge of the wider field of development economics, or will be taking a parallel course in that subject. Thus our primary purpose is to illuminate the economic problems of agriculture in less developed countries (LDCs) and to analyse policy alternatives by applying well-known tools of economic analysis. However, as in other fields of applied economics, the student of agricultural development economics must acquire an adequate knowledge of the relevant institutional framework. Thus, an important secondary objective is to provide readers with a basic knowledge of the institutions through which reformist policies for agriculture in LDCs are commonly conducted. These institutional aspects include land reform, the provision of credit and marketing services, price stabilisation, and the advancement of agricultural research and education. We envisage that the book might serve either as the main text for a full-length course on agricultural development economics, or as a supplementary text for less specialised courses in the economics of development or agriculture.

The analytical sections of the book consist primarily of verbal arguments, supplemented in some cases by simple diagrams. Our sparing use of mathematics (mainly algebra) and statistics is confined to tools of analysis forming an integral part of virtually all modern economics degree courses.

In common with most textbooks, this one is largely an amalgam of many people's ideas. We accordingly gratefully acknowledge our considerable intellectual debt to the many authors cited in our references, as well as to our numerous unnamed professional colleagues. SG wishes to thank David Pyle and Clement Ayisa in particular for their help and co-operation. A special word of thanks is due to Edward Elgar of Wheatsheaf Books for his help and encouragement in publishing the book. We also thank Olwen Bradley, Jane Dewick, Gill Limbery, Dorothy Longsdon, Janet Wimperis and Marian Wearmouth for typing our manuscript with so much patience and skill. We are also indebted to our students for the stimulus they have given to our teaching and indirectly to the writing of this book, by their many pertinent observations and critical comments over a period of years. But we naturally accept full responsibility for all such errors, omissions, inconsistencies and other defects that may remain in the text.

1 Introduction

In this book, an attempt has been made to provide an up-to-date and comprehensive account of the interaction between agriculture and the economic development of the less developed countries (LDCs). We give a detailed account of both theory and techniques to help the readers to evaluate policy implications. The plan of the book is as follows:

First, we consider how developed countries (DCs) and LDCs differ in respect to the structure and organisation of agriculture and the behaviour of agricultural producers. Some features of traditional agriculture receive special attention and a few important peculiarities in the behaviour of peasant farmers in LDCs are emphasised. Then some comments are made on such peculiarities of the farming environment and its effects on the behaviour of peasant farmers.

In chapter 3, the role of agriculture in economic development is thoroughly investigated in the light of the Kuznets's analysis. Agriculture provides both food and raw materials to the rest of the economy; a growing agricultural sector provides an enlarged market as it expands aggregate demand; it also provides labour for employment in the industrial sector; and agriculture is often a principal source of capital for investment elsewhere in the economy. Exports from the agricultural sector are important to earn foreign exchange which is critical for imports of capital goods and other equipment for rapid industrialisation and economic growth.

In chapter 4, we analyse the concept of agricultural surplus and the theory of economic rent. The interchangeability of product, land and labour surpluses is mentioned, and the pattern of distribution under alternative tenure systems is noted. Alternative systems of land tenure are described and attention given to their role in the agricultural systems of LDCs.

Next, in chapter 5, the dual economy models and the role of 'surplus' agricultural labour is evaluated critically, particularly in the light of the works of Lewis, Fei and Ranis, Jorgenson and Kelly, and Williamson and Cheetham. Results of the 'tests' of some of these models are stated to derive policy implications.

Chapter 6 is an analysis of the important issue of efficiency in resource use in underdeveloped agriculture. In this context, Schultz's 'poor but efficient' hypothesis is examined. The behaviour of poor farmers in the presence of risk and uncertainty is also analysed. The important relationship between farm size and efficiency is examined in this context to formulate land policies. Next, technical change in underdeveloped agriculture is discussed in detail. Here, we have analysed some characteristics of 'modern' agricultural technology and the impact of their adoption in underdeveloped agriculture. The barriers to adoption of new technology is analysed with the distributional consequences of their adoption. We argue that dangers of dualistic development within the agricultural sector of LDCs should be properly emphasised.

Chapter 7 deals with the problem of 'supply response' in LDCs. The impact of economic incentives on agricultural production and supply are investigated. Both the theoretical and empirical issues in the analysis of the problem of supply response are thoroughly discussed. In this context, supply response of both annual and perennial crops is mentioned. Finally, the role of agricultural price policy is considered for mobilising 'surplus' from the agricultural sector.

Chapter 8 is an investigation of the important institutional constraints on agricultural development. Here, we turn to land-ownership structure and the case for land reform. The problem of capital and finance in underdeveloped agriculture is discussed. The structure of agricultural credit market is analysed at length and the problem of 'high' interest rates in such markets examined in detail. Credit policy recommendations, including loan security and interest rate structure, are formulated in the light of our analysis.

Since market imperfections present considerable problems in LDCs, the restructuring of agricultural markets, and the necessity to follow sensible price policy, are advocated. Here again, the case for institutional reform has been specifically emphasised. In chapter 9, the relationship between population and food supplies is discussed. In chapter 10, the role of agriculture in the international trade is analysed from the point of view of LDCs. Problems such as adverse movements in terms of trade, primary price fluctuations, and so on, are examined, with special attention to the welfare gains and losses from primary product price stabilisation; objectives and achievements of compensatory finance schemes are discussed at length.

In conclusion, we discuss the macro- and micro-planning of agriculture. Under macro-planning, we discuss the Soviet 'tribute' model, the Chinese model, and their relative merits and demerits. Next, we argue the merits and demerits of uni-modal and bi-modal agricultural strategies. It has been argued that, from the standpoint of LDCs, perhaps a uni-modal strategy might be superior to a bi-modal one. Then the input–output analysis for agricultural planning is discussed. Under micro-planning, we show how techniques such as cost–benefit analysis can be used for agricultural project evaluation in LDCs. Special emphasis is given to the UNIDO method of project appraisal including a practical example of its application to the agricultural project. At the same time, readers are made aware of the limitations to such exercises.

2 Structure and Characteristics of Agriculture in LDCs

In this chapter we shall consider how developed and developing countries differ with respect to the structure and organisation of agriculture, and the behaviour of agricultural producers. In the first, major section we discuss the nature and characteristics of the 'traditional' agricultural sector in LDCs and some peculiarities of the behaviour of peasant farmers. Then we consider inequalities in the distribution of agricultural land and access to capital and their consequences for the majority of farmers. Finally, we remark on some peculiarities of the farming environment and its effects on the behaviour of peasant farmers.

1 Traditional Agriculture

In developed countries, where the process of economic and social integration between agriculture and other sectors of the economy is virtually complete, farming is a business and farmers behave like businessmen. They are both commercially oriented and technically well-informed. They have at their disposal the services of an array of financial, marketing, advisory and research institutions, both public and private. Nowadays, despite the image of poverty which their political representatives have a vested interest in projecting, the majority of developed-country farmers are not notably worse-off than people in other occupations. Indeed, when account is also taken of their relative independence, the amenity value of country living and, for many farmers, the social advantages and economic rewards of being a rural landowner, it might be considered that in *developed* countries farmers as a class are rather better-off than their non-farming compatriots. But in *LDCs* it is otherwise.

The term 'traditional agriculture' is used to describe the characteristic farming type in countries where agriculture is the dominant employer (including those who are self-employed), though not necessarily the largest sector in terms of contributions to GNP. But before examining the attributes of traditional agriculture, let us be clear that in most LDCs the agricultural sector is not uniform, but marked by considerable economic and social diversity. Typically, a 'traditional' sub-sector of agriculture, which is large in terms of employment but smaller in terms of production *for the market*, co-exists with a 'modern' agricultural sub-sector, which is relatively small in terms of employment, but much larger in terms of its contribution to the country's marketed output of agricultural products. This is referred to as 'economic dualism'. Alternatively if, instead of applying the dual economy concept to the agricultural sector alone, we apply it to the economy *as a whole*, we can view the traditional sub-sector of agriculture as but one element in a larger traditional *sector* (which also includes other categories of self-employment using little capital) whereas the modern agricultural sub-sector belongs to a larger 'modern sector' (including other 'non-traditional' industries).

1.1 Agricultural production unit: form and function

We shall now proceed to examine the main attributes of traditional agriculture. The first point to emphasise is that peasant agriculture is typified by the *small, family* farm. As a production unit this is characterised not only by its small size – whether measured in terms of the volume of resources employed or the volume of output – but also by its high degree of self-sufficiency. The main factors of production are labour and land; few purchased inputs are employed. The farm workforce consists principally of family labour. Despite the possibility of a direct correlation between farm size and family size, at least where land is relatively abundant, some hiring-in and hiring-out of labour may occur amongst neighbouring farms according to the differing circumstances of individual farmers, including seasonal patterns of labour demand. Where the correlation between farm size and family size is not direct but *inverse*, the fragmentation of small farms exacerbates inequality in the distribution of land and agricultural income. On the output side, the emphasis on self-sufficiency implies that a proportion of farm output is not sold but retained on the farm for consumption by the farmer's household. In a purely subsistence agriculture farm households produce *only* for their own consumption.

In a semi-subsistence agriculture farmers produce partly for their own consumption and partly for the market. The amount of the marketed portion of total output may either be planned or unplanned. If it is planned the farmer may grow one or more cash crops specifically for the market; if it is unplanned it is likely to consist of food which is surplus to the farm household's subsistence needs. Unplanned food surpluses may be confined to years when the harvest is unusually good. Pure subsistence farmers, and semi-subsistence farmers of both types described, may co-exist in the same locality. Whereas a purely subsistence farmer is not necessarily fully self-sufficient, a fully self-sufficient farmer may also produce a marketable surplus. However, if no market exists, or if the farmer does not desire a cash income, the surplus may be reabsorbed by the farm family which consequently increases its consumption to match the larger-sized crop – Engel's Law does not apply to a subsistence economy.

The specification of the peasant farmer's objective function has been the subject of much debate amongst economists. Whereas some have held that his primary concern is to maximise profit (like the modern sector farmer) others have argued that, above all, he seeks economic security or the minimisation of risk. The pure subsistence farmer is concerned only with managing a household, not a business, so it is to be expected that his primary motivation will be food security for himself and his family. The semi-subsistence farmer is both a household manager and a businessman, so that he may be motivated by the goal of profit maximisation, but subject to the constraint of withholding sufficient resources from cash crop production to provide for the subsistence of the farm household. But since, for reasons to be discussed, his cash income is typically very low, it would not be surprising if risk-aversion took precedence over profit maximisation in deciding which crops to grow and how to grow them. However, allocating resources in a way which trades a marginal profit for a marginal gain in security does not signify economic irrationality, since such behaviour is fully consistent with utility maximisation. Moreover, the results of empirical studies show that peasant farmers who do produce for the market are not unresponsive to changes in the relative prices of production alternatives.

1.2 Resources structure and labour reward system

Since land and labour are the dominant factors of production in traditional agriculture, the level of production tends to be limited by the availability of those factors. The labour input includes labour embodied in farm-produced capital inputs, such as simple buildings, drainage channels and land reclamation. Although, in principle, the static constraints on production imposed by available supplies of land and labour can be relaxed by dynamic improvements in the quality of both inputs, this consideration is of little relevance in the *short term* either to individual farmers or to those concerned with government policy for agriculture. Moreover, for reasons yet to be explained, there are formidable institutional and other obstacles to improving the quality of the inputs used by farmers in the traditional sector.

Although the scarcer factor may be either land or labour it is more frequently land due to population pressure and limited non-agricultural employment opportunities, combined with landownership concentration. (Restrictions on access to land in LDCs are discussed in the next section.) The combination of scarce land and plentiful labour favours labour-intensive systems of agriculture giving detailed attention to crops in order to secure higher yields. The relative scarcity of land is signified by the rent – characteristically 'high' under labour-intensive agricultural systems – whereas the relative abundance of labour is signified by 'low' wages. But given sufficient land, and freedom of access to it for farmers, labour may be the scarcer factor. This favours a land-intensive system of agriculture which places more emphasis on output per worker than on output per unit of land. There is consequently much less incentive to grow crops requiring detailed attention in order to achieve an economic yield. Moreover, it is feasible to vary the cultivated area according to changes in the available labour supply. In this situation the rent of land is characteristically low and, indeed, individual property rights in land may be ill-defined or even non-existent, but agricultural wages may be relatively high, at least by the standards of the country or region concerned.

The peculiarities of the labour reward system in traditional agriculture derive from distinctions between hired labour (the modern sector norm) and family labour. Whereas in the modern sector, under competitive conditions, the firm's equilibrium level of employment is where the marginal productivity of labour equates with a wage which is exogenously determined by market forces, family enterprises in the traditional sector are characterised by work- and income-sharing. Work-sharing explains why open unemployment is absent from traditional agriculture. Because the farm family shares the work-load, the family (farm) income is also shared amongst its members. Thus the individual worker's implicit 'wage' is set endogenously by the average productivity of the family rather than by the marginal productivity of his labour. There has been much discussion in the literature about the implicit wage of family workers in the traditional sector and a minimum *subsistence* wage. It has been held both that agricultural wages cannot fall below the minimum subsistence level and that, due to Malthusian pressures, they are unlikely to rise above that level. However, both these arguments are oversimplistic. In the one case, no allowance is made for farm family workers supplementing their incomes from outside sources, such as part-time or casual employment on modern sector farms. In the second case, it is unrealistic to suppose that the ratio of labour to land is equally unfavourable throughout traditional agriculture because the norms of perfect competition, including the perfect mobility of resources, are violated in practice. Where, for a variety of reasons, the ratio is relatively favourable, either for an individual family, a village, a region or some even larger social group, the expectation must be that the implicit family farm wage will be above the minimum level for physical survival.

1.3 Labour market dualism

The different processes of wage determination in the traditional and modern sectors give rise to 'labour market dualism'. Although the implicit wage of workers in the traditional sector (based on their average productivity) is not *necessarily* lower than the modern sector wage level (based on marginal productivity), it generally is so, due to the low opportunity cost of labour in the traditional sector. Large numbers of workers remain in traditional agriculture despite low wages, due either to ignorance of better opportunities outside agriculture, or to their inability to obtain a modern sector job despite wishing to do so, or to the costs of moving being unacceptably high (including the cost of giving up the relative security of remaining at home) in relation to the expected wage premium. An important element of a well-known and influential theory of the rural–urban migration process in LDCs is that potential migrants discount the higher modern sector wage by the probability of remaining unemployed (Todaro, 1969).

A possible counter-argument to labour market dualism is that modern sector employers, including large-scale farmers, behave monopsonistically in the labour market. Thus, it may be predicted from the theory of monopsony that the *volume* of modern sector employment will remain below the level of the social optimum, with modern sector wages consequently being depressed below the social opportunity cost of labour. However, proponents of this theory, such as Griffin (1979), concede that due to pressures on modern sector employers to increase wages the modern/traditional sector wage gap is unlikely to be entirely eliminated. Certainly, there is ample empirical evidence from LDCs that for those traditional sector migrants who are lucky enough to obtain a modern sector job a substantial wage premium exists (e.g. Ritson, 1973).

Some obvious consequences of price distortions in the market for agricultural labour ensue. The cheapness of *labour* in the traditional sector causes it to be used extensively there. That is, extra labour is employed to perform tasks which would be unprofitable at the modern sector wage rate. Moreover, cheap labour favours the retention of labour intensive *methods* of production, such as cultivating by hand rather than mechanically. Price distortions in the markets for land and capital (see section 2 below) accentuate the intensive use of labour. Now let us assume that the technology known to, and employed in, traditional agriculture is given and fixed (a not unrealistic assumption for reasons to be explained below). Then, if access to additional farm land, or the acquisition of capital inputs such as new tools or even fertiliser, are not feasible options due to either their physical inaccessibility or their too high price, the application of additional labour may be the only means left to the peasant farmer who wants to produce more. With the number of workers comprising the family labour force fixed, in the short-run at least, any increase in work *hours* must augment the total product to be divided amongst the family, provided only that the marginal product per *man hour* is positive.

It follows from this argument that, provided the opportunity cost of labour time is zero, it is rational to intensify labour use to the point of zero marginal productivity. But there are two reasons why labour time may have a positive opportunity cost, even in peasant agriculture. First, supplementary employment may be available off the family farm as, for example, on modern sector farms that employ hired, part-time or casual workers. Second, because the marginal cost of extra work time is leisure forgone, it is quite possible – or even likely – that the equivalence of the marginal utility of extra income and the marginal disutility of extra work will be reached at a lower level of labour intensity than that corresponding with *zero* marginal productivity of labour time. We may reason, *a priori*, that for leisure to be valueless, even at the margin of consumption, a society must either be extremely poor or surfeited with leisure time. Though either of these conditions could occur as special cases in backward agrarian societies, there is no compelling argument or evidence for treating them as general case conditions applying to all traditional agriculture. (We return to the trade-off between work and leisure, for deeper analysis in chapters 3 and 6.)

1.4 Access to information and technological choice

Having dealt with the factor reward system in traditional agriculture we now turn to the questions of access to information and choice of technology. To observers from developed countries, or even to those from the modern sector of LDCs, farmers in traditional agriculture appear to be ignorant of modern farming methods and the technology they do use appears to be primitive. There is little reason to doubt that these are accurate observations, at least from a western viewpoint. Although it would be naive to equate knowledge and use of modern agricultural technology with economic and social well-being without qualification, there is ample evidence that peasant farm families in developing countries do aspire to a higher material standard of living. So, enquiring into the reasons for technological backwardness in traditional agriculture, and its economic consequences, are questions of some importance.

Some analysts regard technological stagnation as the definitive characteristic of traditional agriculture, for example, 'Farming based wholly upon the kinds of factors of production that have been used by farmers for generations can be called traditional agriculture.' (Schultz, 1964). This raises the three interrelated issues: (1) the type of technology used; (2) why it does not change; and (3) the consequences for agricultural labour productivity and farm incomes.

1.4.1 Characteristics of Traditional Agricultural Technology
The technology itself is simple, even primitive to western eyes, though well-tried and reliable through its use by many generations. It is also labour-intensive and self-sufficient, with little reliance on outside suppliers of materials or services. Thus, for example, the digging-stick or hoe is used instead of the plough, and human draught-power, or perhaps animal-power, is used instead of the internal combustion

engine. Any capital inputs which are employed, like simple implements and containers for transport or storage, are home-produced – either on the farm itself or in the village. There are, of course, different degrees of technological backwardness, according to systems of cultivation, the natural environment and the extent of any outside influences on farming methods. Few rural societies, nowadays, are entirely cut off from the outside world, but some are more isolated than others. Thus, other things being equal, the plough is more likely to be used where farmers are settled in one place than where shifting cultivation is practised: technological change is more likely to penetrate where cash crops are grown, perhaps with some for export, than where production is only for subsistence. Moreover, because, even in a traditional society, individual farmers differ in their resource endowments, and some are better informed or more interested in farming success than others, different methods of performing the same task may co-exist in the same locality.

1.4.2 Reasons for Technological Stagnation

The reasons for technological stagnation can be grouped into four main categories:

(1) the lack of an appropriate alternative technology;
(2) farmers' ignorance of 'better' methods;
(3) farmers' lack of motivation – the risks and costs of adoption; and
(4) barriers to adoption due to other market failures.

Lack of appropriate alternatives. Could the technological backwardness of traditional agriculture be due to the absence of alternative technologies with a higher production or profit potential which are also properly adapted to its conditions of plentiful labour, very scarce capital, very small scale of production, and limited literacy and knowledge of modern technical skills amongst farmers? Much has been written, in a dual economy context, about the 'inappropriateness' of attempting to transfer modern sector technology to the traditional sector: the transfer of technology from DCs to LDCs, by multinational companies and through trade and aid, has also been condemned. So, for example, it may be feasible to evolve tractors, power-tillers, harvesters, irrigation pumps, and other items of farm equipment which are better adapted than large-scale, expensive and labour-displacing 'Western models' to the requirements of small-scale producers in LDCs. The hallmarks of a more appropriate agricultural technology might include technical simplicity, low cost and capacity to raise farm productivity without displacing labour.

The case for fostering the development of more appropriate alternative technologies which, if successful, would help to remove economic dualism by reducing and ultimately closing the present gap in per capita income in favour of the modern sector, has been well made (Stewart, 1978). A number of specialised institutions, such as the Intermediate Technology Group, have been established to collect and exchange information, to encourage research and to foster the actual creation of more appropriate technology, for use in both industry and agriculture, in other ways. Yet to carry this argument to the point of saying that there can be *no* technological advance in traditional agriculture without the invention of 'new tools' is unconvincing. By no means all technical improvements involve substituting capital for labour, or call for a high order of technical skill or literacy in their application. Nor need they be particularly expensive to introduce. For example there may be scope for raising crop yields by changing seed varieties or sowing dates. Or it may be feasible to raise farm income by producing a crop which has not been grown before. This line of argument suggests that farmers in the traditional sector may not be fully informed about *existing* alternative techniques which are appropriate and have an output and income-raising potential.

Lack of knowledge: barriers to spread of technological information. In seeking to explain why farmers may be ignorant of better methods it is tempting to blame their low levels of literacy and formal education. Although this notion cannot be entirely discredited, a number of other factors also need to be considered. However receptive people may be to new knowledge and new ideas, actual technological innovation is dependent on effective channels of communication between the sources of new knowledge and its recipients. In agriculture there are two major impediments to the dissemination of new information. First, in most countries much of the farm population is located in relatively remote and inaccessible areas. However, their 'remoteness' and 'inaccessibility' are, to a considerable degree, expressions of the under-development of the rural communications infrastructure, particularly in terms of the number and quality of roads and railways. The underprovision, by developed country standards, of other means of communication such as postal, telegraph and telephone services add to the relative isolation of the rural population. Newspaper circulation is low, partly because of widespread illiteracy but also because of the difficulties of distribution in remote areas. A comprehensive analysis of the reasons why the rural communications infrastructure is deficient extends well beyond the scope of our subject-matter, except to make two points: first, the per capita cost of providing services tends to be higher in rural areas than in towns and cities because of their relative geographical isolation and lower population density (this occurs in developed as well as in developing countries); and secondly, the problem is exacerbated by the relative poverty of the rural population. In most LDCs, although poverty is not confined to the rural sector it is concentrated there: farm incomes are substantially below non-farm incomes on a per capita basis. The relative poverty of farmers is the second of the major barriers to the spread of information about new farming methods, and technological stagnation can be viewed as an aspect of the vicious circle of poverty. Farmers fail to adopt better methods because they cannot afford to travel to find out about them – they may even be unable to afford a radio.

But is it necessary for farmers to travel in search of new information? Can it not be conveyed to them at home by their professional advisers? In developed countries farmers get advice from both public and private sources. Advice is available (usually free) both from government advisers and from the suppliers of the many purchased inputs used in modern agriculture. In LDCs agricultural advice is much less readily available for several reasons. Dealing first with the government side, there are two major problems. First, there is the difficulty of recruiting and training adequate numbers of agricultural extension personnel. Suitable recruits need to be of better-than-average education and also able to command the respect and confidence of farmers. They should also be motivated towards working in agriculture and living in rural surroundings. However, due to agriculture's inferior status as a vocation in most LDCs, it is often difficult to find adequate numbers of people with this motivation amongst those with sufficient education to compete in the urban professional job market. Whereas university degree courses in medicine and law are invariably over-subscribed, would-be university entrants in LDCs rarely express a preference for specialising in agriculture. The second problem is the high cost of disseminating advice to large numbers of *small* farmers, many of them in remote parts of the country and/or in areas where access to villages and individual farmers, using modern methods of transport, is impeded by infrastructural underprovision as already discussed. In particular, all-weather roads (either main or feeder roads) are frequently lacking. Thus, handicapped as they are by inadequate numbers of trained personnel, and subject also to budgetary constraints on the cost of travel and transport, it is small wonder that government agricultural extension services in LDCs tend to concentrate their meagre resources on the bigger farmers and those in the more accessible areas. Farmers in the modern sector – small in number but large in terms of their contribution to marketed agricultural output – tend to hold the advantage with respect to both farm size and farm accessibility. Farmers in the traditional sector tend to lose out on both counts.

The second potential source of agricultural advice is the private sector where advice complements trade. Profit-motivated commercial firms tend to concentrate on giving advice to large-scale commercial farmers in the modern sector who are their major customers. Attempting to expand trade and accompanying advice into the traditional sector is fraught with much greater risks and uncertainties. The initial demand for modern inputs in the traditional sector must inevitably be small and the costs of servicing it correspondingly high. The high costs of travel to remote areas, and the difficulties of reaching farmers where there are no proper roads, are unlikely to be appreciably less daunting to private enterprise than to the government. In order to make such business profitable it might be necessary to raise the prices of goods supplied above the level that peasant farmers would be willing to pay, or the cost of recovering debts might be prohibitive. We conclude that in considering whether to offer services to small-scale farmers in remote areas the private sector is confronted with the same high costs as government, and the benefits are very uncertain. The difference is that, being profit-oriented, the private sector is even more likely to reject the option.

Inadequate motivation for adoption. We now consider whether technological stagnation could be due not to ignorance, but to farmers' inability to perceive the benefits of adopting new technology. The perception of adequate benefits provides the motivation for adoption. The main barriers to perception are the risks and costs of adoption. These are interrelated. The risks of adoption include the potential costs of making mistakes due to inexperience. Although traditional methods of agriculture suffer from low productivity their margins of error also tend to be low, because the traditional methods have been refined and improved by use over many generations. Mistakes are less likely to be made in repeating familiar tasks than in applying new and unfamiliar ones. For example, to an inexperienced user of chemical herbicides that method of weed control would be more risky than hand weeding, in the sense that the risk of damaging the crop would be higher. Similarly, in adopting a new crop variety which, under experienced management, is capable of out-yielding more traditional varieties, an inexperienced farmer might suffer a *lower* yield, or even a complete crop failure, due to mismanagement. These are examples of production risks but farmers embarking upon commercial enterprises also face *marketing* risks as discussed in section 3.

But why should risk deter farmers in traditional agriculture from adopting new methods? It is not so in the West where the acceptance of risk is regarded as a normal consequence of being in business. Indeed, under capitalism, the function of profit is to reward enterprise and risk-bearing. But, as already discussed in this chapter, peasant farmers are not western-style business entrepreneurs. Their objective function is different: some produce only for their own subsistence; others produce partly for subsistence and partly for the market. Most are poor, so poor in fact that they cannot afford to take unnecessary risks. A crop failure could result in hunger or even starvation for members of the farm household, so it is safer to grow relatively low-yielding crops by well-tried methods than to hazard diverting resources to a new enterprise or technology with a higher expected yield but with a higher risk of failure also. Whether the unfamiliar technology concerns a subsistence crop or a cash crop makes little difference. If it is the former, then a crop failure curtails the family food supply; if it is the latter, then the family's cash income to buy food and other essentials is reduced.

The extent to which the adoption of new technology does in fact expose the adopter to additional risk is, of course, arguable. To suppose that the degree of risk varies according to both the particular technology and the individual adopter is a plausible hypothesis. Some modern agricultural technologies are ostensibly designed to reduce risk; irrigation to offset the effect of drought, for example. But all this is really a matter for empirical verification, which we return to in chapter 6. For our present purpose the important point is that farmers commonly believe that some, though not necessarily all, new methods are more risky than traditional techniques. In a sense, contrary empirical evidence is irrelevant unless they can be persuaded to change their minds, but this is bound to take a long time.

We now turn from the risks to the costs of adopting new technology. Even though a farmer may be convinced that a new method of production is of potential benefit to himself, he may still be unable to fund its adoption. If, in reality, adoption is without risk this seems perverse. At positive odds, there can be no scarcity of backers for a 'sure winner'. It is understandable that the farmer may lack sufficient savings to raise the stakes himself, either because he is too poor to save or because, in the past, he has lacked an adequate incentive to save. This could be explained by, for example, a very low marginal productivity of capital using traditional methods of agriculture.

But this does not explain a peasant farmer's inability or unwillingness to *borrow* in order to fund the adoption of new technology offering a favourable marginal return on investment. However, such a state of affairs can be explained by capital market imperfections. Outside investors may over-estimate the risks of lending to agriculture due to imperfect knowledge. Hence, they may either be unwilling to lend to agriculture at all, or only at a risk premium that farmers are unable or unwilling to pay. Even without misinformation about the true risks of lending to agriculture, the price of credit may be biased against small farmers for other reasons. Expressed as a proportion of the value of the loan, the transactions cost of advancing credit to small farmers in small packages must tend to be higher than the cost of making large advances to large-scale farmers. Therefore, small farmers in the traditional sector may either be refused institutional credit or have it offered to them only on terms they find unacceptable. Their only alternative source may be the private moneylender whose transactions costs are probably lower (because of his proximity to the borrower) but whose price may include an element of monopoly profit. These, then, are the capital market imperfections which restrict the supply and raise the price of capital to farmers in traditional agriculture. Credit restrictions hamper the adoption of new technology and therefore help to explain technological stagnation.

Other barriers to adoption. Let us now suppose that a farmer in the traditional sector knows about better methods of production, desires to adopt them and is able to fund adoption, either from his own savings or by borrowing. His adoption of new technology may still be blocked by bottlenecks in the supply of physical inputs such as seeds, fertilisers, pesticides and irrigation equipment. These bottlenecks may be caused by inadequate domestic manufacturing capacity, by import constraints, or simply by an under-developed distribution network's inability to cope with an increased demand for its service. Moreover, in situations of excess demand caused by supply bottlenecks, farmers in the modern sector may have preferential access to available supplies. Not only do modern sector farmers buy in larger lots, they also tend to have greater influence with both distributors and government. Political influence is of particular value to farmers if the distribution of modern agricultural inputs is directly controlled by government, which commonly occurs in LDCs. In contrast to the strong bargaining position of modern sector farmers in competing for scarce input supplies, farmers in the traditional sector are commonly found to be in a position of relative economic and political weakness. The market failures underlying these weaknesses are the fourth and last of our explanations of technological stagnation in traditional agriculture.

1.4.3 Consequences of Technological Stagnation: Labour Productivity and Farm Incomes

Although traditional methods of agriculture are 'safe' and well-adapted to the relative scarcities of land, labour and capital in the sector, they suffer from the drawback of low productivity. Where traditional crop varieties are grown by traditional methods the expected output per unit of land area is usually lower than for modern crop varieties grown by modern methods. This explains the low productivity of land in traditional agriculture. But because of labour market dualism (section 1.2, above) the productivity of labour is depressed as well as the productivity of land. A low level of farm income follows as an inevitable consequence of low labour productivity. Poverty is the price of technological stagnation in traditional agriculture.

The relative poverty of peasant farmers explains their aversion to risk and their unwillingness or inability to innovate and invest, as already discussed. But there are also other side-effects. Extreme poverty can cause malnutrition and ill-health. Moreover, these consequences of poverty may have feedback effects upon labour productivity. Because they are underfed or suffer from debilitating diseases, peasant agricultural workers may lack sufficient energy to intensify their work effort during periods of peak labour demand. Thus, the sowing of crops may be delayed beyond the optimum date or necessary weeding may remain uncompleted, with inevitable adverse consequences for crop yields and labour productivity. To western eyes the ratio of working to non-working time in traditional agriculture may appear to be low. For example, it has been reported that on small farms in Africa 'adult family workers typically work only twenty to forty hours a week on their land, even during peak months' (Lele, 1975, p. 24). Although it may be suspected that part of the non-working time may be enforced rest due to an inadequate diet and poor health, there is unfortunately insufficient systematic evidence to confirm or reject this hypothesis.

Poverty in agriculture is also a cause of other social ills, such as high infant and neonatal mortality. On the other hand, the fertility rate in poor rural populations tends to be high, possibly in compensation for high infant and child mortality. In a society where every newborn baby is seen as a potential addition to the family labour force, and where parents are anxious to ensure that their offspring survive to care for them in their old age, parents may tend to 'over-insure' against the risk of losing their children by premature death. Hence the problem of poverty in traditional agriculture is exacerbated by excessive population growth. Poverty is perpetuated by the pressure of population on the land in seeming confirmation of the Malthusian doctrine.

We return to the subject of population and food supplies in chapter 8. In this chapter we confine ourselves to describing traditional agriculture and its inherent problems. All discussion of policy options for reforming and modernising traditional agriculture, in order to ameliorate its economic and social ills, is deferred to later chapters.

2 Access to Non-labour Resources

In many developing economies, pronounced inequality in the distribution of resources – particularly land – is a major characteristic of the rural sector. The distribution of land amongst families dependent on agriculture for their livelihood is important for two reasons. First, since land tends to be the scarcest factor of production in traditional agriculture, there is a close and direct correlation between farm size (land area) and family income. In other words, the economic well-being of the farm household depends, to a large extent, on the amount of land to which it has access. Although there tends to be an inverse relationship between farm size and land-use intensity, diminishing marginal returns to non-land inputs, particularly labour, prevent farmers from fully compensating for the handicap of small farm size by means of more intensive cultivation. Regardless of their efficiency in terms of output per unit of land area, many traditional sector farmers are thus condemned to a low income by the inadequate size of their holdings. Second, in many LDCs land distribution coincides with the distribution of political and economic power. Feudalism is characterised not only by the high concentration of landownership but also by government by the landowners. Thus although the principal motivation for land reform may appear to derive from a mass desire for greater equality in the distribution of land, a shift in the locus of political power away from the traditional large landowners is usually also implied.

Empirical evidence of inequality in the distribution of agricultural land is available from a wide cross-section of LDCs. But, due to incomplete or defective land registration in many countries, most of the evidence relates not to the distribution of *landownership* rights *per se*, but to the distribution of cultivation rights. Although it has been held that the only way of ensuring access to land is to own it (Griffin, 1979, p. 18) this distinction may be judged not to be of major importance. Until more precise statistics become available we can only presume a reasonably close correlation between landownership distributions and distributions of cultivation rights. However, in countries where the landlord-tenant system is widespread, it is to be expected that ownership will be more concentrated than the right to cultivate, because large estates are typically sub-divided into many separately tenanted holdings.

Statistics on the distribution of cultivation rights in India reveal that, in 1970-71, approximately half of all farm occupiers were on holdings of less than 1 hectare. But such holdings accounted for less than 10 per cent of the total holdings areas of the country. At the upper end of the holding size scale, approximately 15 per cent of occupiers on holdings of 4 or more hectares controlled more than 60 per cent of the total area (See Table 2.1). These statistics exclude landless rural households, estimated at 22 per cent of all rural households in India in 1959. Similar statistics on the size distribution of cultivation rights are available for Kenya, Pakistan, Bangladesh, Nigeria (Kano state), Colombia and Ecuador (ILO, 1972, Table 2; 1977, Tables 7 and 53; Morris, 1971, Table 3; Griffin, 1974, Table 5.7;

Table 2.1: *Size and distribution of operational holdings in India, 1970-1*

Holding size (hectares)	Number of holdings %	Holdings area %
Less than 1	50.6	9.0
1-2	19.0	11.9
2-4	15.2	18.2
4-10	11.3	29.7
10 +	3.9	30.9
All sizes (000s)	70 493	—

Source: Compiled from B. Dasgupta (1977), Table 6, who cites the *Indian Agricultural Census*, 1970-1.

1979, Table 4.4). Statistics on the distribution of land-ownership include India (Tamil Nadu State), Morocco, Gautemala and Turkey (ILO, 1977, Table 44; Griffin, 1974, Tables 2.4, 4.6 and pp. 243–4).

It is important to recognise that the degree of concentration in the ownership and control of agricultural land is not uniform amongst countries. The highest concentrations appear to exist in those Latin American and Asian countries where feudalism in the countryside still prevails.[1] Concentration is lower where redistribution has already taken place through land reform. Although land reform tends to be associated with socialism it has also occurred in non-socialist countries such as Taiwan. Land concentration is also lower in countries where traditional systems of communal landownership still prevail, as in parts of Africa (World Bank, 1975, pp. 243–8). Contrary to what might be expected, *a priori*, the degree of land concentration does *not* appear to be directly correlated with density of population. Restricted access to land for the majority whose livelihood depends on it is certainly not confined to densely populated countries, as exemplified by the highly concentrated patterns of landownership and control in a number of Latin American countries with quite sparse populations (World Bank, 1975, Annex 1, Tables 1.2 and 1.9).

A further aspect of inequality in access to agricultural land in LDCs is that rents may be distorted in favour of a few large-scale farmers and against large numbers of small-scale operators. Whereas monopsonistic modern sector farmers with preferential access to the land market can rent land relatively cheaply, small-scale farmers in the traditional sector must pay more than the competitive equilibrium rent due to their weak bargaining position *vis à vis* a powerful landlord. A parallel distortion occurs in *capital* markets if, having allowed for differential risks and transactions costs, modern sector farmers can still borrow more cheaply than those in the traditional sector. The contention that the prices of land and capital are in fact raised against small farmers is backed by empirical evidence from a number of LDCs (Griffin, 1979). A theory of 'inter-locking factor markets' has been advanced to explain the dominance of large landowners over small-scale peasant farmers and landless labourers in poor agrarian economies (Bardhan, 1980). The structure and conduct of the markets for agricultural land and capital are discussed at greater length in chapter 8. But due to variation amongst developing countries in the extent of economic and social inequality between farmers and landowners in the traditional and modern

STRUCTURE AND CHARACTERISTICS OF AGRICULTURE IN LDCS 21

sectors, further empirical investigation is needed to establish the extent of price distortions in factor markets.

The implications of restricted access to land and distorted factor prices for small-scale traditional sector farmers in LDCs are clear. Because they start with so little land and the costs of acquiring additional land and capital are so high (even if supplies are available) farm production and farm income must both be low. Thus the problem of low farm income arising from technological stagnation, as discussed in the previous section, is compounded.

3 The Farming Environment: Natural Hazards and Economic Uncertainties

We have already discussed the risks of agricultural production, and the risk-averse behaviour of peasant farmers, in relation to the adoption of new agricultural technology. Before concluding this chapter we refer briefly to the nature and causes of risk and uncertainty in agriculture, and their effects on the behaviour of farmers in a broader context.

Agriculture is profoundly affected by an unstable natural environment as well as by major economic uncertainties. The biological basis of agricultural production, and its exposure to the elements, pose special problems in attempting to forecast yields; these are affected not only by extremes of weather but also by damage caused by pests and diseases against which farmers have only limited defences. Most of LDC agriculture is situated in the tropics where extremes of weather (temperature, wind and rainfall) are greater than in temperate areas. The main economic effect of an unstable physical environment is *yield uncertainty*. Although complete crop failure can occur under the most extreme conditions of wind, flood or drought, the risk of this happening to a particular farmer may not be all that high. But the risk of a low yield is considerably greater. Whether his crop is substantially reduced in size or is lost altogether, the consequences for the subsistence farmer can be extremely serious. Unless he can buy, borrow or beg food from another source he and his family risk starvation. For the semi-subsistence or commercial farmer intending to produce a market surplus, the risk of starvation may be lower, though the loss of money income due to crop failure or a low yield can have serious repercussions for the purchase of food and other necessities. Although production uncertainty can rarely be eliminated at a feasible cost, it can be reduced by the avoidance of unnecessary risks. In view of the uncertainty of their natural environment it is quite rational for farmers in traditional agriculture, who are mostly very poor, to be risk-averse.

A major source of economic uncertainty *per se* is *price* uncertainty. Agricultural market prices are inherently unstable due to short-term shifts in aggregate demand and supply which farmers are powerless to control. The root cause of price uncertainty is that farmers are unable to predict price changes. Although market prices tend to be inversely correlated with levels of aggregate supply – prices being high during shortages and low during gluts – the individual producer has little reason to be confident that the market price will happen to be favourable in those years in which he experiences a bad harvest. The causes of crop failure can be quite localised, so that a bad harvest in one region may coincide with more than normal production elsewhere in the market supply area. So, despite the individual's misfortune, aggregate supplies can be normal or even above normal, with the obvious implication for the level of market price, all other things remaining equal.

Market prices are also affected by shifts in demand. A shortfall in the aggregate supply of a particular product will not necessarily cause its price to rise, provided adequate supplies of an acceptable (to consumers) substitute product are available. Cross-elasticities of demand are likely to be relatively high for non-staple foodstuffs, such as fruit, vegetables and some livestock products, though possibly less so for staples such as rice, millet, maize and wheat. A further complication is that in an open economy market prices are affected by shifts in imported supplies and export demand, as well as by domestic market forces. This adds to the price uncertainty facing the producer. Price variability and price uncertainty are exacerbated by poor communications. Due to the lack of a cheap and efficient communications system agricultural markets in LDCs tend to be narrow, particularly in the remoter regions. Farmers are aware of local prices, but it is difficult for them to get reliable information about prices in more distant markets and, in any case, it may not pay to send supplies there. Transport and handling costs, including wastage in transit, are uncertain. Thus there are serious obstacles to reducing local gluts and shortages of produce, and to the smoothing of price disequilibria between markets by engaging in arbitrage.

Because of agriculture's competitive structure – many predominantly small producers producing mostly perishable and relatively homogeneous products – farmers are price takers. Unlike nonagricultural producers in oligopolistic industries, they are unable to set their selling price. Moreover, due to the perishability of many agricultural products and the burden of storage costs, there is little scope for accumulating stocks to bridge periods of low prices.

Price stabilisation schemes to damp price variability and price uncertainty exist in some LDCs for some commodities. Such schemes are usually operated directly by government or by a producers' marketing board with delegated compulsory powers. But for reasons of administrative convenience and economy, minimum prices tend to be fixed at the wholesale or retail level rather than at the farm level. Thus small producers and those without direct access to wholesale or retail markets for other reasons, such as geographical isolation from urban areas, may be denied the benefits of price stabilisation due to their weak bargaining position *vis à vis* the intermediaries to whom they sell. Moreover, price stabilisation is generally restricted to a small number of commodities, particularly those produced for export. A more extended discussion of price stabilisation schemes appears in chapter 10.

Price uncertainty compounds yield uncertainty and is an additional reason for risk-aversion amongst traditional sector farmers producing for the market. The expectation that peasant farmers might tend to prefer enterprises with low yield and price variances to more risky but possibly more profitable enterprises, is confirmed by empirical evidence (Wolgin, 1975).

4 Summary

Poverty is the outstanding characteristic of traditional agriculture. The causes of poverty are resource constraints and technological stagnation, both of which limit production. Whilst labour is generally abundant, with work-sharing the norm on family farms, land is extremely scarce due to population pressure and the unequal distribution of cultivation rights. Moreover, access to capital is very restricted due to low internal saving (linked with low farm income) and capital market imperfections. Farming methods and equipment are primitive by western standards, but technological progress is hampered by a lack of appropriate small-scale alternatives, barriers to the spread of technological information, the risks of adoption and input supply bottlenecks.

The scarcity of land encourages the intensive use of labour, especially family labour, to obtain high crop yields: but the productivity of *labour* is reduced. The problem of low labour productivity in traditional agriculture is exacerbated by technological stagnation. Although wages in the modern sector tend to be higher, due to labour market dualism, the number of workers in traditional agriculture remains abundant due to rural isolation and constraints on rural-urban migration, such as the risk of prolonged unemployment in the urban sector. A low level of *per capita income* in traditional agriculture is an inevitable consequence of low labour productivity.

The distribution of land in LDCs tends to be distorted in favour of a minority of landlords and large-scale farmers. Most rural households are restricted to only a very small plot of land, or none at all in the case of landless labourers. The distribution of wealth and income in the rural sector is directly correlated with the distribution of land. A majority of peasant farmers are further handicapped in the capital market because of large farmers' preferential access to cheap credit. These distributions in access to non-land resources greatly exacerbate low labour productivity and farm income for the majority of households in traditional agriculture.

Due to agriculture's biological basis, and an unstable natural environment, forecasting agricultural production is particularly hazardous. Yields are therefore very uncertain, particularly at the level of the individual producer. Farmers producing for the market are also confronted by major *price* uncertainties. Because of their poverty, farmers in traditional agriculture may tend to place security before profit maximisation. Those who depend on agriculture for their survival are exposed to numerous *inescapable* hazards, such as drought, flood, hurricane and damage by wild animals and pests: it is wholly rational should they choose not to expose themselves to further *avoidable* risks such as growing an unfamiliar crop by an untried method for an unknown market.

References

Bardhan, P.K. (1980), 'Interlocking factor markets and agrarian development: a review of issues', *Oxford Econ. Papers*, **32** (1).

Dasgupta, B. (1977), *Agrarian Change and the New Technology in India*, UNRISD, Geneva.

Eicher, C.K. and Witt, L.W. (1964), *Agriculture in Economic Development*, McGraw-Hill, New York.

Griffin, K. (1976), *Land Concentration and Rural Poverty*, Macmillan, London.

Griffin, K. (1979), *The Political Economy of Agrarian Change*, Macmillan, London (2nd edn).
International Labour Office (1972), *Employment, Income, and Equality: A Strategy for Increasing Productive Employment in Kenya*, Geneva.
Griffin, K. (1977), *Poverty and Landlessness in Rural Asia*, Geneva.
Morris, J. (1981), 'Agrarian structure: implications for development – a Kano case study', *Oxford Agrarian Studies*, X, pp. 44–69.
Norman, D.W. (1977), 'Economic rationality in traditional Hausa dryland farmers in the north of Nigeria', in Stevens, R.D. (ed.) *Tradition and Dynamics in Small-Farm Agriculture*, Iowa State Univ. Press, Ames, Iowa.
Ritson, C. (1973), 'A framework for analysing the contribution of the agricultural sector to economic development', *J. Agric. Econ.* XXIV(1).
Schultz, T.W. (1964), *Transforming Traditional Agriculture*, Yale University Press, New Haven.
Stewart, Frances (1978), *Technology and Underdevelopment*, Macmillan, London, 2nd. ed.
Todaro, M. (1969), 'A model of labour migration and urban development in less developed countries', *Amer. Econ. Dev.*, pp. 138–48.
Wolgin, J.M. (1975), 'Resource allocation and risk: a case study of smallholder agriculture in Kenya', *Amer. J. Agric. Econ.*, 54(4)
World Bank (1975), *The Assault on World Poverty*, Johns Hopkins University Press, Baltimore, pp. 187–261 (Land Reform).

Notes

1. The term 'feudalism' is admittedly ambiguous. Following Warriner, in Eicher & Witt (1964), we use it loosely merely to denote any system of large land ownership giving rise to rigid social stratification and economic inequality in favour of the land-owning class and detrimental to the economic and social welfare of a large subservient non-landowning class.

3 Role of Agriculture in Economic Development

1 A Framework of Analysis

Following the classic analysis of Kuznets (1961) the agricultural sector in LDCs may be seen as being potentially capable of making four types of contribution to overall national economic growth and development.

(1) Expansion of the non-agricultural sector is strongly reliant on domestic agriculture, not only for a sustained increase in the supply of food, but also for raw materials used in manufacturing products such as textiles. Kuznets terms this the 'product contribution'.

(2) Because of the strong agrarian bias of the economy during the early stages of economic growth, the agricultural population inevitably forms a substantial proportion of the home market for the products of domestic industry, including the market for producer goods as well as consumer goods. This is termed the 'market contribution' by Kuznets.

(3) Because the *relative* importance of agriculture in the economy inevitably declines with economic growth and development, agriculture is seen as a principal source of capital for investment elsewhere in the economy. Thus the development process involves the transfer of surplus capital from agriculture to the non-agricultural sector. Similarly, development also entails the transfer of surplus *labour* from agriculture to non-agricultural occupations, especially over the long term. Kuznets terms these agriculture's 'factor contributions'.

(4) Domestic agriculture is capable of contributing beneficially to

the balance of overseas payments either by augmenting the country's export earnings or by expanding the production of agricultural import substitutes. This 'foreign exchange contribution' is not explicitly identified by Kuznets but is implicit in his market contribution.

2 Product Contribution

An expression derived from Kuznets (1964) shows how the agricultural sector's share of GDP growth is related to the product of agriculture's initial share of GDP and the relative rates of growth of agricultural and non-agricultural net products. Define:

P_a = agricultural net product
P_n = non-agricultural net product
P = total national product

Then:

$$P = P_a + P_n \ldots \quad (1)$$

and

$$\delta P = \frac{\delta P_a}{P_a} P_a + \frac{\delta P_n}{P_n} P_n \ldots \quad (2)$$

Writing r_a for $\delta P_a/P_a$, r_n for $\delta P_n/P_n$:

$$\delta P = P_a r_a + P_n r_n \ldots \quad (3)$$

$$\therefore \quad P_a r_a = \delta P - P_n r_n \ldots \quad (4)$$

and

$$\frac{P_a r_a}{\delta P} = 1 - \frac{P_n r_n}{\delta P} \ldots \quad (5)$$

Substituting for δP on the RHS of equation (5) from equation (3):

$$= 1 - \frac{P_n r_n}{P_a r_a + P_n r_n}$$

$$= \frac{P_a r_a + P_n r_n - P_n r_n}{P_a r_a + P_n r_n}$$

$$= \frac{P_a r_a}{P_a r_a + P_n r_n}$$

$$= \frac{1}{(P_a r_a + P_n r_n)/P_a r_a}$$
$$= \frac{1}{1 + P_n r_n/P_a r_a} \ldots \quad (6)$$

Kuznets's formula expressing an inverse relationship between agriculture's share of GDP growth ($P_a r_a/\delta P$) and the product of the ratio of sectoral shares of GDP (P_n/P_a) and the ratio of sectoral growth rates (r_n/r_a), is given by equation (6).

In Table 3.1 empirical estimates of $P_a r_a/\delta P$ in selected countries at different stages of economic development are presented. These have been derived from data published by the World Bank (1980), using Kuznets's formula. Separate estimates are shown for 1960 and 1978 corresponding with the available data on sectoral shares of GDP (P_a, P_n) for those two years. The data on sectoral growth rates (r_a, r_n) are trend estimates based on the periods 1960–70 and 1970–78. Overall growth rates during these periods are shown by countries on the extreme RHS of the table.

The empirical evidence on the A sector's share of the growth rate ($P_a r_a/\delta P$) is in broad agreement with *a priori* expectation. Thus the evidence confirms that:

(1) the share declines over time as a consequence of economic growth and development, as shown by the comparative values of $P_a r_a/\delta P$ in 1960 and 1978 in particular countries; and
(2) the share is inversely correlated with the level of economic development measured in terms of GNP per capita, as shown by cross-sectional comparisons of $P_a r_a/\delta P$ amongst countries at different stages of development at a particular time, such as 1960 or 1978.

So, for example, between 1960 and 1978, the A sector's share of the growth rate fell both in Bangladesh, a low income country, from 33 to 22 per cent, and in Japan, an industrialised country from 5 to 1 per cent. In Egypt, a middle income country, the share fell from 20 to 11 per cent over the same period. Although there is clearly considerable diversity amongst countries, both in the rate at which the A sector's share of the growth rate declines over time and in the size of the share, even amongst countries at a similar level of GNP per head, average values for low income, middle income and industrialised countries (shown at the top of the table) suggest that these relationships are reasonably systematic.

Table 3.1: Agriculture's contribution to the rate of economic growth

Countries (ranked in ascending order of GNP per head ($US) in 1978)	P_a^1		P_n^2		r_a^3		r_n^4		P_n/P_a		r_n/r_a		$\dfrac{P_a r_a}{\delta P}^5$		$\dfrac{\delta P}{P}^6$	
	1960	1978	1960	1978	1960–70	1970–78	1960–70	1970–78	1960	1978	1960–70	1970–78	1960	1978	1960–70	1970–78
Low income	0.50	0.38	0.50	0.62	0.025	0.020	0.050	0.044	1.0	1.6	2.0	2.2	0.33	0.22	0.039	0.026
Middle income	0.22	0.16	0.78	0.84	0.034	0.031	0.065	0.063	3.5	5.3	1.9	2.0	0.13	0.09	0.060	0.057
Industrialised	0.06	0.04	0.94	0.96	0.012	0.010	0.054	0.036	15.7	24.0	4.5	3.6	0.04	0.01	0.051	0.032
Bangladesh	0.61	0.57	0.39	0.43	0.027	0.016	0.046	0.051	0.6	0.8	1.7	3.2	0.50	0.28	0.036	0.029
India	0.50	0.40	0.50	0.60	0.019	0.026	0.053	0.046	1.0	1.5	2.8	1.8	0.26	0.27	0.036	0.037
Sri Lanka	0.34	0.35	0.66	0.65	0.030	0.023	0.053	0.037	1.9	1.9	1.8	1.6	0.23	0.25	0.046	0.034
Pakistan	0.46	0.32	0.54	0.68	0.049	0.019	0.079	0.057	1.2	2.1	1.6	3.0	0.34	0.14	0.067	0.044
Indonesia	0.54	0.31	0.46	0.69	0.025	0.040	0.071	0.099	0.9	2.2	2.8	2.5	0.29	0.15	0.035	0.078
Egypt	0.30	0.29	0.70	0.71	0.029	0.031	0.052	0.100	2.3	2.4	1.8	3.2	0.20	0.11	0.045	0.078
Thailand	0.40	0.27	0.69	0.73	0.055	0.056	0.098	0.084	1.5	2.7	1.8	1.5	0.27	0.20	0.082	0.076
Morocco	0.23	0.18	0.77	0.82	0.047	0.001	0.040	0.077	3.3	4.6	0.9	77.0	0.25	0.003	0.042	0.064
Colombia	0.34	0.31	0.66	0.69	0.035	0.049	0.058	0.063	1.9	2.2	1.7	1.3	0.24	0.26	0.051	0.060
Mexico	0.16	0.11	0.84	0.89	0.038	0.021	0.077	0.054	5.3	8.1	2.0	2.6	0.09	0.05	0.072	0.050
Italy	0.13	0.07	0.87	0.93	0.028	0.005	0.056	0.030	6.7	13.3	2.0	6.0	0.07	0.01	0.053	0.028
UK	0.04	0.02	0.96	0.98	0.023	0.008	0.029	0.020	24.0	49.0	1.3	2.5	0.03	0.008	0.029	0.021
Japan	0.13	0.05	0.87	0.95	0.040	0.011	0.113	0.055	6.7	19.0	2.8	5.0	0.05	0.01	0.105	0.050
Norway	0.09	0.05	0.91	0.95	0.001	0.023	0.052	0.049	10.1	19.0	52.0	2.1	0.002	0.02	0.049	0.047
USA	0.04	0.03	0.96	0.97	0.003	0.009	0.047	0.032	24.0	32.3	15.7	3.6	0.003	0.009	0.043	0.030

[1] P_a = A sector (= agriculture, forestry and fishing) share of GDP
[2] P_n = non-A sector share of GDP
[3] r_a = Average annual growth rate of A sector product
[4] r_n = Average annual growth rate of non-A sector product
[5] $\dfrac{P_a r_a}{\delta P}$ = Ratio of A sector growth to GDP growth (derived from P_a, P_b, r_a, r_b using Kuznets formula)
[6] $\delta P/P$ = Average annual growth rate of GDP
Source: The World Bank, *World Development Report 1980*, Appendix Tables 2 and 3.

The magnitude of any change in the ratio of non-A sector to A sector product is determined in part by the difference in sectoral rates of growth, thus:

$$\Delta\left(\frac{P_n}{P_a}\right) = \frac{P_n^1}{P_a^1} - \frac{P_n^0}{P_a^0} \dots \quad (7)$$

where the superscripts relate to time

$$= \frac{P_n^0(1+r_n)}{P_a^0(1+r_a)} - \frac{P_n^0}{P_a^0} \dots \quad (8)$$

$$= \frac{P_n^0[(1+r_n)-(1+r_a)]}{P_a^0(1+r_a)} \dots \quad (9)$$

$$= \frac{P_n^0(r_n - r_a)}{P_a^0(1+r_a)} \dots \quad (10)$$

Two major deductions follow from equation (10). First, the *magnitude* of the change in sectoral shares depends upon:

(i) the initial ratio (P_n^0/P_a^0);
(ii) the agricultural growth rate (r_a);

and (iii) the difference in sectoral rates of growth $(r_n - r_a)$.

Other things being equal, $\Delta(P_n/r_a)$ will be 'large' ('small') if either:

(a) P_n^0/P_a^0 is large (small)
or (b) r_a is small (large)
or (c) $r_n - r_a$ is large (small).

Second, the *sign* of the change in sectoral shares depends only upon the sign of $r_n - r_a$: $r_n - r_a > 0$ signifies an increase in the non-A sector's share of the total product, whereas $r_n - r_a < 0$ signifies a decreasing share.

The highest rates of decline in the relative importance of agriculture in the economy tend to be associated with the *combination* of a relatively high initial product shares ratio in favour of the non-A sector, a relatively low rate of agricultural growth and a relatively high rate of non-agricultural growth (making for a large positive difference in favour of the non-A sector). A specific example of this type is Japan where the A sector's share of GDP declined from 13 per cent in 1960 to only about 5 per cent in 1978 (Table 3.1). Conversely, the lowest rates of decline in the A sector's relative importance tend to be associated with a low initial ratio in favour of the non-A sector combined with a relatively high rate of agricultural growth and a low rate of non-agricultural growth. This situation is exemplified by Sri Lanka where the A sector's share of GDP marginally *increased* from 34 per cent in 1960 to 35 per cent in 1978.

In a developing economy, where per capita incomes are rising, growth in the A sector can be expected to lag behind non-A sector growth for three reasons. First, the demand for food and other agricultural products is generally less income-elastic than the demand for non-agricultural products, due to the Engel effect. Second, due to scientific advances and associated technological innovations in agriculture farmers become increasingly reliant on inputs purchased from the non-farm sector of the economy: this is termed the changing resources structure of agriculture effect. Third, because the demand for off-farm marketing services – distribution, storage and processing – is more elastic than the demand for agricultural products at the farm gate, the farmer's share of food expenditure at retail prices declines with time (the urbanisation effect).

In recognising the development trends making for the declining relative importance of the agricultural sector in the long term, it is necessary to avoid the trap of overlooking the critical importance of domestic agriculture's product contribution to the maintenance of an adequate rate of economic growth in the short term. This is the trap that a number of developing countries have fallen into in opting for a strategy of rapid industrialisation without parallel development in agriculture.

There are two basic reasons why, in most LDCs, development based on structural diversification of the economy is constrained by the rate of growth in the marketed output of domestic agriculture. The first reason is that the domestic farm sector is an important source of raw materials for use in industries such as textiles and food processing, as well as being the principal source of food for consumption by growing numbers of non-food producers employed in industry. Secondly, as agriculture becomes more and more closely integrated with other sectors of the economy – due to the changing resources structure and urbanisation effects – the multiplier effects of increased agricultural production and incomes assume an increasing importance in relation to the growth in demand for the products of domestic industry and the associated demands for labour and other industrial inputs. The multiplier effects of domestic agricultural expansion are discussed later under agriculture's 'market contribution' where the inter-industry linkage effects of the adoption of modern agricultural technology in LDCs are analysed. In the meantime we undertake a closer examination of the direct food and raw materials contribution.

2.1 Food contribution

In most developing countries the domestic agricultural sector is the principal source of food for consumption by non-agricultural workers. Diversification of the economy is therefore contingent upon domestic food producers producing a *surplus*, in excess of their own subsistence, which is large enough to feed a growing number of non-food producers. Although, in principle, shortfalls in domestic food supplies can be made good by expanding food imports, in practice such imports are frequently severely constrained by the scarcity and high cost of foreign exchange. Unlike imports of capital goods, food imports are consumed and do not augment the capital stock. So, where a choice has to be made between importing food and importing capital goods, the opportunity costs of food imports may be very high in terms of lower investment and a consequently reduced rate of economic growth.

The marginal opportunity costs of domestic food supplies are derived from the opportunity costs of the additional agricultural resources employed. In countries where the principal agricultural inputs are land and labour, rather than capital, and where alternative employment opportunities for agricultural workers are very scarce, these opportunity costs will tend to be 'low'. Thus, in LDCs, there is typically a marked contrast between the high opportunity cost of food imports and the relatively low marginal opportunity of costs of domestic food production. However, an adequate rate of domestic agricultural expansion is often difficult to achieve in practice due to institutional and policy constraints, as already mentioned chapter 2, and discussed at more length in subsequent chapters. Meanwhile, we consider how industrialisation and urbanisation affect the demand for food and possible consequences of failing to match the increased demand with a corresponding increase in supply.

As industrialisation and urbanisation proceed, the rate of increase in the off-farm demand for food tends to exceed the rate of growth in industrial employment for two reasons: first, because the real earnings of industrial workers are usually higher than those of agricultural workers and, secondly, because in LDCs the income-elasticity of demand for food is relatively high. So the migration of workers from agriculture to higher paid industrial employment may cause a relative scarcity of food unless the productivity of those remaining in agriculture rises fast enough to provide the migrants with a *higher* level of per capita food consumption than previously. Failure to expand food supplies quickly enough to meet the growing demand inevitably results in higher prices, if market forces are permitted to operate, or rationing in some other form if government intervenes to prevent prices from rising. Whether rising food prices can actually cause general price inflation is a question to which we shall return shortly. But, whatever the answer is, there is no doubt that due to the primacy of food as a wage good, as reflected by the high ratio of food expenditure to total household expenditure in LDCs, rapid food price inflation frequently leads to serious social and political instability which is inimical to economic growth. Apart from the untoward indirect effects of social and political unrest on the rate of industrial development, the principal *economic* effect is that, because rising food prices increase the pressure for employers to concede higher wages which are unrelated to productivity, industry's terms of trade deteriorate. Hence, due to falling profits and lower reinvestment there is a drastic decline in the rate of industrial development.

Where consumer incomes are relatively high, there is scope for substituting less costly types of food for more expensive ones as food prices in general rise. However, in LDCs the majority of consumers are relatively poor, their diet consisting mainly of the cheapest possible starchy foods (grains or tubers). The price elasticity of demand for these foods is typically very low and if their price rises, the quantity demanded is little affected due to the absence of less expensive substitutes. Thus, in an LDC, a substantial rise in the price of a staple foodstuff can cause a substantial loss of real income for large numbers of non-food producing consumers. Conversely, if staple prices fall, due to a technologically induced shift in supply, the real incomes of food consumers will rise. Thus, in contrast to the consequences of lagging food production, success in closing the gap between production and the growing demands of industrial and urban consumers will remove a major obstacle to progressive industrial development. Moreover, lower food prices will leave relatively more in the pockets of urban consumers for the purchase of industrial goods, as well as making such goods more competitive on external markets.

2.2 Food supplies and inflation

There are a number of arguments linking lagging food supplies with inflation in LDCs. First, because in LDCs supply bottlenecks are more characteristic than underutilised productive capacity due to deficient demand, the effect of income augmenting policies, such as employment creation, may be excess demand and inflation, rather than higher real output. In countries where food consumption accounts for a high proportion of total household consumption, a correspondingly high proportion of excess demand is likely to relate to food. Second, because agricultural product prices are largely determined by market forces, and because the demand for most food products is relatively price-inelastic, excess demand for agricultural and food products generally results in steeply rising prices. Thirdly, although industrial product prices tend to be 'administered' by firms rather than being determined by the market, they are usually inflexible downward (due to trade union resistance to wage reduction, for example). However, because higher food prices are likely to result eventually in higher industrial wages, industrial product prices can also be expected to rise after a time-lag. With food and industrial product prices both higher, a rise in the general price level (i.e. inflation) is virtually inevitable. Thus a process of cost-push inflation may be visualised in which lagging supplies and rising prices of food provide the initial impetus, but other prices follow after a time-lag (Maynard, 1961; Chakrabarti, 1977). Because of the time-lag, agricultural prices may appear to 'lead' other prices at one stage of the inflationary process whereas non-agricultural prices appear to be leading at a later stage. If farmers purchase substantial amounts of industrial inputs, such as fertilisers and machinery, and in countries where they can also claim compensation from the government for increased costs of production (as under schemes of minimum guaranteed prices for farm products), this gives a further twist to the inflationary spiral (Hathaway, 1974).

However, even in LDCs it is naive to attribute inflation to a single cause, such as rising food prices. Theoretical considerations and empirical evidence both point to the conclusion that inflation is due to a number of causes, including government policy variables. In particular, it seems likely that a country's rate of inflation is affected by macro-economic policy decisions on exchange rates, interest rates and changes in the money supply. Although food supplies and prices may play a significant role in determining the rate of inflation, in practice their effect may be masked by these and other factors (Edel,

1969; Chakrabarti, 1977). Such complications may explain why comparatively few studies of the influence of food supplies on inflation in developing countries have been attempted. But two such studies are worth mentioning.

A study based on a number of Latin American countries in the 1950s and 1960s reached the conclusion that 'although agricultural improvement alone would not automatically curb inflation singlehandedly... eliminating pressure from food prices would lessen the number of factors which can set off inflationary spirals' (Edel, 1969, p. 95).

This conclusion is based on a country-by-country comparison of the difference between actual rates of growth in domestic agricultural supply (symbolised by a) and the so-called autonomous rate of agricultural growth required for price stability (a^*). The autonomous rate of growth is derived from a structural model of food pricing and inflation, with certain parameter restrictions imposed. The model postulates inflation if $a < a^*$, and general price stability if $a \geqslant a^*$. Empirically, it was found that whereas some countries had experienced rapid inflation despite an 'adequate' rate of agricultural growth, in others the rate of inflation had been lower despite failure to achieve a^*. Hence the above quoted conclusion to the effect that at least in Latin America, rising food prices alone fail to explain inflation. Although this is an interesting study it also contains a number of conceptual weaknesses. First, inflation is defined simply as an increase in the relative price of food without formal behavioural justification. Second, the empirical application of the structural model is limited by a number of questionable simplifying assumptions, including a closed economy and a constant level of population.

A study of price behaviour in India during the period 1962-6 was based on a more complete model than the Latin American study (Chakrabarti, 1977). It was postulated that the overall rate of inflation is functionally related not only to changes in the prices of both agricultural and non-agricultural goods, but also to the rate of growth in the money supply.[1] The empirical findings point to the conclusion that, in India, rising domestic food supplies and falling prices have exerted a downward pressure on inflation (as expressed by a general price index), whereas rising industrial production and, at best, static prices had the opposite effect. Given the generally low price-elasticity of demand (or the high coefficient of price flexibility) for food grains, which apparently exists despite the extent of poverty in India, any increase in aggregate industrial output and incomes is bound to result in steeply increasing agricultural prices, all other things being equal. The results of this study point to the conclusion that given the upward movement of prices built into industrial expansion, a continuous fall in the relative price of basic foodstuffs is necessary to permit overall price stability to prevail in the long run. Thus, given a static or rising demand for food, real food prices will show a declining trend only in response to a sufficiently rapid upward trend in output. This implies that, where containing inflation is a major policy objective, inducing the agricultural sector to produce more for the market, despite declining prices, must be a central objective of government agricultural policy. The attainment of this goal may well demand measures to enhance productivity in agriculture, especially of labour, so that farm incomes may be maintained, or even improved, despite falling product prices. Since, for numerous reasons discussed elsewhere in this book, it may not be feasible to achieve an adequate rate of agricultural productivity gain in practice, a government intent on a cheap food policy may need to resort to redistributional measures such as food price subsidies.

Like all partial analytical studies, the study of Indian price behaviour is unrealistic in the sense that it concentrates on but one aspect of a more general problem. In practice, the achievement of price stability may not always be of paramount importance to government policy-makers. If a measure of inflation is the price that has to be paid for attaining a target rate of economic growth, or even for achieving the more limited objective of 'getting agriculture moving', it may be decided that this price is worth paying, at least for a limited period.

There is also some empirical evidence supporting the view that inflationary expectations are influenced by historic rates of food price inflation (Holden and Peel, 1977). This evidence comes from the UK, a developed country: it might be expected that in developing countries, where food accounts for a higher proportion of the consumer's total budget, this association would tend to be even more pronounced. To the extent that there is a direct link between actual rates of inflation and inflationary expectations, the policy implication is that anticipatory measures to contain food price rises may contribute towards the achievement of greater general price stability.

2.3 Raw materials contribution

In many LDCs the early stages of industrialisation are marked by a predominance of industries based on the processing of agricultural raw materials. One possible measure of the relative importance of a country's agro-industries is food and agriculture's share of value added by manufacturing industry. Statistics published by the World Bank show that, within a group of 17 'low income' countries (< US $390 GNP per capita in 1978) in the mid-1970s, this share ranged from 95 per cent (Madagascar) to 17 per cent (Sri Lanka) with a median value of 46 per cent (Pakistan); within a group of 43 'middle income' countries (US $390 to US $3500 in 1978) the equivalent shares were 92 per cent (Nigeria), 7 per cent (Singapore) and 41 per cent (Morocco). In contrast, within a group of 18 'industrialised countries' (US $3500 to US $12,100) the comparable figures were 31 per cent (Ireland), 8 per cent (Japan) and 14 per cent (UK, Canada and Norway). However, for the purpose of compiling these statistics, only the food manufacturing, beverage and tobacco industries (ISIC Major Groups 311, 313 and 314) were defined as agro-industries. Because industries like textiles (ISIC Major Group 321) and footwear (ISIC Major Group 324) are excluded, these figures underestimate the relative importance of industries using agricultural raw materials. Actual figures to indicate by how much the inclusion of these additional categories increase the relative importance of industries that are importantly dependent on agriculture for their raw materials are regrettably unavailable. However, it is clear that in some LDCs, notably those with industries based on the domestic production of natural fibres, hides and skins, the exclusion of these categories results in serious underestimation of the relative importance of agro-industries.

2.4 Production linkages between agriculture and other industries

As discussed at the begining of this chapter, there is a 'natural' long-term tendency for the A sector's relative share of GDP to decline as real GDP per capita rises. However, because the income-elasticity of demand for the value added to agricultural raw materials by agro-industries is usually higher than for the raw materials themselves, the much broader based 'agri-business' sector, including the food manufacturing, clothing and footwear industries as well as farming *per se*, declines at a much slower rate.

This broader view of agriculture's product contribution via related industries leads on to an application of the concept of inter-industry linkage (which describes inter-sectoral inter-dependence). Formally expressed, inter-sectoral production linkage measures the affect of an autonomous increase in final demand for the product of a given industry not only on the output of that industry and the outputs of industries supplying the first industry with inputs, but also the outputs of yet other industries 'supplying the suppliers' in a second round of transactions. The second round leads to a third, and so on, for an indefinite number of further rounds. The first round of linkage effects are sometimes termed the *direct* effects to distinguish them from the *indirect* effects represented by the second and subsequent rounds.[2]

Distinctions may be made between backward, forward and total linkages. Backward linkage measures the ratio of intermediate input purchases from other industries to a particular industry's total value of production; forward linkage measures the ratio of intermediate output sales to other industries to a particular industry's total sales (including sales to the final consumer); total linkage is the sum of backward and forward linkages (Yotopoulos and Nugent, 1976, ch. 16). Thus a hypothetical industry which purchased nothing from other industries and sold the whole of its output directly for final consumption would register zero coefficients for its backward, forward and total production linkages. In practice, even in developing countries, virtually all industries add value to some quantity of purchased inputs, as well as selling a proportion of their output in the form of intermediate output. Hence the direct linkage effects are non-zero and positive: the sum of the direct and indirect effects is even greater, due to the multiplier effect of successive rounds of inter-industry trade.

It has been held that, as a primary industry, agriculture lacks *backward* linkages by definition: also that because much agricultural production goes directly for home consumption or for export, without intermediate processing, especially in LDCs, *forward* linkages are likely to be weak (Hirschman, 1958, ch. 6). However, even in LDCs, the 'modernisation' of agriculture entails the growing usage by farmers of industrial inputs such as fertilisers, pesticides and simple mechanical aids. Thus, in practice, LDC agriculture is rarely so primitive as to be entirely without backward linkages with input suppliers. Moreover, on the downstream side of LDC agriculture, it is a common observation that industries based upon the processing of agricultural products such as grain milling, textile manufacture and fruit and vegetable canning, are often pioneer industries in the industrialisation process. Empirical evidence exists to support the view that, although some other industries may have stronger inter-industry production linkages, those exhibited by agriculture are not negligible. For example, total inter-industry linkage coefficients estimated for various industries in Taiwan, in 1966 were as shown in Table 3.2:

Table 3.2: *Production linkage coefficients*

Industry	PL
Agric. production for food	1.90
Agric. production for raw material	1.50
Working capital for agriculture	2.46
Food processing	2.16

Source: Yotopoulos and Nugent, 1976, Table 15.4.

These data indicate that although a marginal unit of agricultural production (food or raw material) generated less extra production in linked industries than that generated by a marginal unit of production in the agricultural supply or food processing industries, the production linkage coefficients of agriculture as such are nevertheless substantially in excess of unity (PL = 1 would signify the complete absence of inter-industry linkage effects). Although the agricultural supply and food industries may have somewhat stronger linkages with other industries (including agriculture), the development of industries directly linked with agriculture is clearly dependent on parallel development in agriculture itself, apart from the possibility of procuring raw materials from abroad or producing agricultural requisites for export. However, these latter alternatives may not be feasible or economically attractive to producers due to the higher risk and for other reasons (Hirschman, 1958, pp. 99–100).

Inter-industry *production* linkages are complemented by linkages with respect to both *employment* and *income generation*. We shall refer to these again under agriculture's factor and market contributions.

Empirical estimates of inter-industry linkage coefficients are general based on an application of the technique of input – output analysis. In adopting this approach, the usual assumption is that production, employment and income generation linkage effects occur in response to autonomous changes in *demand*. Despite its convenience, this assumption rules out the initiation of growth from the supply side. Another limitation of the technique is that it does not embrace the effects of supply constraints such as input price inflation.

3 Market Contribution

Consider, hypothetically, a closed, single-sector agrarian economy on the threshold of sectoral diversification. Even though the per capita incomes of those employed in the 'new' industries may be expected to be somewhat above those of farmers, the farm sector, because of its sheer size, must initially be the major market for domestic industrial products. Farmers' expenditures on industrial goods – both consumer goods (clothes, furniture, household utensils, building materials) and producer goods (fertilisers, pesticides, tools and implements) – represent one aspect of agriculture's market contribution to general economic development (through sectoral diversification). But, as farmers' purchases of industrial goods have their counterpart in inter-sectoral sales of *agricultural goods*, the A sector's market contribution also includes the sale of food or other farm products to the non-A sector. In discussing these two aspects of the market contribution, Kuznets (1964) describes the first as 'marketization of the production process' and the second as 'marketization of agricultural net product'. Both are accelerated by the adoption of new agricultural technology. Although this does not *necessarily* entail increased farm expenditure on industrial inputs, many agricultural innovations, some mechanical others chemical, are of this type. On the income side of the agricultural sector account, because of the favourable effect of the adoption of new agricultural technology on the productivity of the farm resources (land and labour) adoption almost invariably results in higher agricultural output and a large marketable surplus of farm products.

The effect of dropping the restrictive closed economy assumption is to break the circular flow of income between domestic agriculture and domestic industry: domestically produced goods can, in principle, be sold abroad instead of on the domestic market. This applies to both agricultural and non-agricultural goods. In a perfectly competitive world, competitive forces would ensure that the extent and pattern of each country's external trade conformed with comparative advantage. A country specialising in manufactures could export these in exchange for food imports produced by countries specialising in agriculture: it could, in principle, meet the *whole* of its domestic demand for food with imported supplies. By the same token, a country specialising in agriculture for export, as well as to meet domestic market demand, might buy all its manufactured goods abroad. In reality, of course, international trade is not conducted in conformity with the norms of perfect competition: competition is distorted by numerous restrictions upon both the exchange of goods and the mobility of resources. Moreover, because of the risks of being wholly dependent on overseas sources for the supply of food or other 'strategic' goods, it is not surprising that most countries choose to be self-sufficient in some degree, regardless of the extent of specialisation indicated by considerations of comparative advantage alone. Our over-simplified hypothetical examples serve to show, nevertheless, that agriculture's market contribution remains in the open economy, but its form is modified.

In a closed economy, the ultimate benefits of sectoral diversification would include the diversification of *consumption*, i.e. consumers have a wider choice of goods. In an open, single-sector agrarian economy exporting agricultural products in exchange for manufactures, the sectoral diversification is by-passed (or postponed), but the diversification of consumption remains. The example is artificial because, in the real world where practically all national economies are open in some degree, production is never so specialised as to prevent the diversification of consumption being derived from both forms of agriculture's market contribution. The importance of agriculture's market contribution in an open economy clearly depends on the relative importance of the agricultural sector itself, which is typically 'large' in LDCs, though generally declining as economic development proceeds. In order to find a country where agriculture's market contribution 'does not count' one would have to look for a country with no agriculture, apart from farmers producing exclusively for their own subsistence. Despite the existence of a few small, open economies specialising in the production of light manufactures, such as Hong Kong and Singapore, it may be doubted whether such a country exists anywhere in the world.

All other things being equal, the effect of affording protection to domestic manufacturing industries will be to reduce agriculture's market contribution. Because protection of infant industries shifts the terms of trade against agriculture, agricultural production, income, savings and investment may all decline. With agriculture both selling less to and buying less from other sectors, its market contribution must inevitably decline (Nicholls, 1963). Further aspects of the benefits of agricultural trade are discussed in a subsequent section, under agriculture's 'foreign exchange contribution'.

Because, due to inter-industry linkage, the expansion of production and income in one sector generates increased production and income in other sectors, production linkages (as previously discussed) are complemented by income-multiplier or income-generation linkage effects. Moreover, these may be regarded as proxy measures of the sectoral market contribution to economic growth and development. Further results of the empirical study of inter-industry linkages in Taiwan showed agricultural production of both food and raw

Table 3.3: *Income linkage coefficients*

Industry	YL
Agricultural production for food	1.64
Agricultural production for raw materials	1.80
Working capital for agriculture	1.14
Food processing	1.09

Source: See Table 3.2.

materials exerting stronger income-generation linkage effects than either the agricultural input supply or food processing industries. The estimated coefficients are shown in Table 3.3.

Agriculture's high ranking in terms of income generating capacity outside the sector reflects the relatively high labour component of agricultural value added. Additional net output means additional employment which, in turn, generates additional income, a proportion of which (depending on the marginal propensity to consume) is spent on the products of other industries. Because per capita income in agriculture tends to be lower than elsewhere in the economy, the agricultural population's MPC might tend to be relatively high.

So far in this section, we have implicitly assumed that agriculture is fully integrated with the remainder of the economy. This overlooks the distinction between farmers who produce primarily for the market and those who produce either primarily or even exclusively for their own subsistence. Although the agricultural sector is partially commercialised in virtually all LDCs, a large subsistence or semi-subsistence sub-sector is characteristic of many. Although the commercialisation of agriculture is promoted by general economic growth and development, the rate of commercialisation is inevitably slow where the linkages between agriculture and other sectors are weak. Very weak linkages are characteristic of the 'dualistic economy', where the bulk of domestic agricultural production is for subsistence, where farmers have a minimal cash income to spend on non-agricultural products and where the urban population relies upon imports for the bulk of its food supplies. Clearly, agriculture's market contribution can only be small while dualism prevails, whereas for it to be large agriculture must be fully integrated with the rest of the national economy. Most LDCs are to be found between these extremes and accelerating the commercialisation of agriculture is a prominent agricultural policy objective, at least implicitly, in many. In other words, a major short-term objective is to expand the absolute magnitude of agriculture's market contribution. In the *long term*, although the absolute magnitude of the market contribution may continue to grow, its relative importance in fostering economic growth and development naturally decreases as agriculture's share of national product declines.

4 Factor Contribution

Whereas agriculture's product contribution derives from agricultural production *per se* (other than subsistence production), and the market contribution derives from trade with other sectors, the factor contribution derives from resource transfers to other sectors (Kuznets, 1961). The resources transferred are capital and labour, including human capital. We deal first with the transfer of capital.

4.1 Capital contribution

Before proceeding to discuss alternative means of transferring capital from agriculture to other sectors and criteria for use in judging the most appropriate amount or rate of transfer, let us briefly consider why the net transfer of capital from agriculture is a credible means of development. The main argument *against* inter-sectoral capital transfers, if they are compulsory, are based on considerations of equity. Is it not unfair that farmers should be deprived of part of their wealth in order to fund developments in other sectors from which they derive no direct benefits? The main arguments in favour of transferring capital out of agriculture, compulsorily if necessary, are fourfold. First, even assuming that capital-output ratios in agriculture and non-agriculture are identical, the incremental demand for capital in the non-agricultural sector may be higher in a developing economy because the demand for non-agricultural products and services is generally more income-elastic than the demand for food and other agricultural commodities. In effect, then, the transfer reflects the declining relative importance of agriculture in the economy. Second, incremental capital-output ratios in LDC *agriculture* may, in fact, tend to be lower than in LDC *industries*. Certainly, scope apparently exists for raising productivity in agriculture by using methods requiring only a moderate outlay of capital, such as adopting higher yielding crop varieties and improved strains of livestock, and intensifying the use of fertilisers and pesticides.

Another example is the adoption of simple mechanical aids to supplement, rather than displace, human labour (intermediate technology). Third, as the dominant sector in the economy of an LDC, agriculture is virtually the sole *domestic* source of savings and investment during the initial stages of development: foreign private investment and overseas aid are only supplementary and not the principal sources of investment capital, with the possible exception of a few of the poorest countries in the world. Fourth, farmers are likely to benefit *indirectly* from non-agricultural-type investments such as the improvement of communications and provision of public utilities. Except in a dualistic economy, with a wholly uncommercialised agriculture, indirect benefits should also derive from growth and development in the industrial sector. If sectoral diversification is beneficial to the nation, by raising average living standards, it should also benefit those making their livelihood in agriculture (or any other sector), provided inter-sectoral resource mobility and income distribution are not unduly impeded by market imperfections and discriminatory government policy interventions.

Although realism compels us to recognise that market imperfections are prevalent in most countries, including LDCs, and that government intervention is sometimes inimical to the interests of farmers (although this is not always intended), there is no convincing evidence pointing to the acceptance *as an LDC norm* that the agricultural sector is excluded from deriving *any* indirect benefits from incremental investment in the non-agricultural sector.

Having argued a case for some transfer of capital from agriculture to other sectors, we proceed to a brief discussion of alternative means of transfer. The broad choice for governments lies between relying on the voluntary decisions of private investors in a free market to effect the transfer, or resorting to compulsion. The conditions governing the free market transfer of capital from the agricultural to the non-agricultural sector have been neatly summarised by Griffin, as follows:

(i) Farmers must sell part of their output outside their own sector, i.e. a market surplus of agricultural products must exist.
(ii) Farmers must be *net savers*, i.e. they must consume less than they produce.
(iii) Farmers' savings must exceed their investment in agriculture.

'If these three conditions are satisfied agriculture will have a "balance of payments" surplus with the rest of the economy' (Griffin, 1979, p. 109). The same author draws attention to the possible effect of changing the land tenure system on the amount of agriculture's market surplus and hence upon the inter-sectoral transfer of capital. Whereas, under the conditions of peasant agriculture, *owner farmers* may not produce a market surplus (being content with their lot as subsistence farmers) farm landlords exact a rent (in cash or in kind) from their tenants. Thus *tenant farmers* are obliged to produce a surplus in order to pay the rent. If the rent is paid in kind, and provided it exceeds the landlord's own subsistence requirements (which it surely must do save in exceptional cases), the landlord will sell part of his share of the surplus to urban-industrial consumers. If the rent is paid in cash a proportion is likely to be saved and invested outside agriculture. Agricultural landlords as a class are often credited with a high marginal propensity to save (due to their relative wealth) and are considered to have 'city connections'. The crux of the argument, then, is that the landlord–tenant system cannot only generate an agricultural surplus (which might not otherwise exist), but also *enforce* a net transfer of capital from agriculture to the non-farm sector. It follows that *ceteris paribus* land reform – whereby former tenant farmers become owners and so cease to pay rent – may lead to a fall in agricultural output and so also in the size of the agricultural surplus corresponding with the inter-sectoral transfer of capital. The policy implication for avoiding or reducing this 'hidden cost' of land reform is clear: farmers must have adequate economic incentives, adequate resources and, if necessary, technical assistance to maintain or even increase output when the changeover from tenant to owner-occupier takes place.[3]

Without the complication of land reform, relying on the market to transfer capital from agriculture at any required rate – through voluntary savings and investment – means giving the agricultural population adequate incentives to accumulate and hold savings in forms that can be used to fund non-agricultural type investments. For example, an agricultural credit co-operative might receive deposits from farmers for reinvestment in industrial enterprises, with appropriate provision for liquidity: this has worked in Japan. However, in most countries the direct purchase by farmers of industrial securities must generally wait until a relatively advanced state of development, when joint stock companies and an organised securities market are both well established.

Having adopted the device of national economic planning, the governments of LDCs generally prefer to intervene in the inter-sectoral allocation of capital and other resources, rather than relying on the aggregate outcome of the decisions of large numbers of independent private investors and entrepreneurs. On the particular issue of transforming agricultural savings into non-agricultural investment capital, government may consider that it is in a better position than farmers, or even agricultural landlords as well, to perceive the long-term benefits of sectoral diversification: left only to the voluntary savings and investment decisions of farmers and landlords, the rate at which capital was transferred out of agriculture would be 'too slow'.

Having chosen to intervene in the inter-sectoral transfer of capital, the broad choice for government lies between indirect and direct methods of control. *Indirect* methods, such as price controls, indirect taxes and exchange rate manipulation, have the common objective of changing the inter-sectoral terms of trade ratio (i.e. the farm: non-farm product price ratio). If the objective is to extract capital from agriculture by this method, then the inter-sectoral terms of trade must be shifted *against* agriculture – by raising duties on imports of manufactured goods, for example. Some possible dangers and disadvantages of this include:

(1) The creation or, more likely, aggravation of rural–urban income inequality conflicting with broader objectives of income distribution.
(2) The creation of 'infant' industries which impose a burden on the economy by continuing to require protection from outside competition, even in the long term.
(3) Stagnation or even a decline in the amount of the agricultural market surplus, due to the reduced purchasing power of agricultural products at the lower price level; the corresponding decline in agriculture's product and market contributions could eventually lead to industrial collapse as well.

Direct methods of control include the direct taxation of farmers and landowners (usually based on income or property values); compulsory deliveries of agricultural products to the state (usually at less than the prevailing market price); product-input barter exchange schemes with government acting as the monopoly supplier of inputs, such as seeds and fertilisers.

Direct and indirect methods of control are not mutually exclusive; they can be successfully combined as exemplified by what happened in Japan during the late nineteenth and early twentieth centuries. The Japanese experience suggests that it is feasible to extract a substantial surplus from agriculture, without resort to measures as drastic as physical confiscation, provided certain conditions are satisfied. First, agriculture itself needs to be technologically progressive and on a rising plane of prosperity. Second, an appropriate framework of financial institutions is needed to channel agricultural savings into industrial development, without unacceptable losses of security and liquidity. Third, a *land tax* is a feasible and very effective method of forcibly extracting wealth from agriculture to fund sectoral diversification, provided government has the will and ability to levy and collect the tax. Over 80 per cent of central government taxation in Japan during the last two decades of the nineteenth century came from this source (Kuznets, 1964, p. 115). The relative importance of the land tax and other factors in explaining Japan's first industrial revolution is open to debate: evidence exists that a proportion of land tax revenues was 'wasted' on military expenditures rather than being used for industrial investment (Bird, 1977). But the fact that the tax was successfuly collected to form a high proportion of the government's total 'tax take' at that time is not in dispute.

Compared with other forms of taxation in LDCs, a land tax is thought to possess several advantages (Lewis, Jr, in Southworth and Johnston, 1967, pp. 464–7):

(1) It is relatively easy to collect and difficult to evade (providing landowners hold legally valid titles which are officially documented).
(2) It discourages the speculative holding of idle land (earning zero rent).
(3) It induces farmers to *market* more of their output – due to their higher cash requirements to pay the tax (even if output itself remains unchanged).

But the political problem remains of getting the landowning classes to vote for legislation increasing their own tax burden. For example, an empirical study of the relative tax burdens of agriculture and non-agriculture in India concluded that the agricultural sector was 'undertaxed' (Southworth and Johnston, p. 482). Although the study was confined to the incidence of direct taxation, thus ignoring the inter-sectoral distribution of indirect taxes, this finding is consistent with the widely held view that in India tax evasion is rife in the rural sector.

India is a mixed economy. However, the socialist solution of abolishing private landownership in order to mobilise surplus agricultural savings has not been notably more successful. In Russia, although the peasants obtained independence and private ownership of their land for a brief period following the 1917 Revolution, land reform was followed, within a few years, by forced collectivisation and compulsory deliveries to the state. Although the squeeze on Soviet agriculture in the late 1920s and 1930s was intended to accelerate industrial development, especially of heavy industry, the *unintended* effects on agricultural output, especially in the livestock sector, were disastrous and long lasting. In China, where the communist revolution came later, the lessons of Russia's adverse experience probably influenced the choice of a different development strategy affording a better balance between industry and agriculture. Although some of the familiar devices have been employed to extract a surplus from Chinese agriculture – especially rigid price controls and compulsory deliveries – agriculture's terms of trade have been allowed to improve in recent years through policy adjustment, such as:

(1) Raising the prices paid to agricultural producers relative to the prices charged to urban food consumers.
(2) Allowing the real prices of manufactured goods to decline in response to improved productivity.
(3) Keeping direct taxation of agriculture at a comparatively 'low' level and exempting *agricultural* productivity gains from tax altogether, by basing tax liability on *historic* rather than current production (Reynolds, 1975, pp. 21–2).

Thus in capitalist and socialist countries alike, government intervention to influence and control the transfer of capital from agriculture to the remainder of the economy has been fraught with many difficulties. The choice between intervention and reliance on the market and voluntary transfers is difficult to make. Whereas the benefits of sectoral diversification and improvement of the economic and social infrastructure are more apparent to government than to private investors, the 'costs' of intervention cannot be ignored. In addition to those already discussed, these include:

(1) The inflexibility of intervention measures in response to changing conditions.
(2) The extra administrative burdens of operation and policing.
(3) The wide scope for corruption.
(Griffin, 1979, p. 134)

4.2 Labour contribution

4.2.1 The Turning-point in the Growth of the Agricultural Labour Force

The absolute size of the agricultural labour force cannot decline until the growth rate of the non-A labour force exceeds the growth rate of the total labour force. The conditions underlying this turning point are of interest.

Define: r_T = growth rate of total labour force
r_A = growth rate of A labour force
r_N = growth rate of non-A labour force
W^A, W^N = sectoral weights, $0 \leqslant W \leqslant 1$

Then, $r_T = r_A(W_B^A W_C^A)^{\frac{1}{2}} + r_N(W_B^N W_C^N)^{\frac{1}{2}}$

where the subscripts B and C on W^A and W^N denote base and current periods, respectively.

Suppose that $r_A = 0$; then

$$r_T = r_N(W_B^N W_C^N)^{\frac{1}{2}}$$

Therefore, since $(W_B^N W_C^N)^{\frac{1}{2}} < 1$

$$r_N > r_T$$

Two deductions follow. First, if W_B^N and $W_C^N - W_B^N$ are both 'small', $(W_B^N W_C^N)^{\frac{1}{2}}$ is also small, accentuating the degree to which r_N must exceed r_T to be consistent with $r_A = 0$. For example, given $W_B^N = 0.15$, $W_C^N = 0.20$, $(W_B^N W_C^N)^{\frac{1}{2}} = 0.17$. That is, r_N must exceed r_T by a factor of more than 5. But given $W_B^N = 0.6$, $W_B^C = 0.8$, $(W_B^N W_C^N)^{\frac{1}{2}} = 0.69$ and the factor is reduced to only 1.4. It is unnecessary here to analyse the determination of r_T except to make the obvious link with population growth. Similarly with r_N, which must be closely linked with the growth in demand for non-A products. The important first deduction is that it is inherently much easier for a country to achieve zero growth of its agricultural labour force when the non-A sector has already captured a relatively large *share* of the total labour force, than when agriculture is still the dominant sector in terms of employment. Thus, in most LDCs, the turning point is unlikely to be reached until a relatively 'late' stage of their development.

The second deduction is that, *ceteris paribus* the higher the exogenously determined value of r_T, the higher the 'required' value of r_N to be consistent with $r_A = 0$. In other words, the higher the rate at which the total labour force is growing, the more 'difficult' it is to reach the turning point. By implication, the higher the population growth rate the more difficult the task. Our empirical estimates of rates of growth in the agricultural labour force, the non-agricultural labour force and the total labour force in a number of countries over the period 1960–70 are presented in Table 3.4. The countries are the same as those shown in Table 3.1, dealing with sectoral contributions to GDP, except that Bangladesh is omitted due to missing information, and Nepal is included instead. The present table also contains estimates of average labour force growth rates within two large groups of countries, namely, 38 low income countries and 52 middle income countries (again, corresponding with groups shown in Table 3.1). The method of estimation is explained in the appendix.

Amongst the 38 low income countries, the agricultural labour force was still growing at an average rate of 2 per cent per annum over the observation period. Within this group, amongst the five countries shown in the table, the rate ranged from a low of 0.9 per cent (Indonesia) to a high of 2.9 per cent (Pakistan) (Table 3.4). Even amongst the middle income countries, the average rate was 1.4 per cent per annum. However, amongst the nine individual countries, shown in the table, although several exceeded the average rate, the rate was negative in four, namely, Algeria, Argentina, Yugoslavia and Spain. This signifies that these countries had reached the turning point by 1960 or before. Moreover, all are amongst the 'better-off' countries in the middle income group. Thus empirical observation confirms our earlier *a priori* conclusion that few countries reach the turning point until a relatively late stage in their economic development.

4.2.2 Transfer of Labour from Agriculture to Industry

The feasibility of accelerating development through the transfer of labour from agriculture to industry is subject to the following constraints:

(1) The size of any reservoir of 'redundant' agricultural labour, i.e. $MP \leq 0$ (or the rate at which labour can be made redundant by productivity enhancement).
(2) The 'quality' of rural migrants as potential industrial workers (or the costs of industrial training).
(3) The supplies and prices of non-labour inputs and other components of the demand for industrial labour.

Availability of surplus agricultural labour. Although realism compels us to recognise that in many LDCs the rate of rural exodus to towns

Table 3.4: Employment in LDCs: Shifting sectoral shares and relative growth rates, 1960–78

Countries (ranked in ascending order of GNP per capita ($US) in 1978)	Working population (millions)		Percentage distribution of labour force				Size of labour force (millions)				Compound growth rate 1960–78 (% per annum)		
			A		non-A		A		non-A				
	1960	1978	1960	1978	1960	1978	1960	1978	1960	1978	A	non-A	Total
38 low income countries	468.3	711.6	77	72	23	28	360.6	512.4	107.7	199.2	2.0	3.5	2.4
Nepal	5.2	7.5	95	93	5	7	4.9	7.0	0.3	0.5	2.0	2.9	2.1
India	246.6	360.6	74	74	26	26	182.5	266.8	64.1	93.8	2.1	2.1	2.1
Sri Lanka	5.3	8.3	56	54	44	46	3.0	4.5	2.3	3.8	2.3	2.8	2.5
Pakistan	22.3	39.4	61	58	39	42	13.6	22.9	8.7	16.5	2.9	3.6	3.2
Indonesia	51.9	76.2	75	60	25	40	38.9	45.7	13.0	30.5	0.9	4.9	2.2
52 middle income countries	290.3	480.0	58	45	42	55	168.4	216.0	121.9	264.0	1.4	4.4	2.8
Egypt	14.2	22.3	58	51	42	49	8.2	11.4	6.0	10.9	1.8	3.4	2.5
Thailand	13.5	23.6	84	77	16	23	11.3	18.2	2.2	5.4	2.7	5.1	3.2
Morocco	6.1	9.5	62	53	38	47	3.8	5.0	2.3	4.5	1.5	3.8	2.6
Colombia	7.1	14.3	52	30	48	70	3.7	4.3	3.4	10.0	0.8	6.2	4.0
Algeria	5.7	8.6	67	30	33	70	3.8	2.6	1.9	6.0	−2.1	6.6	2.3
Mexico	17.8	33.4	55	39	45	61	9.8	13.0	8.0	20.4	1.6	5.3	3.6
Argentina	12.8	16.6	20	14	80	86	2.6	2.3	10.2	14.3	−0.7	1.9	1.5
Yugoslavia	11.7	14.5	63	33	37	67	7.4	4.8	4.3	9.7	−2.4	4.6	1.2
Spain	19.3	23.4	42	18	58	82	8.1	4.2	11.2	19.2	−3.6	3.0	1.1

Sources: World Bank, *World Development Report, 1980*, Appendix, Tables 1 and 19.
United Nations, *Statistical Yearbook, 1961*, Table 1.

and cities is currently higher than the growth of urban and industrial employment – as signified by very high rates of urban unemployment and underemployment – the economic implications of removing labour from agriculture, under the limiting assumptions of various models, has been a popular topic of research and a source of controversy amongst development theorists. We therefore undertake a brief survey of the relevant theoretical issues, and some empirical evidence.

Several development models (discussed at more length in chapter 4) are based on the mobilisation of redundant agricultural workers for socially-productive employment outside agriculture. Labour redundancy in LDC agriculture is supposed to take the form of *disguised unemployment*. This term has been defined in several ways. First, there is a definitional distinction between redundancy in the sense that

(1) the productivity of the marginal agricultural worker is zero $MP_L = 0$), and
(2) marginal productivity is positive but below average consumption or the subsistence wage $(0 < MP_L < SW)$.

Second, a distinction can be made between two forms of redundancy depending on whether the length of the working day in agriculture is fixed or variable.

With the existence of disguised unemployment conditional upon $MP_L = 0$ and, all other things being equal, some workers could be permanently withdrawn from the land with no consequent fall in aggregate agricultural output (Ragnar Nurkse, 1953). But, given that $0 < MP_L < SW$, and a fixed agricultural work norm, a marginal but permanent contraction of the agricultural labour force *would* cause a decline in output and, *ceteris paribus*, an improvement in the agricultural sector's terms of trade (Lewis, 1954; Ranis and Fei, 1961).

Given a flexible agricultural work norm (i.e. a variable length of working day subject to a maximum constraint), and $0 < MP_L < SW$, then permanent contraction of the agricultural labour force need *not* cause a reduction in output (Sen, 1975 ch. 4). These arguments are clarified in Figure 3.1 where TP is a total product function of agriculture with labour, the sole variable input, measured on the horizontal axis in *hours*. The number of *workers* is measured on the vertical axis, reading down from the origin. Supposed that TP is maximised at Op_2 and that, initially, the labour time input is Ol_3 hours. The MP of labour at this point is zero ($MP_T = 0$). Suppose

ROLE OF AGRICULTURE IN ECONOMIC DEVELOPMENT 53

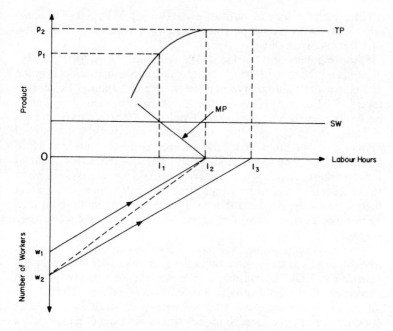

Figure 3.1: *Disguised unemployment*

further that the total number of workers is initially Ow_2. Then the agricultural work norm is Ol_3/Ow_2. But, since the labour time input could be reduced from Ol_3 to Ol_2 without loss of output, and since Ol_2 labour time could be provided by Ow_1 workers without altering the work norm ($Ol_3/Ow_2 = Ol_2/Ow_1$, given that w_2l_3 and w_1l_2 are parallel) $Ow_2 - Ow_1$ workers are disguisedly unemployed in Nurkse's sense. However, even if initially the labour time input was only Ol_2, but the number of workers was again Ow_2, the number of disguised unemployed in Sen's sense would again be $Ow_2 - Ow_1$, provided that agricultural workers were prepared to raise their work norm from Ol_2/Ow_2 to Ol_2/Ol_1.

Suppose now that equilibrium exists where the marginal productivity of labour time equates with the subsistence wage ($MP_T = SW$). corresponding with a labour time input of Ol_1 and a total product of p_1. Suppose further that the work norm given by dividing the initial number of agricultural workers into Ol_1 is lower than the maximum acceptable level. Then, the number of workers can again decline without loss of output, even though $MP_T > 0$. This case combines disguised unemployment in Lewis's sense with Sen's notion of a flexible work norm.

Choosing between Nurkse ($MP_L = 0$) and Lewis ($MP_L \leqslant SW$) depends on the reward system in agriculture. Neither model is in the neo-classical tradition in which wages are determined by market forces. Nurkse's model fits a society in which farmers are prepared to retain completely non-productive 'employees'. This seems inherently implausible, except possibly in a highly paternalistic traditional agricultural system with a labour force consisting exclusively of family labour. The Lewis model also precludes the employment of hired workers according to commercial criteria for two reasons. First, the subsistence wage is determined institutionally and not by market forces. Secondly, although the marginal productivity of labour may be positive it cannot exceed per capita consumption (the subsistence wage).

Sen's model has aroused substantial controversy due to its inherent assumption that, up to some limiting input value, the labour supply function in LDC agriculture, at the subsistence wage, is *perfectly elastic*; above the limiting value it is completely *inelastic*. The limiting value of labour input is set by the *minimum* number of hours in a day (or other time period) needed for eating, sleeping and other non-work activities, including leisure. Although no reward, however large, is sufficient to induce exceeding the limit, no extra inducement is needed to move up to the limit from some lower level of working hours. Thus, provided the limit on individual working hours is not exceeded, it may be feasible to obtain *either* extra output from a given-sized labour force, *or* a given output from a smaller labour force. The second of these alternatives signifies Sen-type disguised unemployment.

Sen's model contrasts with neo-classical wage theory which postulates that the supply price of labour is a positive function of work effort, as measured by the length of the working day, for example. Since workers need a 'bribe' to induce them to work longer hours, the supply function is upward sloping, i.e. it is less than perfectly elastic.

The crux of the argument between Sen and adherents to the neo-classical model concerns the peasant agriculturalist's marginal valuation of leisure. Is leisure a superior good, as assumed by the neo-classicists? Or does it alternate between worthlessness above a critical level (corresponding with the limit on labour input) and pricelessness below that limit, as postulated by Sen?

The theoretical issues raised by notions of disguised unemployment have been neatly summarised and clarified by Berry and Soligo (1968), who attempt to demonstrate that, *ceteris paribus*, aggregate agricultural output *cannot* be maintained in the face of a decline in the stock of agricultural workers, except in three special cases which they identify. Berry and Soligo formulate the theoretical problem posed by the concept of disguised unemployment as follows: given a representative family farm worker's production function and the average product $(\overline{AP})$ per *man hour* corresponding with a given total farm output $(\overline{P})$ and total man hours per farm $(\overline{W})$, find the conditions under which total output (and total man hours) can remain *constant*, despite the loss of one (or more) members of the farm labour force.

In Figure 3.2,

$$Bw_1 \times \text{slope of BE} = OS$$

where the slope of BE measures $\overline{AP}$, and OS is the subsistence age.

At point C on BE,

$$MP \text{ per man hour} = 0$$
$$\text{Total farm output } (\overline{P}) = OS \times n$$
$$\text{Total work hours } (\overline{W}) = Bw_1 \times n$$

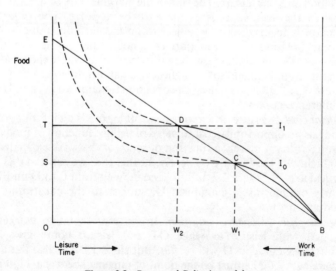

Figure 3.2: *Berry and Soligo's model*

with n signifying the number of workers comprising the farm labour force.

Suppose that one worker leaves the representative farm permanently: the production function shifts up from SCB to TDB. To maintain total farm output at $\bar{P}$ (and total work hours at $\bar{W}$) it is necessary to move up BE from C to D, where BE intersects TDB,

At point D on BE,

$$\text{MP per man hour} = 0$$

ensuring that total farm output remains at $\bar{P}$

$$Bw_2 \times \text{slope of BE} = OT$$

where OT is the daily wage

$$\bar{P} = OT \times (n-1)$$
$$\bar{W} = Bw_2 \times (n-1)$$

So, provided D is the new equilibrium point on TDB, rather than some lower point corresponding with a shorter working day, $n-1$ workers will produce the same output $\bar{P}$ as was formerly produced by n workers. But what *demand* conditions need to be fulfilled to ensure that the new equilibrium *is* at D?

Consider a general utility model of the from $U = f[F, L]$, where F = food and L = leisure. The sign of the marginal rate of substitution between the arguments is assumed to be negative. The objective function is to maximise U subject to an income constraint, where income (= food) is a function of labour time (or leisure time forgone). The maximand on the utility surface is where the production function is tangent to the highest attainable indifference curve. In Figure 3.2, TDB is the relevant production function and I_0 and I_1 are indifference curves.

Special case 1: leisure an inferior good. If leisure is an inferior good, food has no opportunity cost. Instead of being a function of food and leisure, utility is merely a function of food (or the work expended in producing it) i.e. $U = f[F]$. Clearly, in this case we can re-label the vertical axis of Figure 3.2 'utility', and moving from C to D on BE is clearly optimal as it maximises U, subject to the constraints on production.

Special case 2: leisure satiation. If the representative worker is satiated with leisure at wage OT and leisure hours Ow_2, his indifference curve at D will be *flat*. But since TDB is also flat at D (where MP = 0), D must be the point of tangency between TDB and I_1, the highest attainable indifference curve.

ROLE OF AGRICULTURE IN ECONOMIC DEVELOPMENT

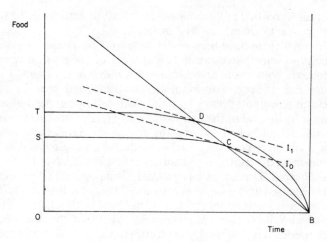

Figure 3.3: *Sen's model*

If Ow_2 does *not* represent leisure satiation, the point of tangency will be on some sloping portion of TDB to the right of D. Thus work day length Bw_2 will *not* be reached and total output will *fall*.

Special case 3: food and leisure perfect substitutes (Sen's model). If food and leisure are perfect substitutes, with a constant marginal rate of substitution regardless of the consumption ratio, the indifference curves will be linear, as shown in Figure 3.3.

It can be shown that if I_0 is tangent to SCB at C, I_1 must be tangent to TDB at D. So, even if $MU_w < 0, MU_L > 0$, the required shift from C to D will occur provided F and L are perfect substitutes.

It is worth nothing that, in all three special cases, extending the length of the working day raises the daily wage above the original subsistence level OS. This violates a commonly made assumption that real agricultural wages remain constant even when the number of agricultural workers is falling (Ranis and Fei, 1961). But provided they cannot earn more by leaving agriculture, it is theoretically feasible to hold down the wage at level OS by imposing a *poll tax* of TS on all agricultural workers.

From their analysis, Berry and Soligo conclude that 'only under rather special circumstances does output remain constant when labour is withdrawn from agriculture.' But are the limiting conditions of the special cases ever met in practice? This is clearly a question which can be answered only in the light of empirical evidence. In particular information is needed to determine the form of the peasant farmer's utility function.

Although there have been several attempts to measure disguised agricultural unemployment in LDCs, some of the results obtained fail to carry conviction. Some studies have been of the *post hoc ergo propter hoc* variety. For example, Schultz found that in India agricultural output declined in the year following the influenza epidemic at the end of the first world war. He also observed that flu epidemic fatalities had reduced the size of the agricultural population. Had there been disguised agricultural unemployment *before* the epidemic, the reduction in manpower would not have affected the level of output. The fact that output *had* fallen pointed to the rejection of the disguised unemployment hypothesis (Schultz, 1964 ch. 4). This conclusion is open to several criticisms. First, although the winter crop which Schultz observed was smaller than usual, the *summer* crop which preceded it (but not the epidemic) was of normal size. Second, the size of the winter crop may have been affected by low rainfall. Third, the smaller size of the winter crop may have been in response to a shift in demand (caused by the higher death rate). Fourth, whereas epidemic fatalities 'prune' a population unselectively, the rural–urban migration process may tend to remove the less productive members of the farm population. Fifth, even if Schultz's conclusion carried conviction for India in 1918–19 it would not necessarily be valid now (Mehra, 1966; Sen, 1975 ch. 4; Thirlwall, 1978 ch. 3).

An important distinction exists in agriculture between peak season unemployment and unemployment at other times of year. Unless a worker is unemployed during the peak season, as well as at other times, output is likely to decline if he permanently leaves agriculture. The results of a rural employment survey in Egypt appeared to discredit the existence of peak season unemployment of agricultural workers there (Hansen, 1966). But, in reaching this conclusion, insufficient weight may have been given to institutional rigidities which hinder transfers from farms with surplus labour to farms with a labour deficit, where the former are generally 'small' and the latter 'large' farms (Mabro, 1967).

A different approach was adopted in another study of disguised agricultural unemployment in India (Mehra, 1966). Instead of attempting to measure marginal productivity per man day or per man hour, Mehra sought to observe the number of working hours per capita of the agricultural population. The *actual* number of working hours was then compared with the *maximum possible* number, subject to the constraints identified by Sen. The *minimum* number of workers needed to accomplish the agricultural task represented by current output was found by dividing the maximum/actual hours ratio into the actual number of workers employed. By subtracting the minimum number of workers from the actual number, the estimated number of disguisedly unemployed workers, judged according to the criterion proposed by Sen, was found.

Mehra had at her disposal comprehensive farm management survey data on average working hours per capita on farms classified by land area. The *largest* farms were assumed to have no surplus labour time, i.e. all workers on such farms were assumed to be putting in a full working day. We adopt Sen's symbol x^* to denote a full working day. Assuming that x^* is the same for all workers on all farms irrespective of their size. Mehra then calculated the numbers of workers *required* on all other (smaller) farms. The number of 'surplus' workers in each farm size class was calculated by subtracting the required number from the actual number, and a national total of surplus agricultural workers was obtained by aggregation. This total, expressed as a proportion of the actual number of people employed in Indian agriculture, yielded the required measure of disguised unemployment. Because of the discrete numbers of people working on farms, maximum and minimum estimates of labour requirements and disguised unemployment, respectively based on rounding up and rounding down to the nearest whole number of workers, were obtained. The resulting estimates of the disguisedly unemployed proportion of the total agricultural workforce in India in 1956–7 (the data period) ranged from 29 per cent (maximum) to 6 per cent (minimum), giving a mean value of approximately 17 per cent. This suggests that, at the time of the survey, more than one Indian agricultural worker in six was redundant, in the sense defined by Sen. The results also revealed substantial inter-state variation in disguised unemployment, with mean values for individual states ranging from nearly 40 per cent to zero. However, despite its virtue of thoughtful planning and painstaking execution, this study also is open to several criticisms. First, although the assumption of zero surplus labour on the largest farms has a logical basis, and is certainly convenient, it is nevertheless arbitrary. If disguised unemployment exists, its causes include a scarcity of non-farm jobs as well as inequality in the distribution of agricultural land. Second, in this as in other contexts, averages can be misleading by concealing the variance of empirical observations. Differences in working hours per capita amongst *individual farms* must inevitably be greater than the differences between the means of size classes. Thus, the use of class means causes the labour surplus to be *underestimated*. Third, although farms with surplus labour might have sold part of the surplus to farms with a labour shortage, it was not feasible to account for such transfers in the study. Thus, to the extent that, in fact, family members on small farms went to work as part-time labourers on large farms, Mehra's method of measurement *overestimated* the overall labour surplus in the system (Sen, 1975, p. 13). Fourth, further errors could have arisen from sources such as differences in attitudes to work and leisure based on a social class, culture or, of major significance to economists, income. There is also the failure to distinguish between differences in work effort intensity due to variations in the land: labour ratio, and those due to factors such as ill-health and malnutrition, choice of technology and the quality of non-labour inputs (including land) – any or all of which may be correlated with farm size (Thirlwall, 1978, p. 103). Although, *a priori*, these various sources of error are not uniformly positive or negative, it requires an act of faith to assume that their *net* effect is zero. Thus the interpretation of Mehra's findings on disguised agricultural unemployment in India, and the policy implications of her results, remain difficult.

The results of another empirical study of the use of labour in Indian agriculture suggested that, whereas marginal productivity is generally positive on farms where hired labour is employed, it is zero, or even negative, on farms without hired labour. Moreover, farms of both types were found to co-exist in the same geographical area though not in the same village (Desai and Mazumdar, 1970). This suggests, that at the micro-level underemployment in agriculture can be due to imperfections in the market for *agricultural* labour, as well as to more familiar causes such as labour immobility between agriculture and other sectors.

A further problem arises from the difficulty of distinguishing between people who are involuntarily unemployed or underemployed and those who prefer not to have a full-time job, or even not to work at all. In the West, the idea of *voluntary* unemployment (or even underemployment) is a largely alien concept. But the results of a recent study of unemployment in rural Bangladesh show that in at least one LDC voluntary underemployment is of major significance (Ahmed, 1978).

In this study the rate of unemployment was defined as the sum of three components. Expressed as an identity these were

$$U_T = U_{V_1} + U_{V_2} + U_I$$

ROLE OF AGRICULTURE IN ECONOMIC DEVELOPMENT 61

where U_{V_1} = rate of voluntary unemployment
 U_{V_2} = rate of voluntary underemployment
 U_I = rate of involuntary unemployment (including *under-employment*)

All the variables were specified in terms of unused *labour days*, without regard to season. Hence, due to seasonal variation in the demand for labour in agriculture, these unemployment rates do *not* indicate the number of *workers* who could be withdrawn from agriculture without reducing output. Actual work time data were collected from about 300 rural households by sample survey. These data were aggregated to find the equivalent number of eight-hour work days per annum for each individual. Three alternative assumptions were made about the number of working days constituting a man year.

Taken at their face value the results of the survey appeared to indicate that a substantial proportion of the work potential of the rural population remained unused. Under the median assumption of 275 working days per annum, U_T was approximately 42 per cent. But the results also showed that, under all three assumption, voluntary underemployment (U_{V_2}) was the major source of unused labour time. Since the voluntarily underemployed consist of self-employed people who are unwilling to work for anyone else, their unused labour time cannot be mobilised for outside employment in any form, even on a part-time basis. Under two of the three working days in the year norms, the second most important source of unused labour time was the voluntary unemployment (U_{V_1}) of members of village elites who, for cultural reasons, are unwilling to work at all. Thus, the unused labour of this group cannot be counted as part of any kind of mobilisable surplus, short of a revolution in social attitudes. Except under the largest of the working-days in the year norms, involuntary unemployment (U_I) – including those with too little work, as well as those with none – turned out to be the *smallest* source of unused labour time. Under the median assumption of 275 days in the working year, only about 8 per cent of total unused labour time (U_T) derived from this source. This result points to the conclusion that, at least in Bangladesh, the amount of surplus agricultural labour available for alternative employment is a good deal smaller than casual observation might suggest.

A profile of seasonal labour demands in the study area was constructed from information obtained from a sub-sample of households. By this method an estimate was made of the proportion of involuntarily unemployed *workers* on a typical day in the busiest week of the year. This turned out to be fractionally below 4 per cent of the total potential work force, including the voluntarily unemployed who comprised about 8 per cent of the same total. The approximate sum of these two figures (12 per cent) gives an estimate of the proportion of potential workers who could be permanently removed from the rural sector without reducing output. However, this figure is *not* comparable with Mehra's estimate of 17 per cent disguised unemployment in India for two reasons. First, because Ahmed's Bangladesh study is unique in counting the voluntarily unemployed as part of the potential work force. Second, because Ahmed lacked information about how many more working hours the voluntarily underemployed might have been prepared to put in *on their own farms*, in response to the departure of former colleagues. In other words, estimates of disguised unemployment based on Sen's definition are not comparable with estimates based on the assumption that the intensity of work effort is fixed. The overall impression left by the Bangladesh study is that, even in a country which is notorious for its over-population, the degree of disguised agricultural unemployment, as conventionally defined, is not large. Nevertheless, because social and cultural attitudes are resistant to change, a considerable amount of labour in the rural sector appears to be 'wasted'.

Although accurate observation is obscured by problems of both definition and measurement, really convincing evidence of *major* disguised unemployment amongst those working on the land in LDCs is hard to find. Though the evidence is too incomplete to justify an unqualified rejection of the disguised unemployment hypothesis irrespective of place or time, it appears unlikely that large-scale and *permanent* labour redundancy in the agricultural sector is typical in LDCs. Hence, the notion of 'painless' capital formation based on the large-scale mobilisation of surplus agricultural labour may have little or no policy relevance in most countries. This is not to deny that in the *long term* the agricultural sector is bound to surrender labour to the non-agricultural sector, but only to emphasise that the labour transfer must be matched or preceded by other changes which violate the *ceteris paribus* condition, such as the adoption of new agricultural technology.

There may also be scope for mobilising *seasonally* unemployed or underemployed agricultural workers for non-farm work during part of the year. Since workers from farms must be free to return there during peak periods, supplementary employment must derive from activities which operate efficiently despite seasonal fluctuations in manpower. It will also be an advantage if the supplementary employment is actually located in rural areas so that workers recruited from farms do not have to leave home. Local 'public works', such as the construction of roads and dams, might meet these requirements. This happens in China, where industrial workers have assisted on farms during seasonal peaks and agricultural workers have been employed on construction projects during the agricultural off-season. The Chinese have also attempted to integrate agricultural production with the manufacture of simple industrial products, such as building materials, and even consumer goods for outside sale, within the rural commune system (Reynolds, in Reynolds, 1975). Although agricultural workers who work outside agriculture for part of the year are *not* 'disguisedly unemployed', in the strict sense, this may nevertheless be the most feasible *short-term* means of utilising surplus agricultural manpower in LDCs.

Considering the inter-sectoral transfer of agricultural labour in its broadest sense, with the *ceteris paribus* condition relaxed, we now address ourselves briefly to two further aspects. These are

(1) The 'quality' of rural migrants as potential industrial workers and the implicit transfer of 'human capital' and
(2) The availability and prices of complementary factors of production in the industrial sector.

Quality of redundant agricultural labour and the inter-sectoral transfer of human capital. The more obvious costs of transferring workers from agriculture to industry include transport and rehousing, for example. Somewhat less obvious are the cost of training or retraining and the reasons why this is necessary. There is also the question of how these costs are distributed between the two sectors.

Rural migrants will normally require some training in industrial skills before becoming fully-fledged members of the industrial labour force. The costs of training for some types of work may be enhanced by the low formal educational qualifications of many migrants. The farm to factory transfer may also mean a change in attitudes to work. The rural migrant may experience difficulty in adjusting to the standards of punctuality and regular working hours inherent in the factory system. Although some of the consequential adjustment costs may be borne by the newly-arrived workers themselves, perhaps in the form of lower wages, a share may also fall on employers. The rural migrant's productivity in industry may also be affected adversely by a legacy of poor health or weakened physical condition caused by malnutrition. To the extent that the costs of remedying such ills fall on employers, they add to the 'hidden costs' of employing migrants from rural areas, where standards of health and nutrition are almost invariably lower than in towns and cities.

Although industrial employers, or government, bear the costs of training rural migrants in industrial skills, the industrial sector 'inherits' from the agricultural sector the capital invested in rearing and educating them to working age. Schultz has emphasised the complementary roles of material capital and 'human capital' in agricultural development, and the potential benefits of providing better educational and health services in rural areas of LDCs to improve the quality of human resources in agriculture (Schultz, 1964, ch. 12). Kuznets has also made a notable contribution in stressing the size of the investment made by parents of rural migrants in rearing, educating and training them to maturity. He gives an illustrative example, based on 'realistic' assumptions, in which the annual value of this transfer could amount to as much as 6 per cent of a developing country's GNP (Kuznets, 1964, section IV). Although the intersectoral transfer of human capital is closely linked to the transfer of labour, discussing it here, rather than under agriculture's capital contribution, is arbitrary and undertaken largely for convenience.

Since landless labourers are often the poorest section of the rural community, they would appear to have the most to gain from leaving agriculture. However, because of their poverty, they may also be least able to bear the costs of migration.

Supply of complementary factors of production. Industrial development cannot be based on labour alone, capital inputs are also needed, even where industries with a low capital: labour ratio, such as some types of textile manufacturing, are deliberately selected for development. In practice, industrial development is frequently constrained by available supplies of capital and other co-operant factors – including technical know-how and managerial expertise – even though *unskilled* labour may be plentiful. The existence of high *open* unemployment in and around large towns and cities in many LDCs supports the belief that, at least in the short run, industrial development is not constrained by an inelastic supply of labour.

For the immediate future, therefore, an important policy objective in many LDCs might be to increase employment in agriculture, or at least to slow down the rate of out-migration in order to contain, or even reduce, the number of urban unemployed. This is the main policy conclusion of Todaro's celebrated rural–urban migration model (Todaro, 1969).

The case for expanding agriculture in order to create employment is supported by evidence on employment linkage effects. Employment linkage may be defined as the total effect of a unit increase in final demand for the product of a given industry on the derived demand for labour, both in that industry and in all others linked with it in the inter-industry input–output matrix. Like the income linkage coefficient YL, the employment linkage coefficient EL reflects the labour coefficient (in value terms) of sectoral output. $EL = 0$ would signify a zero labour coefficient. Because of the importance of employment as a source of income, there tends to be a close and direct relationship between income linkage and employment linkage coefficients. Thus an industry which is strong on income generation can be expected to come out well in terms of employment creation also. This is borne out by the results of the empirical study of inter-industry linkage in Taiwan referred to above. For the same four sectors, the employment linkage coefficients were as shown in Table 3.5.

These results indicate that agriculture's employment generating capacity is superior to that of both the input supply (working capital for agriculture) and food processing industries. Indeed, amongst a total of twelve industries, the two sub-sectors of agriculture ranked first and third, whilst the mining sector ranked second. In contrast, the agricultural input supply and food processing industries ranked tenth and twelfth, respectively.

Most of the literature on surplus labour in agriculture is concerned with densely populated countries. Moreover, development economists have tended to assume that in sparsely populated countries or regions, where labour is scarcer than land, surplus labour is ruled out *ex hypothesi*. Any transfer of labour from agriculture to another sector is therefore contingent upon investment to advance farm labour productivity. However, Helleiner (1966) has put forward a model postulating a potential labour surplus in a *land* surplus

Table 3.5: *Employment linkage coefficients*

Industry	EL
Agricultural production for food	0.852
Agricultural production for raw materials	0.908
Working capital for agriculture	0.725
Food processing	0.673

Source: as Tables 3.2 and 3.3.

economy. In this model a difference is postulated between the amount the agricultural working population is technologically capable of producing and the amount it chooses to produce under a 'limited material wants' constraint. Assuming that the potential surplus can be mobilised – by taxation for example – it can take the form of either food or labour, but not both. Unlike Sen, who assumes that, until the maximum length of working day constraint becomes binding, leisure is valueless to family farm workers, Helleiner assumes that once their basic subsistence requirements have been met, such workers have a strong preference for leisure. Although Helleiner offers historical evidence from Nigeria to back his hypothesis, more direct evidence backing the limited wants assumption is clearly needed.

5 Foreign Exchange Contribution

The potential benefits of overseas agricultural trade in an open economy have been discussed under the 'market contribution'. In this section we adopt a more pragmatic approach. Given as an axiom that development in most LDCs is constrained by a scarcity of foreign exchange, we examine how agriculture may contribute to relaxing that constraint. But first we consider briefly whether the foreign exchange contribution conflicts with the market and factor contributions.

It has been argued that there is a fundamental contradiction between

(1) the 'balanced' development of agriculture and industry in a closed economy where agriculture aids sectoral diversification through its market and factor contributions; and
(2) the 'unbalance' of an open economy in which agriculture is no longer constrained by the size of the domestic market and where agricultural exports compete with the non-agricultural sector for capital, or even labour (Myint, in Reynolds, 1975).

Food that is sold abroad is not available for domestic consumption in the non-agricultural sector or to help contain inflation. Resources utilised in producing agricultural products for export are not available for transfer to the non-farm sector.

There is some substance in this argument which cannot be dismissed out of hand. In a country with a lagging agricultural sector and a large and unmanageable food import bill it might well make better economic sense to expand food production for the domestic market than to encourage agricultural exports. But once domestic agriculture is able to meet the basic requirements of the domestic market, and the food import bill is reduced to a manageable size, it may be sound policy to exchange agricultural exports – either of food or other agricultural products – for imported goods of the type needed to quicken the pace of development through sectoral diversification. Indeed, exchanging surplus agricultural goods for imports, particularly of capital goods, may be seen as a first step in the industrialisation process (Nicholls, 1963). More pragmatically, in many developing countries which are chronically short of foreign exchange, selling agricultural products abroad is a feasible and practical method of earning foreign currency to purchase 'essential' imports. Alternative sources of funds for buying from abroad, such as official aid, foreign loans and private foreign investment, are all useful for 'topping up' but are rarely available in sufficient amount to meet more than a minor proportion of the bill for indispensable imports of goods, raw materials and fuels. Although the difference between indispensable imports and those which it pays a country to replace with a domestic substitute may be hard to determine in practice, it cannot be seriously doubted that most LDCs intent on a higher rate of economic growth and development *do* have minimum import needs because of their limited range of industrial skills and gaps in their resource endowments.

For a country wishing to increase its export earnings, an expansion of agricultural exports offers a number of potential advantages not shared by exports of manufactured goods. First, an export crop such as coffee, cocoa or cotton, can often be added to an existing, largely subsistence, cropping system, thus avoiding major new investment. Second, to the extent that new investment is needed, the amount of capital needed is often only moderate due to the relatively low capital: output ratios inherent in some types of agricultural product. Farmers are capable of using highly labour-intensive methods to create capital assets such as roads, dwelling houses and other buildings, drainage and irrigation works, and terracing to prevent soil erosion. Much of the labour is low opportunity cost family labour which would otherwise be underemployed during slack periods of the farming year. Such methods of lowcost capital formation in agriculture have been much used in China. Third, because most agricultural products are fairly homogeneous, and the market share of 'new' exporters is usually quite small, the demand schedule facing the individual exporter is typically fairly elastic. This implies that, given freedom of entry to overseas markets, a modest increase in export volume can be undertaken by an individual country without depressing its terms of trade. The terms of trade effect would of course be different (and adverse) were a number of countries to expand their exports simultaneously.

Another potential advantage of agricultural export trade is that export outlets in developed countries may provide capital and technical advice on production and product improvement. The benefits of such advice may spread from the export sector to the production of food for the domestic market as well.

Where the objective is to 'save' foreign exchange by reducing imports, rather than earning more by expanding exports, *agricultural import substitution* may be especially advantageous. First, there is the question of which kinds of imports it is technically feasible to produce at home. Major agricultural imports may score highly on this count, whereas many industrial imports may not. Second, there are the marginal costs of import replacement to be considered. Agriculture will be superior to industry in this respect if domestic agricultural production can be expanded at a lower real cost – by bringing reserves of underutilised land and labour into production, for example. But if expanding domestic agricultural production to replace imports unavoidably entails diverting capital away from industrial uses, where even more foreign exchange could be saved, then the cost of agricultural import replacement will be too high. Although LDCs are very varied in their resource structures, choice of production techniques and the extent of competition between agriculture and industry for the use of resources, a good many appear still to belong to the first of these categories rather than the second.

In this section we have stressed the *short-term*, foreign exchange benefits of agricultural production for export and import replacement. This emphasis is not intended to disguise recognising that, in the *long term*, the relative importance of agriculture's foreign currency contribution is bound to diminish as the economy becomes more diversified, and the relative importance of agriculture itself declines. Open economies with a pronounced comparative advantage in agricultural production face a dilemma. Should they give full rein to their short-term comparative advantage as agricultural exporters? Or, should they divert resources away from agriculture into industrial investment and training in industrial skills in order to shift their comparative advantage, in the long of term, away from agriculture to manufacturing? In other words will a purely static view of comparative advantage suffice, or must it be dynamic? Assuming that sectoral diversification is a long-term policy objective in practically all LDCs, it is logical that dynamic comparative advantage should *not* be disregarded. This is to recognise a trade-off between the short-term gains from holding back sectoral diversification in order to exploit a static comparative advantage in agricultural production and potential long-term gains from hastening the process of diversification into new industries in which a dynamic comparative advantage is thought to exist. Finding the optimum point of trade-off, where short-term gains are equal to the present value of long-term losses (or short-term losses are equal to long-term gains), is undoubtedly difficult due to the dearth of needed information and the many uncertainties involved. But it would be unrealistic to suppose that, in practice, long-term trading policy is ever based exclusively on what is perceived to be a country's static comparative advantage.

6 Conclusions

This chapter points to a number of conclusions about the role of agriculture in economic development. First, and most importantly, in many LDCs the creation and sustained increase of a marketable surplus of agricultural products is a virtual precondition of sectoral diversification and hence of development itself. Although some exporters of mineral oil or other natural resources may be exceptional in this respect, because these are depletable assets it is doubtful whether even they can afford to neglect agriculture in the long-term. There are four major reasons why a growing surplus of agricultural products (in the sense that production exceeds consumption in agriculture itself) is needed:

(1) to increase supplies of food and agricultural raw materials at non-inflationary prices;
(2) to widen the domestic market for industrial goods through increased purchasing power within the rural sector;
(3) to facilitate inter-sectoral transfers of capital needed for industrial development (including infrastructure); and
(4) to increase foreign exchange earnings through agricultural exports.

Although these four contributions of agriculture to economic development are conceptually distinct, they are also interdependent: moreover, they need to be consistent. We have discussed one aspect of consistency in the previous section, where it was concluded that the foreign exchange contribution need not be inconsistent with the market and factor contributions. But what about the consistency of the capital contribution with other contributions? Is transferring capital from agriculture to the non-farm sector consistent with the retention in agriculture of sufficient resources to enable industry's growing food and raw materials requirements to be met without inflation? And is it consistent with sustaining agriculture's important role as a buyer of industrial output? These questions were discussed at some length in section 4.1, where it is concluded that although the capital contribution is not necessarily inconsistent with the other two 'closed economy' contributions, the choice of method for transferring capital out of agriculture, and finding the best rate of transfer, are critical policy decisions.

A second major conclusion of this chapter is that the importance of agriculture as a source of redundant labour waiting to be mobilised for relatively 'painless' industrial development has been much exaggerated. There is a dearth of convincing evidence supporting disguised agricultural unemployment as an empirically valid concept, even in LDCs. This means that, apart from the recruitment of *seasonally* underemployed agricultural workers for *temporary* employment outside agriculture during the slack season, the agricultural sector's ability to yield labour to the non-agricultural sector is contingent upon the achievement of significant and sustained advances in agricultural labour productivity. The supposed importance of agriculture's labour contribution, at least in the short term, is also undermined by a dearth of evidence that, in LDCs, development outside agriculture is constrained by the scarcity of unskilled labour. On the contrary, high rates of urban unemployment in most LDCs point to the conclusion that the demand for industrial labour typically lags a long way behind the supply of those seeking industrial employment. The rate of rural–urban migration is too high, so that an appropriate policy objective might be to attempt to damp it down by providing would-be migrants with incentives to remain in agriculture. These might include better education, health and other social amenities, as well as higher farm incomes.

Our final conclusion is that because most LDCs need to generate a larger agricultural surplus to enable the agricultural sector to play its developmental role more effectively, they must also consider how best to achieve the needed increase in agricultural output. In some areas the mobilisation of seasonally underemployed agricultural workers for supplementary employment in *agriculture* may be a feasible option, particularly where adequate supplies of complementary inputs, such as irrigation to extend the growing season, are available. Elsewhere, securing the required increase in agricultural output may be contingent upon the scope for relaxing any or all of a large number of production constraints. This may require major institutional changes, such as land reform, as well as a battery of measures to equip farmers with the knowledge, confidence and enhanced economic incentives they need in order to use unfamiliar, output-increasing technology. Successful application of such measures will yield higher agricultural output in the short term, and the release of labour to the industrial sector in the long term. As shown in section 4.2, any decline in the absolute size of the agricultural labour force is likely to be delayed until a comparatively late stage of sectoral diversification.

Appendix: Estimation of Labour Force Growth Rates

The following statistical items were obtained from the sources quoted:

1. Total population levels, by countries, in 1960 and 1978.
2. Percentage of population of working age, by countries, in 1960 and 1978.
3. Percentage distribution of the labour force between agriculture and non-agriculture, by countries, in 1960 and 1978.

Derived estimates were then obtained for:

4. Total working population levels (including the unemployed), by countries, in 1960 and 1978.
5. Total numbers employed, or available for work, in agriculture and non-agricultural, by countries, in 1960 and 1978.

Compound rates of growth in the agricultural, non-agricultural and total working populations over the 18-year period were then obtained by substituting the derived 1960 and 1978 population estimates in the compound growth formula $P_t = P_o(1+r)^t$ and solving for r.

References

Ahmed, I. (1978), 'Unemployment and underemployment in Bangladesh agriculture', *World Development*, **6** (11/12).

Berry, R.A. and Soligo, R. (1968), 'Rural-urban migration, agricultural output and the supply price of labour in a labour surplus economy', *Oxford Economic Papers*, **20** (2).
Bird, R.M. (1977), 'Land taxation and economic development: the model of Meiji Japan', *J. Dev. Studies*, **13** (2).
Chakrabarti, S.K. (1977), *The Behaviour of Prices in India, 1952–70*, Macmillan Company of India, New Delhi.
Chakrabarti, S.K. (1978), 'Money, output and price: guidelines for macro-credit planning', in Reserve Bank of India, *Recent Developments in Monetary Theory and Policy*, Papers and Proceedings of the Seminar held at Bombay.
Edel, M. (1969), *Food Supply and Inflation in Latin America*, Praeger, New York.
Griffin, K. (1979), *The Political Economy of Agrarian Change*, (2nd edn.), Macmillan, London.
Hansen, B. (1966), 'Marginal productivity wage theory and subsistence wage theory in Egyptian agriculture', *J. Dev. Studies*, **2** (4).
Hathaway, D.E. (1974), 'Food prices and inflation', *Brookings Papers on Economic Activity*, 1.
Heady, E.O. (1962), *Agricultural Policy Under Economic Development*, Iowa State University, Press, Ames.
Helleiner, G.K. (1966), 'Typology in development theory: the land surplus economy (Nigeria)', *Food Research Institute Studies*, **VI** (2).
Hirschman, A.O. (1958), *The Strategy of Economic Development*, Yale University Press, New Haven.
Holden, K. and Peel, D.A. (1977), 'An empirical investigation of inflationary expectations', *Oxford Bul. Econ. Statist.*, **39** (4).
Kuznets, S. (1964), 'Economic growth and the contribution of agriculture', in Eicher, C.K. and Witt, L.W. (eds), *Agriculture in Economic Development*, McGraw-Hill, New York.
Lewis, W.A. (1954), 'Economic development with unlimited supplies of labour', *Manchester School*, pp. 130–90.
Mabro, R. (1967), Industrial growth, agricultural under-employment and the Lewis model: the Egyptian case', *J. Dev. Studies*, **3** (4).
Maynard, G. (1961), 'Inflation and growth: some lessons to be drawn from Latin-American experience', *Oxford Economic Papers*, **13** (2).
Mehra, S. (1966), 'Surplus labour in Indian agriculture', *Indian Econ. Rev.* (**1**), (reprinted in Chaudhuri, P. (ed.) (1972) *Readings in Indian Agricultural Development*, Allen & Unwin, London).
Myint, H. (1975), 'Agriculture and economic development in the open economy', in Reynolds, L.G. (ed.), *Agriculture in Development Theory*, Yale University Press, New Haven.

Nicholls, W.H. (1963), 'An agricultural surplus as a factor in economic development', *J. Political Economy*, LXXI (1).

Nurkse, R. (1953), *Problems of Capital Formation in Underdeveloped Countries*, Blackwell, Oxford.

Ranis, G. and Fei, J.C.H. (1961), 'A theory of economic development', *American Econ. Rev.* 51, pp. 533–65, (reprinted in Eicher and Witt, 1964).

Reynolds, L.G. (1975), 'Agriculture in development theory: an overview', in Reynolds (ed.), op. cit.

Schultz, T.W. (1964), *Transforming Traditional Agriculture*, Yale University Press, New Haven.

Sen, A. (1975), *Employment, Technology & Development*, Oxford University Press, Oxford.

Southworth, H. and Johnston, B.F. (1967), *Agricultural Development and Economic Growth*, Cornell University Press, Ithaca.

Thirlwall, A.P. (1978), *Growth and Development* (2nd edn.), Macmillan, London.

Todaro, M. (1969), 'A model of labour migration and urban development in less developed countries', *Amer. Econ. Rev.*, pp. 138–48.

World Bank (1980), *World Development Report*, The World Bank, Washington D.C.

Yotopoulos, P.A. and Nugent, J.B. (1976), *Economics of Development*, Harper & Row, New York.

Notes

1. The structural equations of Chakrabarti's model are:

$$\frac{P_1}{P_2} = A Y_1^\alpha Y_2^\beta \ldots \qquad (1)$$

$$P_2 = B m^\gamma \ldots \qquad (2)$$

$$P = C P_1^{\lambda_1} P_2^{\lambda_2} \ldots \qquad (3)$$

where P_1 = price of A sector goods, P_2 = price of non-A sector goods, P = general price level, Y_1 = real A sector output, Y_2 = real non-A sector output, m = *per capita* money supply, and A, B and C are constants. $\alpha, \beta, \gamma, \lambda_1$ and λ_2 are parameters, with *a priori* expectations of $\alpha < 0, \beta > 0, \gamma < 1, \lambda_1 + \lambda_2 = 1$.

The reduced form equation for P is:

$$P = D Y_1^{\alpha \lambda_1} Y_2^{\beta \lambda_1} m^\gamma \ldots \qquad (4)$$

where the constant D involves the constants A, B and C and some parameters. To facilitate parameter estimation, (4) is reformulated as:

$$P = DR^{\alpha\lambda_1} Y_2^{(\alpha+\beta)\lambda_1} m^{\gamma} \ldots \qquad (5)$$

where $R = Y_1/Y_2$
(Chakrabarti, 1978).

2. In countries with adequate industrial statistics, inter-industry linkage effects can be estimated empirically by using input–output analysis. The input–output model specifies that:

$$X - AX = Y \ldots \qquad (1)$$

where X is a vector of sectoral gross outputs (X_j), A is a matrix of intersectoral input–output coefficients (a_{ij}), and Y is a vector of sectoral final demands (Y_j). In matrix notation, the simultaneous solution to (1) derives from:

$$X = (I - A)^{-1} Y \ldots \qquad (2)$$

where $(I - A)^{-1}$ is the inverse matrix of A_1 of which the elements, A_{ij}, are the demand multiplier or linkage coefficients. Subject to the condition $\Sigma a_{ij} < 1$ for all j, i.e. the column sums of the coefficients of the A matrix are all less than unity:

$$(I - A)^{-1} = I + A + A^2 + A^3 + \cdots + A^n \qquad (3)$$

Hence, by substituting (3) in (2):

$$X = (I + A + A^2 + A^3 + \cdots + A^n) Y \ldots \qquad (4)$$

Intuitively, if A is 'large' X is also large relative to Y. Conversely, if A is 'small' X is not much larger than Y. In other words, the A_{ij}'s give a direct measure of the strength of intersectoral demand multiplier or linkage effects.

Readers are referred to the literature for fuller explanations of the input–output model. Heady (1962 pp. 263–5) contains a particularly lucid explanation emphasising agriculture's interdependence.

3. More comprehensive discussions of both landlord-tenant systems in LDCs and land reform are included in chapters 4 and 8 respectively.

4 Theory of Rent and the Concept of 'Surplus'

1 Introduction

In this chapter we shall first discuss the theory of economic rent. Next, we shall discuss the important concept of 'surplus' in agriculture which is critical for capital formation. A clear distinction will be made among a 'product' surplus, a 'labour' surplus and a 'financial' surplus. We shall then discuss some characteristics of landownership in under-developed agriculture. Here, several features of owner-occupation, share-tenancy, ordinary lease will be described. Given the absence of specialisation in underdeveloped agriculture, the case of 'interlocking' of factors will be mentioned under which the landlord may also act as a moneylender and a trader. Finally, the theory of share-tenancy will be examined in detail.

2 Economic Rent

Economic rent can be generally defined as the difference in reward actually accruing to a factor in a particular use and the minimum reward required to keep it in that use (i.e. its minimum supply price or transfer earnings).

It is generally assumed that agricultural land *as a whole* has no alternative use, i.e. its transfer earning is equal to zero. Hence, agricultural landlords consider any return better than no return. But agricultural land in a *particular* use – wheat production, for example – will generally have positive transfer earnings in terms of alternative agricultural uses. Note that land in particular *locations* may also have alternative value, for example, in housing or industrial development. These peculiarities enable the landowner to extract an unearned surplus or rent from the land *user*, provided land is scarce.

3 The Theory of Rent

In a free market economy, rents are determined in the same way as other prices, i.e. by the interaction of supply and demand. Since land is usually fixed in supply, and 'immobile', its price or 'rent' is peculiarly dependent on demand factors, i.e. changes in rent are mainly due to changes in demand.

Changes in demand for land may be explained by a number of reasons:

(a) A number of existing farmers may wish to take more land under cultivation, and

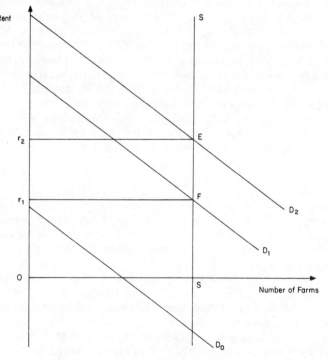

Figure 4.1: *Origin of rent*

(b) many *would-be* farmers may decide to buy more land.

If we assume free entry and exit, then the number in both classes of farmers will tend to vary directly with product prices and farm incomes or product prices and farm income *expectations*. Expectation of *higher* farm income will cause upward shift in the demand curve and an increase in rent and vice-versa. This is shown in Figure 4.1 where we measure rent on the vertical axis and demand and supply of farms on the horizontal axis.

Assuming that all farms are let rather than being farmed by their owners, the initial 'equilibrium' is obtained at r_1 where demand is equal to supply. If the demand for land goes up, the demand curve for land shifts from D_1 to D_2, 'rent' rises by $r_1 r_2 EF$.

Notice that this rise in rent is due to the complete inelasticity of the supply schedule. If the supply of land were completely elastic, no rent would have occurred despite a shift in demand. (Students can verify this point easily by drawing a diagram.) Note also that if the demand for land had been very low (as indicated by the demand curve D_0), land would not have reached its extensive margin of cultivation and no rent would have occurred.

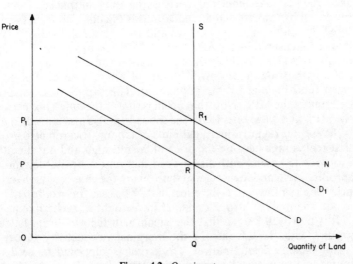

Figure 4.2: *Quasi-rent*

4 Rent and Quasi-rent

It is possible to argue that there are cases where supply of some factors may take some time to react to a change in demand. In such cases, a 'quasi-rent' emerges.

In Figure 4.2, the quantity of factors available in the short run is given by OQ at price OP. No more of this factor is available in the short run. If the demand curve shifts from D to D_1, the price rises to P_1 and a quasi-rent equal to OQR_1P_1 appears. We have implicitly assumed away any variable cost of production. The short period supply curve is given by PRS.

In the long run, the supply curve is given by PRN which is perfectly elastic. The price of factor goes down to the previous equilibrium level ($=OP$) despite a shift in the demand curve to the right and quasi-rent disappears.

Note that in the long run, no rent is earned in excess of cost of production. The short period return to the factor can therefore be regarded as 'quasi-rent'.

5 The Ricardian 'Corn Rent'

The Ricardian corn model has sometimes been used to determine rent as a residual. Assume that corn is the only output that can be produced by available land, and labour is the only input that can be hired. In Figure 4.3, the average and the marginal productivity of land are measured by the APL and the MPL. Assume that the supply of labour is given by the level of population and total 'wages fund' is given by past developments. The rate of wage, w* is then given by wage fund divided by the population. The supply of labour, N* determines the MPL. Note that profit is simply the difference between the MPL and wage rate. Rent (the shaded area) is also a residual as it is the surplus on intra-marginal units of labour. Ricardo also argues that rent is high because the price of corn is high and not the other way round. Adam Smith also argued that rent is a price-determined surplus. 'High and low wages and profit are the causes of high or low price; high or low rent is the effect of it.' The case for taxation of rent is not, then, difficult to understand in the light of such an analysis.

It is now well known that the Richardian theory of rent has lost ground with the advance of the marginal productivity theory of factor income determination. Thus, rent is supposed to be determined by the marginal productivity (MP) of land. However, the MP

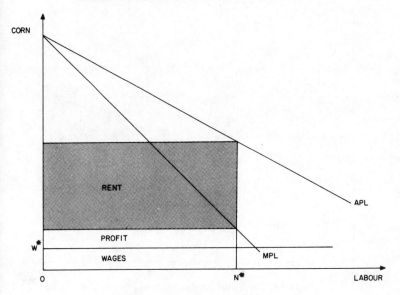

Figure 4.3: *The Ricardian corn rent*

theory explains the demand and *not* the supply of factors. Land is usually a fixed factor in the short run and, as such, it can earn 'quasi-rent'. Also, in the long run, from the stand-point of the society as a whole, the supply of land is fixed.

6 The Rental Market

The important implication of the previous discussion is that the level of rent is a function of the relative profitability of land use. 'The more profitable farming is relative to other comparable occupations, the higher the rents prospective tenant farmers are likely to be prepared to pay and the higher the rents landowners are likely to require to induce them to lease their land' (Currie, 1981). The rental contract between the landlords and the prospective tenants can then be illustrated with the help of Figure 4.4. Let DD stand for the market demand curve and let SS_1S_2 indicate the supply curve for the land (for the sake of simplicity we assume away a 'step' demand and a 'step' supply function (for such analysis, see Currie, 1981). The equilibrium level of rent is then shown by R^e. Units to be cultivated under the

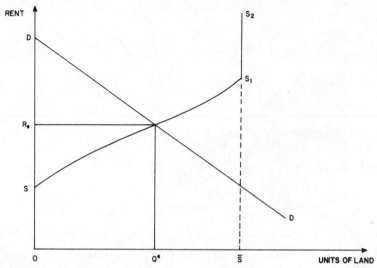

Figure 4.4: *The Rental market with homogeneous land*

landlord–tenant system is given by Q^e and the units that would be farmed by owner–occupier is given by $(\bar{S} - Q^e)$.

6.1 Heterogeneity of land

In the previous analysis, we assumed that all operational units of land are the same in fertility, location, etc. We now drop that assumption and introduce cases of different units of land which are heterogeneous. Suppose there are two different units operating land. There are then those alternatives that a prospective tenant faces: (a) he can rent a type 1 farm; (b) he can rent a type 2 farm; (c) he can accept off-farm employment. The highest rent that will be offered on any type of land will then depend not only on the alternatives to land cultivation, but also on the rent which a tenant can obtain from the other type of land. Hence, for a given rent for one type of land, we can obtain the market demand curve for the other type by 'ordering' the limit rent of the prospective tenants. As Currie argues, that for certain combinations of rent, an owner of one type may decide to lease out his unit and become a tenant in another type. His reservation rent may therefore depend on the market rent for the other type. Therefore, the

THEORY OF RENT AND THE CONCEPT OF 'SURPLUS'

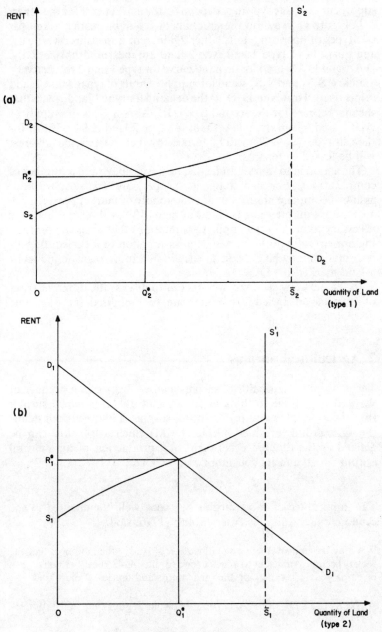

Figure 4.5: *Rent with heterogeneous land*

supply curve of one type may depend on the market rent for the other.

In Figure 4.5, we show the simultaneous equilibrium in markets for two types of operating units. Once again, rent is measured vertically and quantity of type 1 and type 2 land are measured horizontally. D_1D_1 and D_2D_2 land for demand curve for type 1 and 2 respectively. Similary, $S_1S_1^1$ and $S_2S_2^1$ stand for supply of units of types 1 and 2 land respectively. However, D_1D_1 is the demand for type 1 land, given that the market rent for the second type is R_2^e. Also, $S_1S_1^1$ is the supply of type 1 land, given the market rent for type 2 land is R_2^e. Hence, for the i th type, $(\bar{S}_1 - Q_1^e)$ will be farmed by their owners, but the rest will be leased to tenants.

The actual land market in LDCs, however, may pose a number of complications. For one thing, in the previous analysis, we have assumed complete absence of any monopoly or market power. Such an assumption may not be valid in many LDCs. For another, it is necessary to modify the implicit assumptions that costs of contracting are virtually nothing. Also, the assumption of a perfect flow of information may be difficult to sustain in a more realistic model of a land market in LDCs.

In the next few sections, we proceed to discuss the different types of 'surplus' in agriculture which can be mobilised for economic development.

7 Agricultural Surplus

The concept of 'agricultural surplus' can be defined in a number of ways. It is possible to think about agricultural (a) *'product'* surplus (b) *'labour'* surplus and (c) *'financial'* surplus. First, we shall define the agricultural 'product' surplus ($=M$). Such a 'surplus' can be defined as the difference between total production of agricultural output ($=Q$) and its consumption ($=C$), i.e.

$$M = Q - C \dots \qquad (1)$$

The importance of food surplus has been well documented in the economic literature. As Adam Smith (1776) said:

When by the improvement and cultivation of land...the labour of half the society becomes sufficient to provide food for the whole, the other half...can be employed...in satisfying the other wants and fancies of man-kind.

As far as the LDCs are concerned, Nicholls argues that, 'until underdeveloped countries succeed in achieving and sustaining (either through domestic production or imports) a reliable food surplus, they have not fulfilled the fundamental precondition for economic development' (Nicholls, 1963, p. 1). In the same spirit, Kuznets argues on the basis of his studies on economic development: 'an agricultural revolution – a marked rise in productivity per worker in agriculture – is a precondition of the industrial revolution in any part of the world' (Kuznets, 1959).

The emergence of food surplus can be demonstrated with the use of a single diagram. Let food production be measured in the vertical axis and employment be measured in the horizontal axis. Let us assume that labour (= L) is the only factor of production that produces food (= Q). Hence we have:

$$Q = f(L) \ldots \qquad (2)$$

Let us assume that (a) wages remain fixed; (b) agricultural techniques are given; (c) quality of land is homogeneous; (d) labour and land are so combined that any given population maximises its agricultural output. The total production curve is then given by OP as shown in Figure 4.6(a). The wage line is given by OW. It is assumed that the wage level is fixed (subsistence wage). The corresponding marginal (MP) and average product curves (AP) are shown in Figure 4.6(b). Where MP is = AW, the maximum difference between OP and OW is obtained (= QS in Figure 4.6(a) with employment = Ob). The difference between OP and OW can then be regarded as 'surplus' available for reinvestment. Note that such a surplus disappears at a point like E where the total population (Ob_o) consumes the whole of total production and there is no surplus left for capital accumulation. At this point, average product (AP) is equal to the average wage ON_1, which is just enough to keep Ob_o, population alive (see Figure 4.6(b)).

7.1 The concept of 'labour surplus'

It has frequently been pointed out in the literature of economic development that the marginal productivity of labour (MP_{La}) is near or equal to zero in LDC's agriculture, i.e. $MP_{La} \simeq 0$. This implies that in a typical LDC it is possible to remove one or more units of labour without reducing the total output in agriculture. This phenomenon has been branded as 'disguised' unemployment in agriculture by writers including Lewis, and Ranis and Fei (see, chapter 3 of this book; and Lewis, 1954, Ranis and Fei, 1961). It has been argued

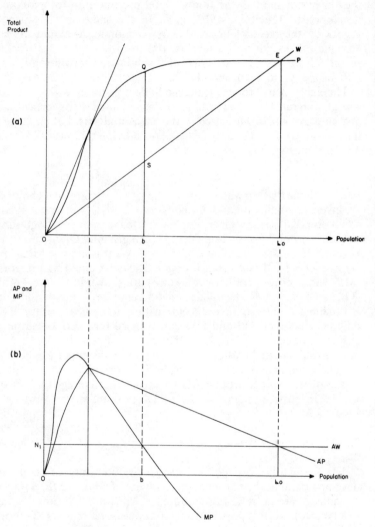

Figure 4.6: *The emergence of an agricultural surplus*

that agriculture can contribute to the growth of the industrial sector by transferring 'surplus' labour from agriculture to industry without any reduction in the level of agricultural output. Some economic models have been developed to show how such surplus labour can aid industrialisation, as will be discussed later.

7.2 The concept of 'financial' surplus

Agriculture has sometimes been regarded as the source of the creation of 'financial' surplus in LDCs. The idea of 'financial' surplus is related to the assignment of money value of the 'physical' surplus generated within the agricultural sector. Thus, if P_a is the price of agricultural goods that have been marketed ($= M$) then $P_a M = S_f$, where $S_f =$ the financial surplus. The behaviour of financial surplus can be postulated by looking at the following relationship.

$$S_f = f(P_a, Q_a, U)$$

where $Q_a =$ the volume of agricultural production, $M = \alpha Q_a$ and $U =$ other variables besides Q and P_a. It is expected that

$$\frac{\partial S_f}{\partial P_a} > 0 \quad \text{and} \quad \frac{\partial S_f}{\partial Q_a} > 0$$

In other words, the higher the level of agricultural prices, the greater will be the supply of agricultural produce. Also, the higher the volume of agricultural output produced, the higher will be the marketable production, i.e. $\partial M / \partial Q_a > 0$.

This phenomenon emerges partly because producers' consumption of agricultural goods reaches an optimum level, and partly due to a rise in the demand for industrial goods by the producers of agricultural goods which, in turn, gives rise to an increased demand for money. Obviously, a positive price policy for agriculture (which implies a relative rise in prices of agricultural goods *vis à vis* prices of industrial goods) and/or a rise in output or both will imply a rise in farm income. A part of this farm income (Y_a) will be consumed ($= C_a$)) and the other part can be saved and invested ($= S_a$). Thus we have:

$$Y_a = C_a + S_a \quad \text{or} \quad Y_a - C_a = S_a$$

Government can play an important role in mobilising the savings ($= S_a$) by a judicious policy of taxation or borrowing (see chapter 3, 4.1).

In view of the above discussion, it is not difficult to understand why economists have paid much attention recently to the problem of mobilising surplus from agriculture for promoting economic growth. The creation and distribution of surplus may vary according to different systems of land tenure in different countries. We shall now discuss some important types of land tenure in underdeveloped agriculture.

8 Characteristics of Landownership in Underdeveloped Agriculture

8.1 Share tenancy

Share tenancy may be defined as a land lease under which the rent paid by the tenant is a contracted proportion of the output produced in a given crop year. Usually, the tenant supplies labour and the landlord supplies land. As Cheung (1969) says:

> Share tenancy is thus share contracting, defined ... as two or more individual parties combining privately owned resources for the production of certain mutually agreed outputs, the actual outputs to be shared according to certain mutually accepted percentages as returns to the contracting parties for their productive resources forsaken.

However, in many cases, tenancy rights are not legally well protected and terms of contract are very complicated.

8.2 Owner–occupier

Under an owner–occupier system, the farmer owns his land and cultivates it with either family or hired labour, or both. Here, the cultivator is a landlord who is free to choose the amount of inputs necessary to realise a certain objective function (e.g. maximisation of output, revenue or profit). In most LDCs, the distribution of land is fairly uneven among the cultivators and as such sometimes a small proportion of farmers may control a very large proportion of land. Such a phenomenon also accounts for the preponderance of small farmers with small or very small plots which are not always viable. Since land is the most important asset that farmers in LDCs have, it is not surprising that most redistributional programmes for agriculture involve land reform. Note that changes in legislation pertaining to landownership would also change allocation of resources and the distribution of wealth.

8.3 Ordinary lease

Under a system of ordinary lease, one observes transfers of ownership rights which may be considered as special cases of land lease. Land lease usually implies the use of another person's agricultural land for the purpose of cultivation by oneself by paying a rent. Sometimes, as in Taiwan, the lessee is not, even with the consent of the lessor, allowed to sub-lease the whole or part of the leased farm land to another person. However, as Cheung has shown, this does not prevent other disguised forms of sub-leasing, such as joint tenancy. Also, the lessor can terminate the lease contract for his own cultivation, and can grant preferential right to lease to the original lessee; there is no constraint on the lessor's right to revise the original terms upon renewal of the contract (Cheung, 1969, pp. 12–13).

8.4 Landlord as the moneylender

The problems of tenants are aggravated by another feature of a semi-feudal agriculture – the absence of a complete specialisation. Landlords frequently act as creditors to their tenants and consumption loans are usually advanced at high interest rates. Hence the tenant leases his land from the same man to whom he is perpetually indebted and this reduces him virtually to the state of a traditional serf (Bhaduri, 1973). Thus the 'semi-feudal' landlord exploits the tenant both through usury and through his 'property rights on land'. It should, however, be conceded that empirical evidence has not always confirmed the interlocking of factors in underdeveloped agriculture (Bardhan and Rudra, 1978; Rahman, 1979; Bardhan, 1980). The problem of usury in the agricultural sector of the LDCs will be discussed later. (See chapter 8 for a detailed discussion.)

There are several other types of complex contracts that may exist between the landlords and the tenants in the LDCs. Here, we shall confine our attention to the analysis of the theory of resource allocation under share tenancy. This discussion will closely follow the arguments first developed by Marshall.

9 The Theory of Share Tenancy

Let us assume that competition exists for the use of the same property resource (land) in an economy. Let us further assume share cropping with *one tenant* first. Later, this assumption can be relaxed without loosing generality in the conclusions drawn. In Figure 4.7 let us measure the amount of land in the horizontal axis and product per unit of land in the vertical axis. Let S denote the total area of land available to a landlord, h indicate the land held by the tenant, and q be the product. The marginal product of land is then given by $\partial q/\partial h$, which falls as h rises. Let the rent ($=r$) for the landlord be 60 per cent of the yield per annum. Hence we can write: $r = 0.6$. The marginal rent curve is then given by $(\partial q/\partial h)r$. The vertical difference between $\partial q/\partial h$ and $(\partial q/\partial h)r$ is the marginal tenant income (Y_{TM}). Thus:

$$Y_{TM} = (\partial q/\partial h)(1 - r)$$

Clearly, the area below $(\partial q/\partial h)r$ is the rent for the landlord, whereas the shaded area is the income received by the tenant $[=(\partial q/\partial h)(1-r)]$, i.e. change in income of the tenant w.r.t. change in land under cultivation by the tenant. If the income of the tenant is as high as, or even higher than, his alternative source of income, he will cultivate the land as long as the MP of land is positive. In order to maximise income, the landlord will raise the rental percentage, (i.e. $\partial q/\partial h \times r$) until the income of the tenant from land equals his income from the alternative source. However, the landlord may divide his land into many parts for *many* tenants to cultivate if such an action results in a higher total rent. When the number of tenants rises, MP of land moves upwards in comparison with the MP of land when there was only one tenant cultivator. Let $(\partial q/\partial h)_1, (\partial q/\partial h)_2, \ldots$ be the MPs of land for each tenant and $(\partial q/\partial h)_1 r$, $(\partial q/\partial h_2 r, \ldots$ be the marginal rent curves for different tenants. The landlord will then optimise the gap between the integral of MPs of land and the integral of the income of tenants (see Figure 4.7). Such an action will maximise the integral of marginal contract rents. It also implies non-farm income of the tenant will be less than his farm income. It follows that with private property rights over land and tenant inputs the terms in a share contract will depend upon the proportion of rent ($=r$) and the ratio of non-land to land input which are consistent with equilibrium (for a mathematical solution, see Cheung, pp. 19–21).

For a long time, economists have discussed the efficiency of resource use under different types of leasing agreements. Traditionally, it has been argued that share tenancy (as well as short period leases) is inefficient compared with the alternatives of cash tenancy and owner cultivation. The economic case against share-tenancy rests on three premises.

(1) The incentive argument. From the point of view of the tenant, share tenancy is inefficient because terms of the contract are such that a

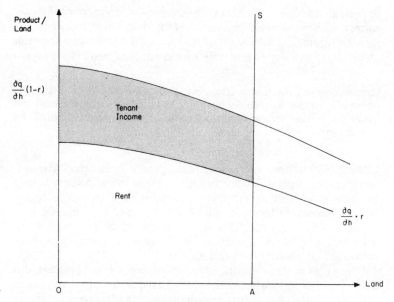

Figure 4.7: *The Marshallian theory of share tenancy*

proportion of output harvested will always be collected as rent by the landlord; hence tenants will be lacking in incentives to produce more only if tenants share is less than the marginal opportunity cost of his labour.

(2) The security argument. Tenants are usually allowed to cultivate land for a short period (typically a year or so). Such a type of contract is not helpful to create a secure atmosphere for the tenants for cultivation of the land owned by the landlord.

(3) The investment argument. Investment under share tenancy is likley to be low because landlords usually charge very high rents from the tenants. Some have considered such rents as 'exploitative' (Bhaduri, 1973). If the tenants are required to pay such high rents, it is obvious that they will not have either the ability or the willingness to invest in agriculture. However, note that the 'exploitation' of tenants may reflect monopoly power rather than a particular tenancy system.

In view of the above arguments, it is easy to understand why land reform has been strongly advocated in many LDCs. Such reforms are regarded as useful on three grounds: (a) protection of tenants from an 'exploitative' mode of production which implies drastic modification or even an abolition of tenancy; (b) promotion of a more egalitarian society with wealth (in this case, land) distributed in favour of tenants; (c) efficient allocation of resources since resource allocation under share tenancy is generally inefficient. As Chen (1961) says:

> Land reform in Taiwan was carried out at just the right time. The time was opportune because by then the landlords had outlived their usefulness and landownership has become an obstacle to further development of agriculture as well as industry.

Cheung (1969), however, argues that different contractual arrangements do not imply different efficiencies of resource allocation as long as these arrangements are themselves aspects of private property rights. Resource allocation will be different and less than optimal if the government interferes in the activities of the market for achieving a rational allocation or if property rights have been controlled or abolished altogether.

Cheung gives the following argument in support of his thesis: it is irrelevant to observe whether the landlord requires the tenant to invest more in land and charges him a lower proportion of rent or whether the landlord invests in land himself and levies a higher proportion of rent from the tenant; the investment will be made as long as this raises landlord's 'rental annuity'. Hence, for any contract, the tenant does not have to possess the required amount of input.

> If what he has is insufficient, the tenant may raise the farming inputs by cooperating with the landowner, by subleasing, by hiring farm hands, by borrowing or by joint tenancy with another family. Moreover, the different tenant input requirements in farms of different land grades and production functions will match tenants with different input endowments accordingly. (Cheung, 1969, p. 26)

However, it is worth noting that all tenant families may not be equally productive, and landlords may fail to discriminate among tenants of different efficiencies. Next, even if all land is of the same quality, the land size per tenant may vary. With difference in production function for the tenants, this will imply different rental percentages for different tenant farms to enable the landlord to maximise the rental annuity from his total landholding. Finally, if the costs of contracting are assumed to be positive, then such costs need to be included in the formulation of optimum contracts.

9.1 The Marshallian theory of share tenancy

In the classical theory as expounded by Smith and Ricardo, rent is not generally included in the cost of production. Also, the marginal analysis has not been used by the classical economists with precision to obtain an equilibrium. Marshall, however, used the marginal analysis to demonstrate clearly the tax-equivalent argument under share tenancy. The exposition is neo-classical and the following diagram is used to explain the basic Marshallian theory.

Let the amount of tenant labour ($=t$) be measured along the horizontal axis and the average and marginal product per unit of t be measured in the vertical axis. Let us assume that t is the only input available to produce output and $\partial q/\partial t$ is the marginal product. In Figure 4.8, the marginal cost of hiring an extra unit of t is given by $\partial(W_t)/\partial t$, where W_t is the current wage level. In a competitive economy, this wage level is supposed to be fixed and hence $W = \partial(W_t)/\partial t$ is a horizontal line. Equilibrium is obtained at a point like B since at this point the marginal product of tenant is equal to its marginal cost, i.e. at B: $\partial q/\partial t = \partial(W_t)/\partial t$. The number of tenants hired will be Ot_2. If we assume owner-cultivation, we derive the same conclusion 'whether the owner works up to t_2 and works elsewhere, or whether he works to less then t_2 and hires extra labourers at W.' Note that at t_2 the rent to landlords is equal to MBD, an area equal to that of a fixed rent contract.

If we assume a tax-equivalent approach to share tenancy than the marginal product minus rent will shift downward. Suppose that the landlord gets r per cent rent on the annual yield of a crop, the tenant will then be left with $\partial q/\partial t \cdot (1 - r)$. Equilibrium is now obtained at a point like A where the marginal cost of a tenant is equal to the marginal gain, i.e. at A $\partial(W_t)/\partial t = (\partial q/\partial t)(1 - r)$. The employment of a tenant is now Ot_1. Out of the total product $ODJt_1$, the landlord will get EDJA and the tenant will receive $OEAt_1$. Note that the tenant now gets an area AEM which is higher than his opportunity cost given by Ot_1 AM. Since, at A, the MP of tenant labour is higher than the marginal cost it is concluded that share tenancy is not an efficient form of cultivation. It can be easily confirmed from the diagram that the area BAJ represents the size of inefficiency in resource allocation.

The implication of the Marshallian analysis has been well summarised by D. Gale Johnson (1950):

Under a crop share lease, if the landlord's share of the crops is half, the tenant will apply his resources in production of crops until the marginal cost of crop

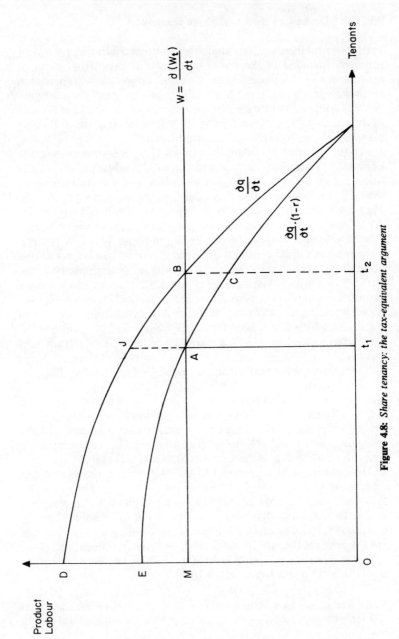

Figure 4.8: *Share tenancy: the tax-equivalent argument*

output is equal to half the value of the marginal output. The same tenant, however, will conduct his livestock operations, where important costs are borne by the landlord and the receipts are not shared with him in the usual manner. The landlord will not invest in land assets unless the value of the marginal product is twice the marginal cost. (p. 111; see also Sen, 1975)

Johnson, however, has not taken into consideration that contracting parties are free only to accept or not to accept a *contract* and that they 'can get by with' only as much as the constraints of competitions permit. Also the emergence of positive transaction costs in contracts will have the following three effects: (a) Existence of positive transaction costs will reduce the number of transactions. This will have adverse effect on specialisation in the process of production and employment; (b) such costs may influence the marginal equalities in the use of resources; and (c) the choice of contracts may also be altered due to the presence of transaction costs.

The last point deserves some emphasis. It is possible to observe three major types of contracts in agriculture: (i) a fixed rent contract, (ii) a share contract, (iii) a wage contract. Given the existence of private property rights, the contracting parties are free to choose a specific kind of contract. The nature of contract varies from country to country, for example, in Japan, a fixed rent contract was the usual phenomenon in agriculture; in Taiwan, before the land reform, a share contracts were more frequent than fixed rents (see Cheung, 1969). The choice of contract is affected by a number of economic and institutional factors. Within the economic costs, we should include negotiation costs, and the enforcement costs of controlling inputs and distributing output. It seems that total transaction costs are likely to be higher under share tenancy than a fixed rent or a wage contract. The terms in a share contract usually incorporate, *inter alia*, the kinds of crops to be grown, the ratio of non-land input to land and the rental percentage and they are generally agreed upon by the tenant and the landlord.

Under a wage contract and fixed rent, usually the landlord (i.e. single party) decides how much of the other party's resources should be employed and the types of crops to be grown. Under a share tenancy, output is usually shared on the basis of actual yield, and hence the landlord must try to monitor the yield. Thus, it is clear that under share tenancy, negotiations and enforcements are quite complicated and more costly than are under either a wage or a fixed rent contract.

If we assume risk-aversion (i.e. given the same expected mean income, a cultivator prefers a lower to a higher variance of income), it can be stated that share tenancy is a way of sharing risk. Note that under a wage contract, the *landlord* bears most of the risk whereas under a fixed contract, the *tenant* bears most of the risk. Hence in a situation of risk-aversion, share tenancy may be preferred to fixed rent or wage contracts. The existence of fixed rent and wage contracts can then be explained by differences in transaction costs between share tenancy and fixed rent and wage contracts.

10 Some Extensions of the Share Tenancy Model

It has been suggested by Cheung that the theory and evidence of inefficiency in resource allocation under share tenancy is wrong. However, he has reached this conclusion only from the standpoint of the landlord who is the maximising agent. Bardhan and Srinivasan (= BS) have used a different model where they have determined the demand side from the point of view of a maximising tenant, just as the supply side is determined from the landlord's maximising decision. Also, in the BS model, the supply of labour is determined by the tenant rather than the landlord. It has been shown in this model that share-croppers stop short of equalising the MP of labour to the wage rate and hence they are not 'efficient'. Given his assumption of a *competitive* market, one would expect r to be exogenously given. It seems that, in fact, Cheung's model can operate only when the landlord has the monopoly power to choose r. Further, Cheung assumes that the landlord can sub-divide his land, leasing out each plot to a different tenant. If this is true, then 'there is no reason why he should ignore a similar decision variable on the part of the tenant, i.e., the tenant can also choose the number of landlords from each of whom he leases in a parcel of land' (Bardhan and Srinivasan, 1971). Indeed, if it is assumed that there are a large number of landlords from whom the tenant leases in a parcel of land, then, in the BS model, it has been shown that the tenant continuously gains from sub-division. This is expected because of the diminishing returns to scale implied by a strictly concave production function.

The BS model explains a common observation in Indian agriculture: when the landlord shares the costs, the rental share he is paid is much higher than otherwise. The Farm Management Surveys in West Bengal show that under the *Bhagchasi* system the landlord does not share in the costs and receives about half of the crop produced by the share-cropper, whereas in the alternative *Krishani* system, the landlord himself covers most of the non-labour costs and usually gets about two-thirds of the crop produced by the share-cropper. It follows that government may fail to implement rent-regulating legislation and at the same time expect to induce the landlords to share more in tenants' fertiliser costs through sheer exhortations.

As regards the superiority of share tenancy over fixed rent and wage contract, the empirical evidence seems to suggest a positive correlation between the relative importance of share-cropping (measured by the percentage of total area under share-cropping) and some measure of production uncertainty. (See Rao (1971) and Bardhan (1977), for India; Cheung (1969), for pre-Communist China.) Such a correlation may, perhaps, be important to understand the nature of risk-aversion in underdeveloped agriculture and its implication for the rate of adoption of new technology. (This is examined further in chapter 6.)

References

Bardhan, P. (1977), 'Variations in forms of tenancy in a peasant economy', *J. Dev. Econ.*, **4**, 2, pp. 105-18.
Bardhan, P. (1980), 'Interlocking factor markets and agrarian development: a review of issues', *Oxford Econ. Papers*, **32**, 82-98.
Bardhan, P. and Rudra, A. (1978), 'Interlinkage of land, labour and credit relations: an analysis of village survey data in East India', *Economic and Political Weekly*, vol. 13, nos. 6 and 7, 367-84.
Bardhan, P. and Srinivasan, T.N. (1971), 'Crop-sharing tenancy in agriculture: a theoretical and empirical analysis', *Amer. Econ. Rev.* **61**, 48-64.
Bhaduri, A. (1973), 'A study in agricultural backwardness in semi-feudalism', *Economic Journal*, **83**, 120-37.
Chen, C. (1961), *Land Reform in Taiwan*, China Publishing Co., Taipei.
Cheung, S. (1969), *The Theory of Share Tenancy*, University of Chicago Press, Chicago.
Currie, J. (1981), *The Economic Theory of Agricultural Land Tenure*, Cambridge University Press, Cambridge.
Fei, J.C. and Ranis, G. (1961), 'A theory of economic development', *Amer. Econ. Rev.* **51**, 533-65.
Johnson, G.D. (1950), 'Resource allocation under share contracts', *J. Pol. Econ.* pp. 156-68.
Kuznets, S. (1959), *Six Lectures on Economic Growth*, Glencoe, Ill.

Lewis, B.A. (1954), 'Economic development with unlimited supplies of labour', *Manchester School of Economic and Social Studies*, **22**, 139–91.

Nicholls, W.H. (1963), 'An "agricultural surplus" as a factor in economic Development', *J. Pol. Econ.*, **71**, 1–29.

Rahman, A. (1979), 'Agrarian structure and capital formation: a study of Bangladesh agricultural with farm-level data', unpublished PhD thesis, Cambridge University.

Rao, C.H. (1971), *Technological Change and Distribution of Gains in Indian Agriculture*, Macmillan, India.

Sen, A.K. (1966), 'Peasants and dualism with or without surplus labour', *J. Pol. Econ.*, **74**, 425–50.

Smith, A. (1776), *A Discovery of the Wealth of the Nations*, Macmillian.

5 Agriculture in Dualistic Development Models

1 Introduction

In most theoretical models of LDCs the relationship between sectoral rather than aggregative growth rates has received considerable attention from the researchers in economic development. It has been argued that while aggregative growth models as postulated by Harrod, Domar, Solow and Kaldor (see Jones, 1976, for a lucid analysis) are useful for analysing the growth problems of the developed countries (DCs), such models are not very helpful in understanding the special problems of the LDCs. In particular, it has been pointed out that the LDCs have special 'dualistic' features (Boeke, 1953). The market structure in LDCs suffers from a great degree of imperfection – for instance, the product market is characterised by a barter system in many rural areas of Asia, Africa and Latin America. In many cases, peasants living within a predominantly agrarian society exchange their crops for industrial goods and as such the extent of monetised economy is reduced. The money and capital markets are far from being homogeneous due to the existence of organised and unorganised credit agencies. It has already been shown in chapter 3 that the labour market also suffers from considerable imperfections. The problem of disguised unemployment clearly shows the lack of a clear relationship between the marginal productivity of labour and wages.

Given such dualism, a number of papers have been written on the theme of dualistic economic growth. Central to the development of such models is the assumption that the whole economy could be divided into two main sectors: (i) an advanced sector which is largely associated with an organised industrial sector; and (ii) a backward sector which mainly comprises the unorganised agricultural or rural sector. A system of interlinkage has been assumed between the two sectors through production relationships, and then the growth of the whole economy has been demonstrated through a growth of the advanced and the backward sector. However, the growth of the agricultural sector has not been discussed in most of these dual economy models. Such models, and their policy implications, clearly rest on a set of assumptions which need to be examined. Since work on the dual economy models started with the pioneering work of Lewis (1954), it is convenient to set out his model first and then to examine later models in turn.

2 The Lewis Model

Lewis has made the following assumptions to develop his model:
(a) The economy is divided into two sectors – a backward, predominantly rural sector, and an advanced well-developed capitalist sector where the market operates reasonably well and exchange takes place. The advanced sector has also been designated as the capitalist sector.
(b) The advanced sector utilises capital stock which is reproducible, and capitalists receive payments for such utilisation. On the other hand, the backward sector utilises non-reproducible capital – for example, land.
(c) The elasticity of the labour supply is infinite in the backward sector because in most LDCs, labourers, particularly those who are unskilled, are available in abundance. This assumption is very crucial for the operation of the Lewis model as it implies that (i) the real wage rate in the subsistence sector is constant; and (ii) the marginal productivity of labour in excess supply must be approximately equal to zero. In other words, if some workers are removed from the paddy or wheat fields of LDCs total output will remain the same. (Such a situation, which has already been discussed in detail in chapter 3, has been termed 'disguised' unemployment.)

However, Lewis has acknowledged the problem of scarcity of Labour, particularly of male labour, in parts of Africa and Latin America.
(d) Due to the differences in the production technologies, it has been postulated that the output per capita is higher in the advanced sector compared with the backward sector. Given the empirical evidence available so far, such an assumption is quite reasonable.

AGRICULTURE IN DUALISTIC DEVELOPMENT MODELS

(e) Although the marginal productivity of labour in the subsistence sector is zero or near zero, wages are not related to marginal product in the backward sector. In fact, wages could consistently be above the MPLa. In the modern, industrial sector, wages are related to marginal productivity. The economy has very little capital. With an unlimited supply of *unskilled* labour, it is very difficult to promote economic growth, as the skill constraint could be very severe. However, such a constraint can be overcome via education and training (i.e. investment in human capital). In co-operation with whatever capital is available, economic growth can take place through a transfer of 'surplus' labour from agriculture to industry. It is argued by Lewis that, at the outset, a rise in the demand for labour in the industrial sector does not raise wages because the supply of labour from the backward sector is infinitely elastic with respect to wages (see, assumption (c) above).

The transfer of 'surplus' labour from the subsistence to the modern sector should be beneficial to both. After the transfer, the backward agricultural sector experiences an improved land: labour ratio; and the modern, industrial sector obtains extra labourers which it needs to increase output. Clearly, the amount of labour that can be transferred will depend upon the amount of capital stock that is available and the numbers of 'surplus' labour. The rate of transfer will depend upon the rate of growth of profits (or 'surplus') within the capitalist sector. Lewis argues that profits or surplus generated in the industrial sector are usually invested by the capitalists. This may not be always true. In fairness to Lewis though, it should be mentioned that he does recognise the possibility of some 'leakages' from profits, but these, like savings by workers, are supposed to be very small.

Fig. 5.1 shows the operation of the Lewis model. The horizontal axis measures the amount of industrial labour while the vertical axis measures the marginal productivity and wages. The MP curve measures the marginal productivity of labour. The fixed subsistence wage is given by OWs. The industrial wage is equal to OWi where OWi > OWs. It is necessary to have OWi − OWs > 0 to offer inducement to the agricultural labourers to be transferred to the industrial sector. Such an argument is, however, open to question.

The industrial sector employs labour up to the point where wages are equal to MPL. Hence P will be the initial point of equilibrium and OM amount of labour employed. The size of profit or surplus within the advanced sector is given by the difference between ONPM and OWiPM (= the total wage bill). Hence profit is equal to $W_i NP$.

Assuming that capitalists will reinvest all their profits, MPL will shift to M_1P_1 in the next phase. Employment now goes up to OM_1 and the size of profit rises to $W_iM_1P_1$. (Note that the real wage has remained fixed at OW_i.)

The process will continue to operate until all the surplus agricultural labour is absorbed into the industrial sector. In Figure 5.1, the point of exhaustion of 'surplus' labour is reached at P_0 with employment at OM_0. After that point, the level of wages will start to rise, indicated by the dashed line. This implies that after P_0, the supply of labour from the agricultural to the industrial sector will be less than perfectly elastic and the agricultural sector will be competing with the industrial sector for more labour. Lewis describes this phase of economic development as the phase of commercialisation of agriculture. It is important to point out that such commercialisation takes place because of the rise of profits in the capitalist sector, and the level of real wages tends to increase.

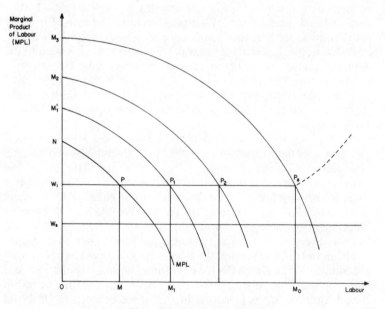

Figure 5.1: *The Lewis model*

2.1 Limitations of the Lewis model

Despite the fact that the Lewis model is one of the pioneering efforts in the development of dual economy models, many criticisms can be levelled against it.

First, the process of exhaustion of surplus labour may come to an abrupt end because of relative change in the distribution of factor shares. For example, the level of real wages may rise instead of remaining constant before the economy has reached the commercialisation point ($=P_0$) because of the operation of minimum wages laws, governmental intervention and/or trade union activities. Sometimes real wages may rise in the subsistence sector due to a rise in productivity. Indeed, 'surplus' labour can be found not only in the agricultural sector, but also in the industrial towns and cities of many LDCs. 'Surplus' labour of the *cities* then, has to be absorbed within the industrial sector and this may effect adversely the process of labour transfer from agriculture to industry. However, Lewis might reply that he never intended to model the rural – urban labour transfer process as such. He specifically stipulates that the backward sector is not confined to agriculture but also includes unorganised urban activities, like domestic service.

Secondly, many have pointed out that it is difficult to accept the view that the marginal productivity of labour is always near or equal to zero in subsistence agriculture. Many empirical studies tend to suggest that the $MPL_a > 0$. (See chapter 3 in this volume; and Mehra, 1966, and Desai and Mazundar, 1970.) If this is true then there will be positive opportunity costs due to a transfer of labour from agriculture to industry since such a transfer will tend to reduce the level of agricultural output. However, some empirical studies have suggested that a fall in labour supply has led to a fall in agricultural output and such studies tend to confirm the Lewis-type hypothesis of a 'surplus' agricultural labour (Schultz, 1964). On the other hand, others have questioned the conclusions of the study of Schultz because (a) the study is not comprehensive enough; (b) MPL_a could be positive in some farms and negative in others; (c) MPL_a could be positive or negative in different *seasons* of agricultural crop production; (d) random elements (e.g. the weather) could probably account for such a fall in output when labour has been displaced (for an elaboration of some of these points, see chapter 3 in this volume; see also Mehra, 1966 and Desai and Mazundar, 1970.

Theoretically, it is possible to argue that the rate of surplus labour absorption will crucially depend upon the reinvestible surplus. Figure

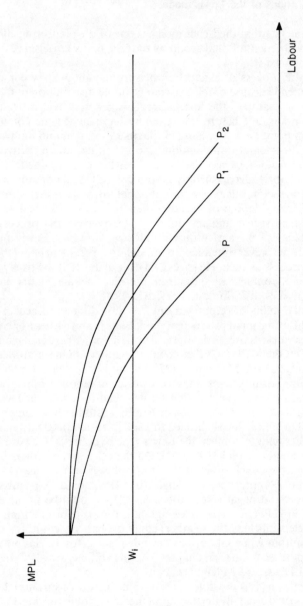

Figure 5.2: *Shifts in the MPL and the size of the surplus*

AGRICULTURE IN DUALISTIC DEVELOPMENT MODELS   103

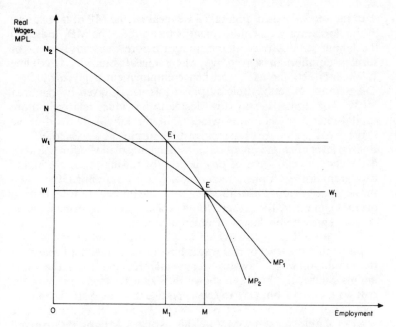

Figure 5.3: *The choice of techniques and the employment effect*

5.1 shows that capitalists reinvest their profit or surplus, i.e. NW_iP and the marginal productivity curve shifts from MP to M_1P_1. It is easy to show that the size of the surplus will be smaller if the shift of the MP curve is not parallel. In Figure 5.2 we show how this occurs. The smaller size of the reinvestible surplus will inevitably slow down the process of labour absorption. Such a situation arises if we assume that the propensity to save and invest by the capitalists is less than unity.

It is also important to demonstrate that the Lewis model operates in a specific way which allows employment to grow with the reinvestment of surplus or profit. But suppose the capitalists decide to adopt capital rather than labour-intensive techniques. Such a situation is not difficult to envisage in large parts of the underdeveloped sector where landlords may try to introduce technical progress by mechanisation (or 'tractorisation') of agriculture. An increase in capital intensity in underdeveloped agriculture will raise the MP per capita and this will shift the MP curve. But such a shift may fail to raise the level of employment: Figure 5.3 illustrates this.

In the vertical axis in Figure 5.3 we measure the MP of labour and in the horizontal axis employment is measured. The MP_1 indicates the labour productivity at the margin before the reinvestment of surplus. Equilibrium is given by E where wages (assumed as fixed) line WW_1 intersects the MP_1 and hence employment is given by OM. The amount of reinvestible surplus is, as usual, given by the area NEW. The capitalist may now decide to introduce relatively more capital-intensive techniques which shift the labour demand curve (MP) to MP_2. The size of the reinvestible surplus now rises to N_2EW. But the level of employment remains unchanged at M. Quite clearly, if the chosen techniques of production for raising profit are highly capital-intensive – a phenomenon which is not rare in LDCs – then the degree of labour absorption will be quite limited. Indeed, instead of employment creation, there could be a labour displacement which would aggravate problems of unemployment.

There are other factors which may raise the level of real wages. Suppose that the level of real wages has been raised either because of the minimum wages law or because of pressures from the trade unions. Such a rise has been shown in Figure 5.3 by a change in the real wages level from OW to OW_1. Notice that the result is a fall in the size of profit from N_2EW to $N_2E_1W_1$ as well as a fall in the level of employment by MM_1. This assumes MP_2 is the demand curve and not MP_1 (the initial position). Such a state may be described as *growth* in the national income due to a rise in capital accumulation without *real* development!

In the light of the above discussion, it is not difficult to see why the Lewis model has sometimes failed in its predictive ability. In the case of Egypt it has been reported that the Lewis model did not perform satisfactorily for the following reasons:

(a) The choice of factor proportions in the Egyptian and Taiwanese industries was much more capital-intensive than envisaged by Lewis for a smooth labour absorption. As such, the level of unemployment in Egypt and Taiwan did not show a significant fall in the 1950s and 1960s (Mabro, 1967, Ho, 1972).

(b) Lewis did not consider the important impact of a rapid rise in population growth on labour supply. Many LDCs have in fact experienced a population explosion between 1950 and 1975 because of a significant fall in the mortality rate without a corresponding fall in the birth rate (Mabro, 1967, Ho, 1972). Note that a high rate of population growth has other effects on consumption, saving, investment and growth.

In view of these criticisms others have tried to modify or extend the basic Lewis model in different ways. One of the major contributions came from Fei and Ranis (FR), which is the subject of our discussion in the next section.

3 The Fei–Ranis (FR) Model of a Dual Economy

The main aim of the dual economy model of Fei and Ranis is to demonstrate that by transferring 'surplus' labour from the agricultural to the industrial sector, an economy can be fully commercialised and developed.

At the outset, Fei and Ranis assume a Lewis-type economy which is characterised by the presence of 'surplus' labour. The level of wages in the agricultural sector is assumed to be fixed by institutional factors. The supply curve of labour in the industrial sector is infinitely elastic at the beginning since the opportunity costs of displacing labour is zero or very small. In the FR model, it has been rightly pointed out that Lewis did not pay much attention to the role of agriculture in promoting industrial and economic growth. Further, FR argue that a labour transfer from agriculture to industry should be preceded by a rise in agricultural productivity.

The FR model of dualistic economic development consists of three stages of economic growth. In the first it is assumed that the labour supply is infinitely elastic and, thus, the supply of labour is invariant with respect to changes in the level of wages. In this stage there is very little difference in the operations of the FR and the Lewis model. 'Surplus' labour with zero or near zero marginal productivity is transferred from the agricultural to the industrial sector. Figure 5.4 illustrates. Figure 5.4 shows that the elasticity of labour supply is infinity, and real wages and marginal product, measured in the vertical axis, remain fixed at a constant institutional wage (CIW). The derivation of the MP of labour is shown by the slope of the production function in Figure 5.4(b). As the slope of the production function is flat between A and D, MP is equal to zero. Figure 5.4 (c) also depicts zero marginal product between O and C. In Figure 5.4 (c) the constant institutional wage is given by OW whereas the average product is given by WVQ. Growth takes place through reinvestment of profit by the capitalist (area d^1st in Figure 5.4(a)) and the transfer of surplus labour is OC = AD. This transfer is completed when the surplus labour is exhausted and the marginal product begins to rise. This is the second stage of economic development. (See DP in Figure 5.4 (b) and CH in Figure 5.4. (c).) But wages remain higher than both average (AP) and marginal product. Since the marginal product in agriculture is positive there would be positive opportunity costs in labour transfer from agriculture to industry (because of forgone output in agriculture) and hence beyond the point t in Figure 5.4. (a) the supply curve of labour ceases to be completely elastic as it begins to rise.

The economy will be fully commercialised at the point where the $MP = CIW$. Such an equality is shown at the point E in Figure 5.4. (c) where $CIW = MP_i$. In Figure 5.4 (b) the corresponding point is indicated by the tangent to the producton function at the point R. Hence, sections DP in Figure 5.4 (b) and CH in Figure 5.4 (c) mark the completion of the second stage of economic development.

From the point of view of the landlord, the logic of such a situation is fairly straightforward. As long as the labour supply exceeds OL_2, it is in his interest to allow the labourer to leave since such a departure will imply a fall in crop yield by an amount less than the level of wages (see Figure 5.4. (a).) but below OL_2 it is in the interest of the landlord to keep the labourer on his farm. In this sense, in the FR model, the points like E, R and L_2 signify points of commercialisation.

It is possible to imagine that, at the outset, the whole population is engaged in agricultural activities. The total output is then given by $\bar{W}L$, where $\bar{W} = AP$ and $L =$ number of labourers. So long as the number of labour taken off the land does not exceed OL_0, total wages fall in proportion to the labour transfer. However, total output does not fall as the MPL is zero (indicated by the flat section of the production function in Figure 5.4(b)). The surplus is $\bar{W}L$, and this amount is proportional to the size of the labour force in urban areas. If we assume along with FR that landlords will sell all the surplus on the market, then the per capita food for the industrial worker remains the same at $\bar{W}$, which is the agricultural subsistence wage. In other words, each agricultural labour goes to industry with the same amount of food for his consumption as before. Hence, as Dixit (1970) argues, a constant industrial product wage and constant terms of trade are a consistent outcome of trade in food.

In Figure 5.4 (a), as the supply of labour in industry is greater than OL_0, a further transfer of labour from agriculture to industry will result in a fall of agricultural output. Hence, beyond this point, 'surplus' for every industrial worker begins to decrease. This will result in a rise in the relative price for food and it is now clear why the supply curve of labour in the industrial sector will begin to rise. In other words, the 'classical' Lewis-type model is valid only up to L_0; after that, such a model no longer functions for the industrial sector

AGRICULTURE IN DUALISTIC DEVELOPMENT MODELS

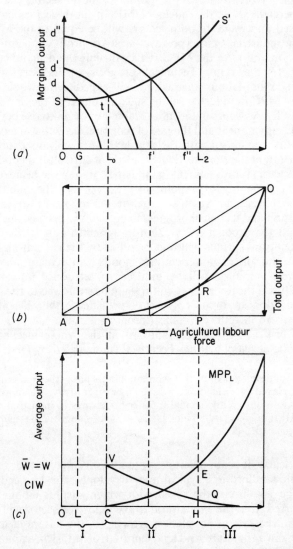

Figure 5.4: *The Fei–Ranis model of labour transfer from the agricultural to the industrial sector*

since 'unlimited' labour supply at zero opportunity costs ceases to be available. Note, that at the point L_2 both sectors have been 'commercialised' as the values of MP_1 in both sectors have been equalised and wages in both sectors will be equal to the MP_1.

In the industrial sector a positive surplus or profit will permit capital formation and raise the MPC. This will shift the demand curve for labour to the right. Technical progress in industry which is neutral/capital/labour augmenting will expedite the whole process and the MPC in industry will be such as to exceed the first turning-point at L_0. As wages in the industrial sector begin to rise beyond the point L_0, investment and the rate of economic growth will tend to fall. Even if it is assumed with FR that landlords will use a substantial proportion of their profits (from the sale of agricultural goods to the urban sector) for investment in industries, it may not be adequate to take the economy to a point like L_2, namely the point of full commercialisation. At this stage, it is important to introduce innovation and technical progress in agriculture because such an act will shift the production function in agriculture upwards. The fall in agricultural surplus will thus be offset by the generation of more output and price of agricultural goods will remain unchanged. However, landlords will be induced to introduce innovation in agriculture if the price of agricultural goods is allowed to rise initially. Subsidisation of input costs may provide further incentives to landlords to introduce technical progress.

The critical role of the landlord in the whole mechanism of economic development has been well summarised by Dixit (1970).

Successful development of this economy hinges altogether too crucially on the role played by the landlord. He should be eager to save. He should sell his surplus to industry, and should transfer his savings to industrial entrepreneurs. He should be eager to innovate, and thereby improve the technology in agriculture. (p. 342)

Quite clearly, it is not always easy to envisage the portrait of such a dynamic landlord in a typical underdeveloped agrarian economy.

In the FR model the speed with which surplus labour can be transferred from the agricultural to the industrial sector will depend upon (a) the rate of population growth; (b) the nature of technical progress in agriculture; and (c) the growth of industrial capital stock, which is determined by the growth rate of profits in industries and the growth of surplus generated within the agricultural sector.

In contrast to the Lewis model, Fei and Ranis argue that a *balanced* growth of both the industrial and the agricultural sector is necessary to avoid stagnation of the rate of economic growth. Hence at the policy level it follows that the growth of agriculture and industry should be balanced since the growth of agriculture is as important as the growth of industry; and the rate at which labour is transferred from the agricultural to the industrial sector must be higher than the overall rate of growth of population.

3.1. Limitations of the FR model

Among many comments that have been made on the assumptions and operation of the FR model, we shall discuss the following major points.

First, it can be argued that FR have not made much attempt to discuss the nature and causes of stagnation that characterise the economies of LDCs. A clear understanding of some important reasons of agricultural backwardness would have been very useful for policy formulation. Second, FR have not made any distinction between wage-based labour and family-based labour. Likewise, a comprehensive analysis of self-sustaining growth in open economies where changes in terms of trade play a crucial role would have been illuminating. Also, the role of money and prices has been neglected. It has been argued that money is not a simple subsititute for physical capital in an aggregate production function. There are reasons to believe that the relationship between money and physical capital could be complementary to one another at some stage of economic development, to the extent that credit policies could play an important part in easing bottlenecks on the growth of agriculture and industry. (See e.g. McKinnon, 1973; Shaw, 1973; Ghatak, 1981.) Note that the investment functions has not been specified in the FR model.

Third, FR assume along with Lewis that MP_L in agriculture is zero in the initial phase of economic development. Such an assumption may be called in question in view of the empirical evidence obtained so far (see chapter 3 in this volume; see also Mehra, 1966; Desai and Mazunder, 1970; Sen, 1975). Furthermore, theoretically it is possible to show that, in the first phase of the FR model, the level of agricultural output will fall except in three special cases.

(a) there is leisure satiation, i.e. people are completely satisfied with leisure and as such additional leisure consumption does not enhance the utility of the individual;

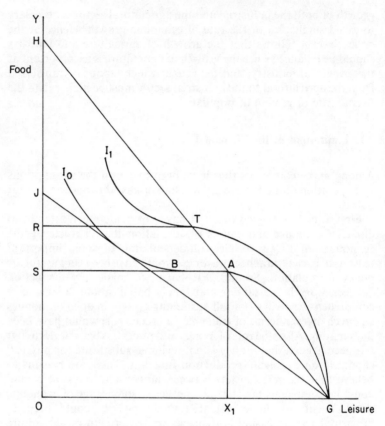

Figure 5.5: *Leisure satiation and change in agricultural output*

(b) leisure is an inferior good and as such people would like to give it up even when income rises;
(c) food and leisure are fully substitutable with one another and as such the marginal rate of substitution of food for leisure will always remain the same.

Such a fixed marginal rate of substitution will imply that the indifferences curves depicting various combinations of income and leisure are no longer convex to the origin (indicating diminishing marginal rate of substitution), but they are linear as shown in the Figure 5.5. (See Berry and Soligo, 1968 for details).

Now if we assume that the MP_L in agriculture is greater than zero, then case (a), which states that there is leisure satiation, is violated. Alternatively, if it is assumed that the $MP_L = 0$, then a perfect substitution between food and leisure is no longer a sufficient assumption for holding the total output constant. The only sufficient condition for a fixed level of output is that leisure is an inferior commodity regardless of the value of the MP_L because under such a condition a transfer of labour will always imply that those who are left on the farm will intensify their work effort (since we have assumed that leisure is an inferior good) to hold output at the same level as before. Figure 5.5 illustrates this point. Let food be measured on the vertical axis and leisure on the horizontal axis. The difference curves show the possible combinations of leisure and food (i.e. $I_0, I_1, \ldots$). The subsistence level of consumption is equal to OS. Since OG measures the maximum leisure hours, a move from G to a point like X_1 indicates a rise in the supply of labour. The transformation of leisure into food is given by the curve SAG which retains its normal properties. At a point like A, the marginal rate of transformation between food and leisure is nil, i.e. the curve is 'flat'. As the indifference curve I_0 is tangent to the curve SAG at the same point, it also indicates a zero marginal rate of substitution between food and leisure. Such a flat section of the indifference curve clearly implies leisure satiation as farmers are willing to give up infinite amounts of leisure for an extra unit of food.

If the wage rate is fixed, the transformation curve is GH. If we now assume that one unit of labourer has been transferred from the agricultural to the non-agricultural sector and his land is equally distributed among those who remain on the farm, then the curve SAG shifts to GTR. Now to obtain the same level of output the indifferences curve at point T must be flat, which will imply that either leisure is a really inferior commodity or complete leisure satiation rules. But without these extreme assumptions, total output will fall. (For a detailed discussion of these points, see chapter 3.)

There are other criticisms of the FR model. It may be argued that FR assume a closed economy model which denies the possibility of importing food and raw materials. Also, if factors like seasonalities are taken into account, then the MP_L in agriculture may not always be equal to zero. Alternatively, 'seasonal unemployment due to seasonality of labour demand may be mistaken for permanent redundancy.' Moreover, FR did not analyse the role of capital in agricultural production – an omission, which many who believe in a significant role of capital will find very difficult to overlook. In fact, some may even argue that the slow growth rate of industries can be explained by slow growth rate in agricultural productivity in LDCs. Under such circumstances, it is very important to raise agricultural productivity and surplus in the first place. As Jorgenson points out, for the expansion of the industrial sector in developing countries it is essential not only to generate an agricultural surplus but also to maintain it through technical progress (Jorgenson, 1961; 1967). The other pitfall of the FR model is that it has completely neglected the role of the *service* sector in the absorption of surplus labour. In many LDCs, the service sector has played a significant role in providing employment.

4 The Jorgenson Model

The Jorgenson model has been considered by some as the application of the neo-classical theory of growth in LDCs. Actually, Jorgenson's model has elements of both the classical and the neo-classical theory. On the one hand, Jorgenson assumes, along with Lewis and FR, that 'surplus' labour may exist in LDCs but not in the sense of zero MP_L in agriculture. The framework of analysis is still a 'dualistic' model which consists of an industrial and an agricultural sector. Total ($= Q_a$) in the agricultural sector is given by land ($= L$) and labour ($= N$). Such a production function is subject to the operation of the law of diminishing returns. It is also assumed that technical progress in agriculture is neither labour nor capital-using, that is, it is neutral. Let this production function to be given by the following equation:

$$Q_a = e^{\alpha t} L^\beta N^{1-\beta} \ldots \quad (1)$$

where $e^{\alpha t}$ measures a change in output due to technical progress. If we assume with Jorgenson that the supply of land is given then equation (1) can be written as follows:

$$Q_a = e^{\alpha t} N^{1-\beta} \ldots \quad (2)$$

Let per capita output be given by $y = Q/N$

$$y = \frac{Q}{N} = e^{\alpha t} N^{-\beta} \ldots \quad (3)$$

If we now differentiate (3) with respect to time ($= t$) and divide the equation by y, we get the following equation:

$$\dot{y}/y = \alpha - \beta \frac{\dot{N}}{N} \ldots \quad (4)$$

If $\dot{N}/N = \eta$ (i.e. rate of growth of labour) then equation (4) can be written as follows:

$$\dot{y}/y = \alpha - \beta\eta \ldots \quad (5)$$

$$\therefore \quad \dot{y} = (\alpha - \beta\eta)y \ldots \quad (6)$$

The time path of output growth is then given by the following equation:

$$y(t) = e^{(\alpha - \beta\eta)t} y(0) \ldots \quad (7)$$

It is clear that to get a positive growth of output, we should have:

$$\alpha - \beta\eta > 0$$

In many LDCs, it is reasonable to assume that the value of β will remain constant as the output elasticity with respect to changes in labour is relatively stationary. Hence, it is important either to accelerate the rate of technical progress (i.e. the value of α), or to reduce the rate of growth of population (i.e. the value of η). If possible, measures can be taken to take care of both problems at the same time to make a significant impact on poverty and underdevelopment. Note, that as long as $\alpha > \eta$ a positive growth of agricultural output per capita will be ensured. But when $\alpha = \eta$ the economy will stagnate at a low level equilibrium trap.

4.1 Evaluation of the Jorgenson model

A number of criticisms have been levelled at the Jorgenson model. We have already noticed Dixit's criticism about the nature of Japanese evidence that Jorgenson has cited in support of his hypothesis. The other important point that has been made is that when Jorgenson claims superiority for the neo-classical model, since he believes that its predictions have been broadly confirmed by the Japanese data for the nineteenth century, such a claim is difficult to accept since short-run predictions of the classical model have been compared with the asymptotic results of the neo-classical theory. Further, Jorgenson, in his description of the agricultural production function, has not included capital as an argument. On the other hand, empirical studies of Nakamura for Japan, Shukla for India and Hansen for Egypt indicate that capital has played a significant role in agricultural development. Also, true to the neo-classical tradition, Jorgenson has emphasised the role of supply factors in economic development. Some argue that demand factors can also play quite an important role in economic development. Finally, both the 'classical' (Lewis and Ranis-Fei) and the 'neo-classical' (Jorgenson) writers have overlooked the important role that service sector plays in promoting overall as well as sectoral growth of both agriculture and industry. Without proper transport and credit facilities, marketing, financial, communication, educational and maintenance services, growth rates in both sectors could be significantly reduced.

5 Kelley, Williamson, Cheetham (KWC) Model

The KWC model embodies a number of features. At the outset it is assumed that the economy is closed. Such an assumption excludes the possibility of analysing the effect of international trade and expansion of intermediate demand on industrial patterns. The economy is 'dual' – it consists of an agricultural and industrial sector each of which produces a single homogeneous commodity. While the industrial good can be consumed or invested, the agricultural output will always be consumed. Thus the KWC model is different from the Uzawa-type two-sector model where the industrial sector produces capital goods and the agricultural sector producers consumer goods.

The production technology in each sector has been described by a continuous twice-differentiable, single-valued function. Let us assume that the industrial sector is sector 1 and the agricultural sector is sector 2. Returns to scale remain constant; joint products or externalities are ruled out. Producers in both sectors wish to minimise cost and maximise profit. Hence, we have

5.1 Production technology

$$Qi_{(t)} = F^i x(t) K_i(t), y(t) L_i(t) \ldots$$
$$(i = 1, 2)$$

Note that $K_i(t) > 0$, $L_i(t) > 0$, which shows the amount of capital (K) and labour (L) currently employed in the ith sector; also $x(t) > 0$ and $y(t) > 0$ are the respective variables of technical progress. Hence, $x(t) K_i(t)$ and $y(t) L_i(t)$ may be regarded as efficiency capital and efficiency labour, respectively. Since it has been observed that the industrial sector in LDCs is more capital-intensive than agriculture, the following constraints have been imposed on the elasticities of factor substitution (σ_1 and σ_2) in industry and agriculture

$$0 < \sigma_1(t) < 1$$

and

$$1 < \sigma_2(t) < \infty$$

The elasticity of factor substitution is usually given by

$$\sigma_i(t) = F_k^i F_L^i / F^i F_{kL}^i$$

where

$$F_k^i = \frac{\partial F^i}{\partial [(x(t)K_i(t)]}$$

$$F_L^i = \frac{\partial F_i}{\partial [y(t)L_i(t)]},$$

$$F_{kL}^i = \frac{\partial^2 F^i}{\partial [x(t)K_i(t)] \partial [y(t)L_i(t)]}$$

These F^i's are really MP coefficients.

The above constraints suggest that the substitution of efficiency labour for efficiency capital is less than 1 in industry and greater than unity in agriculture.

It is important to point out that in contrast to the Jorgenson model, capital has been included in the agricultural production function in the KWC model.

5.2 Technical progress

It is assumed that rates of technical progress are exogenously given, λ_k and λ_1, respectively and they are the same in the two sectors. Note that such an assumption is rather restrictive. Hence:

$$x(t) = x(o)e^{\mu^k t}$$
$$y(t) = y(o)e^{\mu^L t}$$

It is possible to set $x(o) = 1 = y(o)$. The rate of technical progress in each sector is a weighted mean of the rate of factor substitution, the weights being equal to output elasticities of capital and labour. The current rate of technical progress in the i sector is given by

$$R_i(t) = \frac{\partial Q_i(t)}{\partial t} \cdot \frac{1}{Q_i(t)}$$

This can be written as

$$R_i(t) = \lambda_K \alpha_i(t) + \lambda_L[1 - \alpha_i(t)]$$

where

$$\alpha_i(t) = \frac{F_k^i x(t) K_i(t)}{Q_i(t)}$$

i.e. the present elasticity of output w.r.t. capital in the ith sector.

Investment equation: Assume that current net investment ($\dot{K}_t$) is equal to total investment ($I(t)$) net of replacement, which is proportional to the aggregate capital stock [$\delta K_{(t)}$]. Thus

$$\dot{K}(t) = I(t) - \delta K(t)$$

Assume that capitalists behave in a similar way in both sectors, allocate capital goods to both sectors, and $P(t)$ is the present price of industrial goods in terms of agricultural goods and $r_i(t)$ is the present rental rate of efficiency capital in the ith sector. Thus we have

$$P(t)I(t) = x(t)[r_1(t)K_1(t) + r_2(t)K_2(t)]$$

Alternatively, for aggregate investment, we have

$$I(t) = \frac{x}{p(t)} r(t) K$$

5.3 Factor demand conditions

Assume that $W(t)$ denote efficiency wages for labour (L) and $r(t)$ denote efficiency rental for capital. Factor demand is then given by the familiar neo-classical conditions of equality between factor prices ($W(t)$ and $r(t)$) and marginal productivities of inputs (i.e. F_L and F_k). Hence we write:

$$W_1(t) = P(t)F_L^1$$
$$W_2(t) = F_L^2$$
$$r_1(t) = P(t)F_K^1$$
$$r_2(t) = F_K^2$$

where 1 = industrial sector and 2 = agricultural sector.

5.4 Factor supply relation

Assume that $K(t)$ and $L(t)$ are the respective currently available, total stocks of capital and labour. The full employment condition will then be given by:

$$K(t) = K_1(t) + K_2(t)$$
$$L(t) = L_1(t) + L_2(t)$$

The labour force (L) in ith sector is supposed to grow at a given rate, $n_i > 0$. Suppose that $u(t) = L_1(t)/L(t)$, i.e. proportion of the industrial labour force in the total labour force, then

$$n(t) = n_1 u(t) + n_2[1 - u(t)] = \frac{\partial L_{(t)}}{\partial(t)} \cdot \frac{1}{L_{(t)}}$$

It is assumed that $n_2 > n_1$. Since $n(t) = \dot{L}(t)/L(t)$, the above equation can also be written as

$$\dot{L}(t) = \{n_2 u(t) + n_2[1 - u(t)]\} L(t)$$

5.5 Commodity markets

A novel and interesting feature of the KWC model is the introduction of the role of demand relationships. The commodity demand system is described by a set of equations:

$$\frac{D_{1j}(t)}{L_j(t)} = \frac{\beta_{1j}}{P(t)} [y(t) w_j(t) - \gamma]$$

$$\frac{D_{2j}(t)}{L_j(t)} = \beta_{2j} y(t) w_j(t) + (1 - \beta_{2j}) \gamma$$

Where $D_{ij}(t)$ = the total amount of the ith good consumed by the labour force in the jth sector, β_{ij} and γ_{ij} are fixed parameters. Note that γ = minimum subsistence level of per capita consumption of agricultural output. Such demand equations are based on the Stone-Geary system of linear expenditure model which aggregates perfectly over individuals in a group, if each group has the same utility function. In the Stone-Geary system, the utility function for a labour in the jth sector $(U_j(t))$ is given by:

$$U_j(t) = \sum_{i=1}^{2} {}_{ij} \log \left\{ \frac{D_{ij}(t)}{L_j(t)} - \gamma_{ij} \right\} \qquad j = 1, 2$$

The above demand relationships (i.e. $D_{ij}(t)/L_j(t)$) state that given wage income per capita in the jth sector, $yw_j(t)$ and the community price ratio, (P_t), each labour unit consumes at least γ and then he is left with his income which equals $yw_j(t) - \gamma$. This he allocates among goods in the proportions of β_{ij}.

The general form of the Stone-Geary model can be stated as follows:

$$\frac{D_{1j}(t)}{L_j(t)} = \frac{\beta_{1j}}{P_1(t)} y(t) w_j(t) + [1 - \beta_{1j} \gamma_2 \frac{P_2(t)}{P_1(t)}]$$

$$\frac{D_{2j}(t)}{L_j(t)} = \frac{\beta_{2j}}{P_2(t)} y(t) w_j(t) + [1 - \beta_{2j}] \gamma_2 - \beta_{ij} \gamma_2 \frac{P_2(t)}{P_1(t)}$$

Assume $\gamma_1 = 0$, $\gamma_2 = \gamma$, $P_2 = 1$, and we obtain the demand system which has been obtained by KWC.

5.6 Market balancing equations

The overall market balancing equations may now be given in the following sets of equations:

$$Q_1(t) = D_{11}(t) + D_{12}(t) + I(t)$$
$$Q_2(t) = D_{21}(t) + D_{22}(t)$$
$$w_1(t) = w_2(t) = w(t)$$
$$r_1(t) = r_2(t) = r$$

Note that $w(t)$ and $r(t)$ are equilibrium wage and rental rates. The above equilibrium conditions require that excess demand in any market is zero and market prices are not negative. If it is assumed that there is at least one positive value for the terms of trade that will satisfy the equilibrium conditions then it can be shown that equilibrium in the two factor markets and in either commodity market implies equilibrium in the remaining commodity market. Hence, one of the two goods market equations may be neglected; the model then consists of fourteen variables and fourteen equations. Once the equilibrium factor price conditions have been used, the basic model comprises fourteen endogenous variables and four exogenous variables: $L(t) = L$, $K(t) = K$, $x(t) = x$, $y(t) = y$.

Although the KWC model is an extension of the Jorgenson model, there are some differences between them. First, in contrast with the Jorgenson model, KWC assume that both the industrial and the agricultural sector use *capital* and labour. Second, labour supply in the Jorgenson model is exogenously given whereas the KWC, although the labour supply function is exogenous in n_i parameters, endogenously determines the rate of population growth as a weighted function of the differential rates of population growth in the rural and urban sectors. However, with economic growth urban–rural migration reduces the total rate of population growth.

There are other differences in assumptions on parameter values. Whereas Jorgenson assumes a Cobb-Douglas production function for both industry and agriculture, KWC assume that the substitution elasticity between K and L, (σ) in agriculture will be $\geqslant 1$ whereas in industry $\sigma < 1$. Third, Jorgenson assumes demand parameters as the same for both sectors. In the KWC model, such parameters are different in the two sectors. Fourth, in the Jorgensen model, wage rates in the 'traditional' and the 'modern' sectors may not be equal. Income of the peasant in the 'traditional' sector as an owner–operator comprises the per capita product given by returns to both *land* and labour. Such an income need not be equal to the income in the modern sector which consists of a return from *capital* and labour. In the Jorgenson model, the urban wage and farmer's opportunity costs are equal. But in the KWC model, capital returns and labour returns are both equalised between sectors in depreciation equilibrium. Fifth, in the KWC model, the rate of depreciation of capital is positive, whereas in the Jorgenson model it is nil (in the industrial sector, presumably, since A sector does not employ capital).

Thus the KWC model has been extended to analyse dualism not only in production technology, but also in demand behaviour and population growth patterns and as such, it clearly merits serious attention. On the other hand, the model does not discuss in depth different theories of population growth. The overall population growth rate in the KWC model is assumed to be exogenous. Also, enough attention has not been paid to the different patterns of income distribution and their impact on consumption, growth and employment (see Mirrlees, 1975). Finally, the role of institutional factors has not been mentioned as such in the model.

6 Some concluding Remarks on The Dual Economy Models

The advocates of the dual economy models have rightly emphasised the need to examine the role of *both* the agricultural and the industrial sector to promote economic growth in LDCs. Unfortunately, some have neglected the vital role that a *developed* agriculture can play in the promotion of such economic growth. Lewis, despite his seminal contribution, has not paid much attention to the mechanism of development within the agrarian sector. In addition, some of the assumptions of Lewis and FR model have not been sustained in practice. It is not difficult to understand why some of the predictions of these dual economy models have gone wrong. The crucial role of population growth has not *always* been duly emphasised (with the exception of the Jorgenson type of model). Once we include the role of population growth in LDCs, it is easy to understand why the pace of labour absorption has taken place at a very slow rate in *most* LDCs. The process of capital accumulation has been closely related to the choice of techniques in both industry and agriculture. If a technical progress is capital-augmenting (i.e. capital is used relatively more than labour) then the rate of growth of employment in industry will be slower than under a neutral or labour-augmenting technical progress. Similarly, if agricultural development largely means increased mechanisation ('tractorisation') of agriculture in LDCs, then despite some improvement in agriculture (measured by output or productivity), some labour displacement is inevitable. This, again, raises the total supply of labour, unemployment, malnutrition and abject poverty. It is important to emphasise that the failure to relate capital accumulation to the choice to techniques has significantly diminished the predictive power of the dual economy models.

The neglect of the analysis of the development of the agricultural sector has already been mentioned. It has indeed been argued that without rapid technical progress in agriculture, phase 3 of the Lewis, Fei-Ranis-type model would be characterised by hyperinflation rather than self-sustained growth (Watanabe, 1965).

It is worth emphasising that the assumption made by Lewis and Fei and Ranis, that an agricultural surplus already exists to be mobilised, is not always tenable, particularly in the least developed countries. It is, perhaps, more useful to postulate (as Jorgenson does) that the surplus has to be created within the agricultural sector by productivity improvement first.

Further, most writers of the dual economy models postulate a closed-economy model. Such an assumption has some justification in view of the minor role that has been played by trade in the economic development of LDCs. On the other hand, it is possible to argue that some LDCs may actually gain from trade. More specifically, trade in agricultural commodities could play an important role at the initial phase of economic development. In a later stage, apart from trade in agriculture, LDCs may export manufactured goods quite successfully due to the comparative cost they enjoy in the production of relatively more labour-intensive commodities.

The important role that the service sector can play in economic development of LDCs within the dual economy model has never been sufficiently emphasised; nor from the point of view of industrial development, has due allowance for leakage from the wages fund to cover the cost of marketing and distribution been made. It is widely known that the growth of such credit, marketing and distribution facilities aids economic growth considerably.

More generally, most writers of dual economy models do not pay much attention to the crucial problems of *extracting* the agricultural surplus. A more comprehensive investigation of such problems should include, *interalia*, the analysis of terms of trade, inter-sectoral resource flows, taxes and subsidies, the demand functions (as emphasised by Kelley and Williamson and Cheetham) and perhaps the money demand functions. It is no surprise why some writers have taken such a critical view of the dual economy model. Griffin, for instance, argues that the dualistic models are not very helpful because the assumptions on which they are generally based are not always plausible and the predictions of the theory wrong. On the utility of the FR model Griffin states, 'in its country of origin [i.e. Pakistan], it remains pathetically irrelevant' (Griffin, 1969).

As regards the growth of the service sector, it seems sensible to argue that it should increase at the same proportionate rate as the modern sector to avoid bottlenecks in the overall rate of growth. As Dixit (1973) rightly points out: 'If we wish to emphasise the commercialisation aspects of the division, we should classify in the modern sector those services which are available on the market and leave domestic services in the traditional sector' (p. 326).

References

Berry, A. and Soligo, R. (1968), 'Rural–urban migration, agricultural output and the supply price of labour in a labour surplus economy', *Oxford Economic Papers*, **20**, 2, 230–49.

Desai, M. and Mazumdar, D. (1970), 'A test of the hypothesis of disguised unemployment', *Economica*, **37**, 39–53.

Dixit, A. (1971), 'Theories of the dual economy: a survey', in Mirrlees, J. and Stern, N. (eds), *Models of Economic Growth*, Macmillan.

Fei, J.C. and Ranis, G. (1964), *Development of the Labour-Supply Economy: Theory and Policy*, Homewood, Ill., Irwin.

Ghatak, S. (1981), *Monetary Economics in Developing Countries*, Macmillan.

Griffin, K. (1969), *Underdevelopment in Spanish America*, George Allen & Unwin, London.

Hansen, B. (1968), 'The distribution share in Egyptian agriculture, 1897–1961', *International Economic Review*, **9**, 275–94.

Ho, Y.-M. (1972), 'Development with surplus-labour population – the case study of Taiwan', *Economic development and Cultural Change*, **20**, 210–34.

Jorgenson, D. (1961), 'The development of a dual economy', *Economic Journal*, **71**, 309–34.

Jorgenson, D. (1967), 'Surplus agricultural labour and the development of dual economy', *Oxford Economic Papers*, **19**, 3, 288–312.

Kelly, A.C., Williamson, G.G. and Cheetham, R.J. (1972), *Dualistic Economic Development: Theory and History*, University of Chicago Press, Chicago.

Lewis, W.A. (1954), 'Economic development with unlimited supplies of labour', *Manchester School*, **22**, 139–91.

Mabro, R. (1967), 'Industrial growth, agricultural underemployment and the Lewis model: the Egyptian case, 1937–1965', *J. Dev. Studies*, **3**, 4, 322–51.

Mackinnon, R.I. (1973), *Money and Capital in Economic development*, Brookings Institution, Washington, DC.

Mehra, S. (1966), 'Surplus labour in Indian agriculture', in Chaudhuri, P. (ed.) (1972), *Readings in Indian Agricultural Development*, George Allen & Unwin, London.

Nakamura, J.I. (1965), Growth of Japanese agriculture 1875–1920, in Lockwood, W. (ed.), *The State and Economic Enterprise in Japan*, Princetown University Press, N.J.

Schultz, T.W. (1964), *Transforming Traditional Agriculture*, Yale University Press, New Haven.

Sen, A.K. (1975), *Employment, Technology, Development*, Clarendon Press, Oxford.

Shaw, E.S. (1973), *Financial Deepening in Economic Development*, Oxford University Press, New York.

Shukla, T. (1965), *Capital formation in Indian Agriculture*, Vora, Bombay.

Stone, R. (1954), 'Linear expenditure systems and demand analysis: an application to the pattern of British demand', *Economic Journal*, **64**, 511–27.

Watanabe. T. (1965), 'Economic aspects of dualism in the industrial development of Japan', *Economic Development and cultural Change*, **14**, 293–312.

6 Resource Use Efficiency and Technical Change in Peasant Agriculture

In this chapter we examine the scope for increasing agricultural output and incomes in LDCs from three aspects. First, we consider how efficient peasant farmers are in utilising their existing resources and the feasibility of agricultural progress and development without technical change. Second, we consider how agricultural technical change is generated and how it affects agricultural output and incomes, including the distributional effects. Finally, we consider policy implications, particularly for agricultural research and the spread of information, and for farm mechanisation.

1 Efficiency of Resource Utilisation

In this section we attempt to clarify the meaning of various concepts of efficiency, and how these relate to each other, before proceeding to a critique of the hypothesis that peasant farmers are 'poor but efficient'. We also consider how efficiency relates to farm size specifically in the context of LDC agriculture.

1.1 Concepts of efficiency

The term 'efficient' is often used ambiguously, even by economists. So, before proceeding to examine the efficiency of peasant agriculture, we remind readers of how *technical* efficiency differs from *allocative* (*or price*) efficiency, and of how *economic* efficiency combines these concepts (Farrell, 1957; Jones, 1977–8).

Consider a simple production function in which land and labour are combined to produce a single homogeneous product. Thus in Figure 6.1, isoquant SS' is an *outer-bound* production function, i.e. given the current state of knowledge, it is not technically feasible to produce the level of output represented by SS' on an alternative isoquant which is closer to the origin, 0. Thus, any farm lying on SS', such as Q^* or Q_1, is *technically* efficient: but any farm in the space above SS', such as Q_2 or Q_3, is *not* technically efficient.

Allocative efficiency demands that factors be combined in the same ratios as their relative prices. The land: labour factor price ratio is given by the slope of the isocost curve AA'. Thus Q^* is allocatively efficient, whereas Q_1 is not: Q_2 and Q_3 may or may not be allocatively efficient, depending on whether the slopes of the (technically inefficient) isoquants at those points (not shown in the figure) equate with the slope of AA'.

In summary, then, Q_1 is technically efficient, but allocatively inefficient. Q_2 and Q_3 are both technically inefficient, and may be allocatively inefficient as well. Only Q^* is both technically *and* allocatively efficient. Defining economic efficiency as the *combination* of technical efficiency with allocative efficiency, *only Q^** is economically efficient.

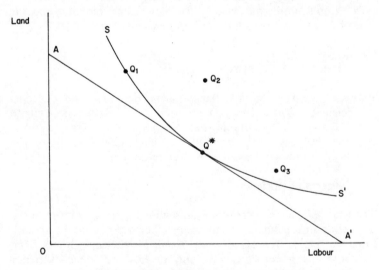

Figure 6.1: *Allocative, technical and economic efficiency*

1.2 Efficiency of peasant agriculture

The 'poor and efficient' hypothesis postulates that peasant farmers are poor, not because they utilise their resources inefficiently, but because of restrictions in the kinds and quantities of the resources they command. It is claimed that the results of a celebrated study of traditional Indian agriculture support this hypothesis (Hopper, 1965; Schultz, 1964).

Hopper fitted a Cobb-Douglas production function to empirical data derived from 43 peasant farms in a village in northern India. The data pertained to four major inputs and four production alternatives (crops).[1] Recall that fitting a C-D function 'imposes' *constant* elasticities of production on the input variables: also that

$$e_{x_i} = \frac{MPP_{x_i}}{APP_{x_i}},$$

where e_{x_i} is factor x_i's elasticity of production. Thus, with knowledge of APP_{x_i} derived from the data, and e_{x_i} from the production function, an estimate of MPP_{x_i} is obtained. By choosing an appropriate numeraire, MPP_{x_i} can be transformed into MVP_{x_i}. Using this method, Hopper derived the MVPs of the four production inputs from the data and production function parameters, together with the implicit prices of the four products (estimated at the mean levels of input and expected output). The results so obtained were then tested against the norms of neo-classical profit-maximising behaviour.

The neo-classical model postulates 3 profit maximising rules. First, that the MVP of each factor must equate with its price, i.e.

$$MVP_{x_i} = P_{x_i} \ldots \tag{1}$$

Secondly, that factors must be combined in the least cost factor combination, i.e.

$$\frac{MPP_{x_i}}{MPP_{x_j}} = \frac{P_{x_i}}{P_{x_j}} \ldots \tag{2}$$

Thirdly, that products must be combined in the highest profit product combination, i.e.

$$MVP_{x_i, y_1} = MVP_{x_i, y_2} = \ldots MVP_{x_i, y_k} \ldots \tag{3}$$

Hopper claimed that, allowing for estimation errors, his empirical estimates of the MVPs of the four production inputs, the prices of the four products, and the relationships between them, were broadly consistent with the model. He went on to make the further inference that, because they appeared to be maximising profits, the case study farmers were allocatively efficient.

The results of another much larger empirical study of Indian agriculture broadly support the Schultz-Hopper hypothesis that, even in traditional agriculture, farmers allocate resources according to the rules of neo-classical profit maximisation. The study covered nearly 3000 holdings, spread over six agricultural regions and was based on accounting records covering three consecutive years (Yotopoulos and Nugent, 1976, ch. 6). Again, a Cobb-Douglas type model was used to test the profit maximisation hypothesis, but in this case, a clear distinction is made in the model between (1) *price* efficiency, concerned with the allocation of variable factors, and (2) *technical* efficiency, as determined by fixed factors or factor endowments. Inter-farm differences in factor productivities can be expected to occur because of differences in factor endowments, such as the qualities of land, labour and entrepreneurship. In fact, Yotopoulos and Nugent found that more than 80 per cent of the observed differences in factor: output ratios amongst farms in the study sample were due to such 'systematic variation'. Less than 20 per cent was left to be explained by 'mistakes in maximisation'. It was thus concluded that 'Indian farmers... seem to be remarkably price efficient' (Yotopoulos and Nugent, 1976, p. 93). The results of this more comprehensive study are more convincing than Hopper's, not only because of the much larger size of the survey sample (and consequent greater representativeness of Indian agriculture as a whole) but also because of the clear distinction made between allocative and technical efficiency. Yet even this evidence is subject to one important caveat, namely, that no alternative resource allocating decision rule to neo-classical profit maximisation was tested. More specifically, the possibility that resource allocation in agriculture is influenced by *risk-aversion* was not investigated.

Before proceeding to criticise the 'poor but efficient' hypothesis and the empirical evidence offered in its support, we refer to its *policy implications*. Schultz and Hopper infer that because peasant farmers are 'allocatively efficient' they cannot materially increase their output (or, by implication, their income) *except through technical innovation*. They are doing the best they possibly can with their existing resources. Thus the only feasible means of enabling them to do better is by equipping them with better quality (i.e. more productive) resources. A 'technical revolution' is essential. More formally, and referring to Figure 6.1, the outer-bound production function must be shifted towards the origin.

1.2.1 Critique of the 'Poor but Efficient' Hypothesis and Supporting Evidence

Recall that the neo-classical model is static and implicitly assumes that producers are endowed with perfect knowledge. These limitations, and others, have been emphasised by critics of the Schultz-Hopper hypothesis that peasant farmers allocate resources efficiently, in the neo-classical mode. The empirical evidence offered in support of the hypothesis has been criticised on the same grounds.

The main criticisms of the 'poor but efficient' hypothesis, and its policy implications, concern: (i) the choice of a neo-classical model to represent the behaviour of peasant farmers; and (ii) the distinction between allocative efficiency and economic efficiency.

Behaviour of peasant farmers under risk and uncertainty. The essence of the first criticism is that the restrictive assumptions of perfect competition underlying the neo-classical model are inapplicable to peasant agriculture. Far from enjoying perfect knowledge and freedom to combine resources and adopt the practices needed to maximise long-run profit, peasant farmers are in fact confronted both by considerable uncertainty and by numerous institutional and cultural restraints. Moreover, being poor, they tend to be risk-averse. Recall that risk-aversion contrasts with risk-neutrality. A *risk-neutral* decision-maker is *indifferent* between accepting: (i) a certain gain of a fixed value; and (ii) an equi-probable gain or loss of the same average or expected monetary value (EMV) as the certain gain.

Given the same choice, a *risk-averse* decision-maker will express indifference between the EMV of the uncertain option and its lower valued *certainly equivalent* (CE). In the jargon of decision theory, CE = EMV with respect to a gamble signifies risk neutrality, whereas CE < EMV signifies risk-aversion. A simple numerical example will serve to illustrate the difference.

Suppose the choice lies between accepting (a) a *certain* gain of £400, and (b) an *uncertain* outcome, with an EMV of £500, where the equi-probable actual outcomes are a £2000 *gain*/£1000 *loss*; i.e. $(2000)(0.5) + (-1000)(0.5) = 500$. If the decision-maker is risk-neutral he must prefer (b), since £500 > £400. But if he is risk averse he will opt for (a), unless the CE of (b) exceeds £400.

If peasants are risk-averse, then their predominant goal is *economic survival*. Adequate stability of output and income, and the avoidance of major short-run losses, take precedence over profit-maximisation

(Lipton, 1968). The results of an empirical study of smallholder agriculture in Kenya support the hypothesis that peasant farmers do trade-off lower risk against higher profits. Or, in other words, they maximise utility rather than income, where utility is a function of the *variance* of income, as well as its average level (Wolgin, 1975).

Referring again to the three profit-maximising conditions of the neo-classical model, it was postulated that if farmers lacked perfect knowledge, and were also risk-averse, condition (2) (least cost factor combination) would still hold, but not conditions (1) or (3). Rather, given uncertainty and risk aversion, condition (1) is revised to:

$$MVP_{x_i} > P_{x_i} \ldots \qquad (1a)$$

The size of the 'safety margin', represented by the positive difference between the MVP of the factor and the factor price, depends upon the degree of risk: the larger the risk the higher the safety margin. Similarly, inequality enters the highest profit enterprise combination condition, since exposure to risk is determined in part by the choice of products and the proportions in which they are combined. Thus, if products $y_1, y_2 \ldots y_n$ are ranked in descending order of risk (i.e. y_1 is the most risky) condition (3) is revised to:

$$MVP_{x_i, y_1} > MVP_{x_i, y_2} > \cdots > MVP_{x_i, y_n} \ldots \qquad (3a)$$

Like Hopper, Wolgin estimated the MVPs of factors (land, labour, capital and purchased inputs) by fitting Cobb-Douglas production functions to farm survey data on alternative crops, classified by ecological zone. The most salient results were:

(1) The MVPs of factors were generally higher than the corresponding market prices of factors thus verifying condition (1a). As expected the safety margin was smaller for subsistence crops than for cash crops, and generally higher for purchased inputs, such as fertiliser, than for non-purchased inputs, such as family labour.
(2) Across crops, the observed ratios of MPPs of factors were broadly consistent with the corresponding factor price ratios, thus verifying condition (2).
(3) The observed ranking of MVPs by crop was found to be closely and directly correlated with the 'marginal increment to risk', a direct function of crop income variance. This observation verifies condition (3a).

A further observation, consistent with the hypothesis of risk aversion, is that most of the 1500 farmers in the survey sample in fact practised multi-cropping. In sum, the results of this study point firmly to the conclusion that if farmers are risk averse, and are obliged to make decisions in risky situations, the neo-classical model of profit maximisation mis-specifies their allocative efficiency.

Subject to the constraints imposed by fixed resource endowments (e.g. farm size), the level and variance of farm income tend to be positively correlated. The enterprises with the highest expected profits tend also to be the most risky enterprises. The risk-averse individual trades off the utility of increasing income against the disutility of increasing risk. The expectations-variance (E-V) frontier formalises this trade-off, as shown in Figure 6.2 where farm income variance, var(Y), is plotted against the expected income level, E(Y). The E-V frontiers of representative farms with 'small', 'medium-sized' and 'large' fixed resource endowments are shown by f_1, f_2 and f_3. Thus, for example, it is technically feasible for a medium-sized farm to be operated *either* at any point on f_2, *or* at any point in the space to the left of f_2. But, due to the fixed resource constraint, operation at any point to the *right* of f_2 is strictly infeasible. A prior condition of reaching the E-V frontier is that the farmer actually operates on the farm's outer-bound production function, i.e. that he is *technically* efficient.[2] However, his preferred position on the frontier will depend on his risk preference. The more risk he is willing to accept, in order to gain more income, the higher up the curve he will be prepared to go: but the greater his risk-aversion the greater the attraction of a position at the lower south-western extremity of the frontier. A neo-classical producer with perfect knowledge is *ipso facto* immune from risk. Such a producer must automatically opt for the highest attainable level of income subject only to the constraint of fixed resources. Thus, in terms of this model, neo-classical *economic* efficiency demands not only that the producer is at his E-V frontier but also that he operates at its north-eastern extremity.

A variant of the E-V frontier model was used to study the choice of cropping patterns, relative to risk, amongst a sample of peasant farmers in a district of India (Schluter and Mount, 1974). The initial step was to estimate, for each farm in the survey, the particular combination of crops likely to minimise the mean deviation of annual net income over six years, subject to various constraints. The most important constraints were annual income level, and available land (classified by productivity). A parametric linear programming model was used to derive these estimates, with the income level parameterised from zero to the maximum attainable level. The second step was to compare each farmer's *actual* position on the E-M map (with E-M used as a proxy for E-V), as determined by his actual

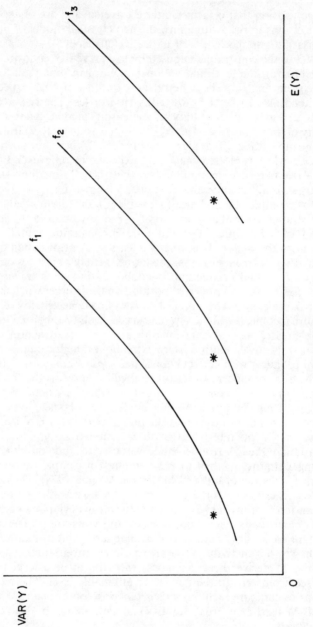

Figure 6.2: *E-V frontier model*

choice of crops, in order to assess: (i) his proximity to the E-M frontier (i.e. his technical efficiency); and (ii) his position relative to its economically efficient NE extremity and its risk minimising SW extremity.

The empirical results were that although most of the cropping patterns actually observed were quite near to their E-M frontiers (i.e. the farmers scored relatively well on technical efficiency) the observed incomes were generally much lower than the maximum corresponding with economic efficiency. Thus it appeared that most of the farmers had deliberately chosen to accept a substantial reduction in income in order to lower their exposure to the risk of income fluctuations. In terms of the simplified model of Figure 6.2, the observed positions of the farmers relative to the E-M frontier were typically as signified by the asterisks. Thus the results of this study provide further evidence that because of risk and uncertainty, peasant farmers in LDCs aim to maximise utility and *not* profit in the neo-classical mode.

If farmers were not exposed to risk (an unrealistic assumption), it would be rational for them to maximise income subject only to the constraint of fixed resources: a farmer who is *risk-neutral* can rationally adopt the same objective function. But for farmers who are *risk-averse* (due to poverty, for example) it is rational to *add a minimum profit constraint*. Such behaviour can be inferred from the results of the study by Schluter and Mount, since opting for a low income *variance* is tantamount to choosing to avoid large reductions in the *level* of income below EMV.

An alternative approach to modelling risk-averse decision-making behaviour directs the primary focus of attention away from maximising EMV to maximising the *minimum* profit at some predetermined probability or confidence level. Two variants of this principle are termed the 'safety-fixed' and 'safety-first' rules (Roumasset, 1974). The rationale of this approach is that, corresponding with each feasible farm production plan there is a frequency distribution of possible profits. The farmer's overriding concern is to avoid low profits. In the presence of risk no level of profit can be attained with certainty. But, given the *distribution* of profits, the probability of falling below some minimum profit level can be determined (from the cumulative frequency distribution of profit levels).

Under the safety-fixed rule, the objective is to maximise d subject to:

$$\Pr(\Pi < d) \leq \bar{\alpha}$$

where Π = profit, d = minimum profit and $\bar{\alpha}$ = probability (confidence) limit. The cumulative frequency distribution of profit is given by:

$$F_\Pi = \int_{x=-\infty}^{\Pi} p(x)dx$$

where x = level of profit and Π = a particular level of profit.

The cumulative frequency distributions of two hypothetical production alternatives, A and B, are depicted in Figure 6.3. In choosing between the alternatives, A maximises d at probability limit $\bar{\alpha}$ (since $d_2 > d_1$) even though $E(\Pi)_B > E(\Pi)_A$ (as signified by the difference in the magnitudes of the two shaded areas between the curves).

Under the safety-first rule, the objective is to maximise Π subject to:

$$\Pr(\Pi < \bar{d}) \leqslant \bar{\alpha}$$

where $\bar{d}$ = a specified 'disaster level' of profit. Suppose that in Figure 6.3 $\bar{d} = d_2$. Under production alternative B:

$$\Pr(\Pi < \bar{d}) = \alpha$$

Since $\alpha > \bar{\alpha}$, production alternative A maximises $E(\Pi)$ subject to the minimum profit constraint $\bar{d}$ at the probability limit $\bar{\alpha}$. Thus, again, because of risk-aversion, A is preferred to B even though $E(\Pi)_B > E(\Pi)_A$.

The parameter $\bar{\alpha}$ is determined subjectively by the decision-maker, but might be set at 0.05 or 0.10. The safety-fixed model was used, with others, in analysing fertiliser application rates, and the use of other cash-intensive techniques, by peasant rice farmers in the Philippines. In this case, the S-F principle did not withstand the test of empirical verification particularly well for two reasons. First, it was found that, generally speaking, yield and profit distributions were such that the goals of profit (EMV) maximisation and 'safety' were *not* in competition. That is, there was no strong positive correlation between average profit (EMV) and risk. There was consequently no strong *a priori* reason for expecting these farmers to show risk-aversion, and, in fact, a *risk-neutral* EMV maximising alternative model fitted the data at least as well, or better, than the S-F model. The second reason why the S-F model failed to carry conviction in

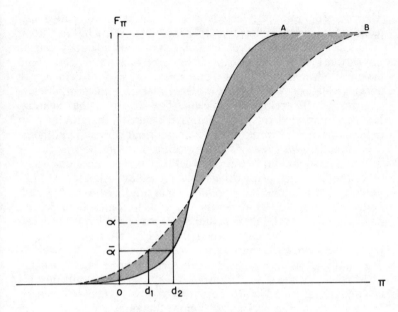

Figure 6.3: *Safety fixed model*

this case was that it was found that, despite the inherent risks of losses due to crop failures or low prices, some poor farmers were prepared to gamble on the use of cash intensive practices in the hope that a favourable pay-off would enable them to get out of debt (Roumasset, 1976, pp. 213–14). Hence there is empirical justification for rejecting the facile assumption that all poor farmers are *necessarily* risk-averse.

Where risk-aversion *is* characteristic of the behaviour of peasant farmers, the actual existence of major risks and uncertainties must constrain agricultural output and incomes in LDCs. Assuming that it is in the national interest to relax that constraint, the obvious policy implication is that measures to reduce the riskiness of farming are needed. So, for example, minimum price guarantees, or better market outlets, or readier access to agricultural credit might be introduced.

Technical efficiency of peasant agriculture. We turn now to the second major criticism of the Schultz-Hopper view of the efficiency of peasant agriculture, namely, that by neglecting the distinction between allocative and economic efficiency, it takes technical efficiency for granted. Recall from the beginning of this chapter that whereas economic efficiency is conditional upon allocative *and* technical efficiency, allocative efficiency is not conditional upon technical efficiency, or vice versa. So, economic efficiency can be proved only by providing explicit evidence of both allocative and technical efficiency. It is not adequate to use explicit evidence of allocative efficiency as implicit evidence that the same firm, or class of firms, is also technically efficient. The allegation that Schultz-Hopper committed this error forms the nub of the argument in criticism of their attempt to prove the poor but efficient hypothesis.

Consider *a priori* reasons for expecting firms to be efficient. A common assumption is that competition will compel efficiency, otherwise firms risk being driven out of business. Another assumption of western capitalist economics is that producers are, above all, motivated by self-interest and the desire to maximise their private profits. However valid these assumptions may be in the West, it can be questioned whether they are equally valid in LDCs, particularly in peasant agriculture (Shapiro, 1977). One argument is that, particularly in a mainly subsistence-type agriculture, competitive forces are likely to be relatively weak. If a farmer gets into economic difficulty his neighbours are more likely to help than to drive him out of business. The same reasoning may apply even to the moneylender who may prefer an extension of credit to foreclosure. A second and more familiar argument is that the maximisation of utility and money profit coincide only so long as the marginal utility of money income exceeds the MU of other satisfactions. In reality, money income competes with other satisfactions such as leisure, fulfilment of social obligations, and the protection of social status. Providing that a farmer or landowner has a 'satisfactory' income, by the standards of his own community, there is no obvious reason why acquiring more income should take precedence over adding to other satisfactions. Bearing in mind that peasant agriculture is labour-intensive, that physical labour is arduous (particularly in the tropics) and possibly socially demeaning also, the notion that farmers in LDCs do *not* typically strive to maximise profits possesses a considerable commonsense appeal. This is so, quite apart from the *risks* of profit maximisation as previously discussed in this chapter. But what does choosing not to maximise profits imply for efficiency? Clearly, non-maximising farmers cannot achieve economic efficiency but can they be allocatively or technically efficient? The reply to this question is that, given the full information they need on MVPs and factor prices, they can achieve allocative efficiency. A farmer allocating his resources efficiently at some level of production or income *below* the outer-bound production function is effectively maximising profit at a sub-optimum level of production (or work-effort intensity) either by choice or due to incomplete knowledge of available production alternatives. This, then, is a rationalisation of allocative efficiency combined with technical inefficiency. The reverse combination – allocative inefficiency combined with technical efficiency is less plausible. A farmer with a full knowledge of technical production possibilities is unlikely to lack the knolwedge and ability to equate real marginal productivities with real factor prices. A study reported by Shapiro (1976) attempted to subject the technical efficiency of Tanzanian cotton farmers to objective assessment.

For the purpose of this study an outer-bound maximum technical efficiency production function was derived by linear programming form the application of the 'best' farm practices in use amongst farmers in the sample. These best practices, relating to fertiliser rates, numbers of sprayings, and similar matters, also accorded with agricultural extension service recommendations. A distribution of technical efficiency scores, ranging from 0–1 (maximum score) was derived by comparing each farmer's actual output with his predicted output on the outer-bound production function. The results showed only a small minority of farms with scores at or very near the maximum, signifying that the majority were *not* operating at the production frontier. The average score was only 0.66 and it was estimated that, had all the farmers modified their practices so as to move to the outer-bound production function, the aggregate output would have been 51 per cent higher. This points to the conclusion that, contrary to the implications of the poor but efficient hypothesis, scope *does* exist in peasant agriculture for improving farm productivity and incomes by persuading farmers to make simple improvements in crop management practices. This conclusion is supported by other evidence (World Bank, 1978, pp. 39–40). But it is facile to suppose that merely by improving the quantity and quality of agricultural extension and farmer education, the performance of every farmer can be raised to the level of the 'best'. Thus the authors of another recent World Bank publication observe:

Best traditional practice is visible to all farmers and to the extent that they are able and willing to emulate the example, they can be expected to do so without extension advice. This suggests that differences in natural ability and assiduity are responsible for part of the observed differences in yields, and differences in land quality, structure of the family labour force, etc., for another part. The part due to information gaps is probably relatively small. (World Bank, 1981, p. 75, fn. 39)

There is an obvious danger in attempting to estimate the production function relating to an individual farm or farmer from a blueprint based on aggregate data or 'average practice'. In agriculture, more than in most other industries, each enterprise is in some sense unique. Farms differ in their physical characteristics and farmers certainly differ in their capacity for farm work and their innate entrepreneurial ability. Although most farmers may be capable of improving their management through learning by experience, there is nevertheless a sense in which each individual is uniquely constrained by his own limitations and those of the farm. Thus, in analysing farming efficiency, there is much to be said for adopting a disaggregated approach which does not implicitly assume either that all farms are capable of being operated at the same absolute level of technical efficiency or that all farmers are capable of reaching the same absolute standard of efficiency. The pitfalls of attempting to apply standardised efficiency norms in agriculture are exemplified by evidence on fertiliser application rates in the Philippines (Roumasset, 1976, ch. 7). The finding that, on average, actual application rates were only about one third of the amounts recommended by the extension service raised the question of why farmers applied fertiliser at less than the recommended rate. The explanation of risk-aversion was ruled out by the poor explanatory power of the S–F model, as already discussed. An obvious alternative is the 'learning lag' explanation. It is unrealistic to suppose that all farmers possess full knowledge of current recommendations, either by contact with the extension service or through 'learning by experience', particularly in LDCs where barriers to the rapid dissemination of information exist. However, the study authors considered that the primary explanation was that the *recommended* fertiliser rates were appropriate only for farmers operating under the most favourable physical and economic conditions. They were *not* appropriate for the majority of farmers in the study who operated in less favourable conditions. In other words, fertiliser recommendations were based on 'idealised' production functions ascribing to farmers higher levels of production intensity and technical efficiency than their fixed resource endowments would permit. A common error is to base extension recommendations on the results of field trials which, for a mixture of technical and economic reasons, commercial farmers are unable to match.

1.2.2 Efficiency and Farm Size
Strictly speaking, the theory of returns to scale is concerned with the relationship between the firm's level of output and its long-run

average costs, *when all factors of production are varied in the same proportion*. In agriculture, scale theory has limited relevance because of factor rigidities and indivisibilities, even in the long run. In seeking to gain profit or utility by changing their level of output, farmers typically *vary* the proportions in which factors are combined. For example, varying the fertiliser rate alters the fertiliser: land ratio. An important potential source of cost saving in agriculture is the fuller utilisation of 'spare capacity' embodied in indivisible factors such as machinery and buildings. The term 'economy of size' embraces not only economies of scale, in the strict sense of constant factor proportions, but also economies derived from using indivisible factors more efficiently.

Indivisibility applies mainly to capital inputs. However, the principal inputs of traditional agriculture are land and labour. Generally these are relatively easy to divide, though we recognise that individual farmers may experience difficulty in gaining access to additional land, and that some farms are too small to provide full-time employment. Thus, *a priori*, traditional agriculture might be expected to show constant returns to scale, i.e. differences in output between farms are broadly proportional to the corresponding differences in the inputs of both land and labour. However, the constant returns hypothesis is subject to two important assumptions: first, the assumption of factor divisibility, as already discussed; and second, the assumption of common factor prices, regardless of farm size. The second assumption requires some elaboration.

If, in fact, the real price of land declines with increasing farm size, whereas the price of labour increases, the land: labour ratio will tend to vary directly with farm size, i.e., *ceteris paribus*, labour will be used most intensively on the smallest farms. Under a labour-intensive system of agriculture, where output per unit of land area is closely and directly related to labour input, the productivity of land is largely a reflection of the degree of labour intensity. In this situation the productivity of land, as reflected by crop yields, for example, will tend to be *inversely* related to farm size. If land is scarcer than labour, which is often the case in LDCs, this argument carries the important policy implication that smaller farms may utilise resources more efficiently than larger farms. Moreover, there may be a direct link between a country's farm size structure and its level of aggregate agricultural production. In an LDC context, a large farm structure may not be consistent either with efficient resource utilisation or with achieving the optimum level of aggregate production.

Due to measurement problems, empirical evidence on differences in the prices of land and labour between large and small farms in LDCs is virtually impossible to obtain. However, there are sound *a priori* reasons for expecting labour to be relatively cheap and land to be relatively dear to the operators of *small* farms, with the obvious implication that the reverse situation of dear labour and cheap land applies to *large* farm operators. The rationale of the wage differential has already been explained, under labour market dualism, in chapter 2, pp. 8–10. The expected land price (and rent) differential is also a consequence of inequality within the agricultural sector. By behaving monopsonistically, large-scale farmers tend to have preferential access to the land market where they buy or rent land relatively cheaply. But small-scale farmers are obliged to pay a higher price or rent because of their relatively weak bargaining position in the market. This is also discussed at greater length in chapter 2, pp. 20–1.

The results of two major empirical studies support the hypothesis that in traditional agriculture agricultural output per unit of land area does tend to be inversely correlated with farm size. In an extension of the production function study referred to earlier in this chapter, Yotopoulos and Nugent (1976, ch. 6) compared the relative economic efficiencies of 'small' and 'large' farms in six agricultural regions of India. Their results appeared to justify the rejection of the null hypothesis of equal efficiency between the two groups of farms and the sign of the farm size dummy variable indicated that, subject to fixed factor constraints, small farms were more profitable (i.e. efficient) than large ones. It was deduced that the observed difference in economic efficiency in favour of the small farms was due to their superior *technical* efficiency, since, as already remarked, *all* the farms covered by the study, regardless of size, appeared to be price efficient. Thus the small farms produced more per unit of scarce resources. But note, following our earlier discussion of the technical efficiency of peasant agriculture, that the results of this study are concerned only with the *relative* efficiency of large and small farms: they throw no light on how the actual performance in either group compared with any *absolute* standard of technical efficiency, based on complete knowledge of available production alternatives and the ability to select and apply the one corresponding with the outer-bound production function.

The second source of empirical evidence on the relationship between the intensity of land use and farm size in traditional agriculture is a set of cross-sectional regression studies of land productivity across farm size groups in six LDCs conducted by Berry and Cline (1979, ch. 4). The countries were Brazil, Colombia, The Philippines, West Pakistan, India and Malaysia. Allowance was made for inter-farm variations in the quality of land, either by using the price of land as a deflator or by the separation of irrigated from non-irrigated land. In all six countries the statistical results appeared to confirm the existence of a negative relationship between farm size and output per unit area of constant quality land. Output per land area unit is only a partial productivity measure. However, where land and labour are the dominant factors of production, as in traditional agriculture, the productivity of land may serve as a reasonable proxy for total factor productivity, especially if the marginal opportunity cost of labour is low. Parallel estimates of total factor productivity across farm sizes in the same group of countries, at estimated social prices of factors, were broadly consistent with this expectation, especially with a zero social cost imputed to labour.[3] As emphasised by the authors of this study, there are at least three components of any change in agricultural output per unit of land area. These are crop yield, cropping intensity and product mix. Thus, in principle, output per unit of land can be changed *either* by changing the yields of specific crops, *or* by changing the frequency of cropping (as reflected by the 'cropping index' which expresses the total area of crops actually grown per unit time period as a proportion of the total area available for cropping), *or* by re-allocating the available land amongst crops with different unit area output values. So, evidence on comparative crop yields alone provides an insufficient basis for measuring the relationships between land use intensity and farm size. Indeed, available evidence suggests that the observed difference in land productivity between large and small farms in countries with a traditional agriculture derives principally from differences in cropping intensity and product mix, rather than from crop yield disparities as such (Berry and Cline, 1979, pp. 13–16).

The empirical observation that, under the conditions of traditional agriculture, output per unit of constant quality land is inversely correlated with farm size appears to remove one of the stock objections to land redistribution, or land reform, in LDCs, namely, the objection that, because small farms are allegedly less efficient than large, land reform is likely to *reduce* the aggregate level of agricultural production. However, even in LDCs, a policy of adopting the small farm as an efficiency norm is limited by two considerations. First, since the superior efficiency of small farms is confined to labour surplus economies, that superiority is destined to disappear as labour becomes scarcer and its opportunity cost rises. Second, and even more importantly, there are sound *a priori* reasons for doubting whether the superior efficiency of the small farm can be sustained in the face of agricultural technical change in LDCs. Virtually all the evidence on the superior efficiency and profitability of small farms, such as that provided by Yotopoulos and Nugent and Berry and Cline, pertains to traditional agriculture and pre-dates the 'Green Revolution'. This represents a major caveat to the findings of these studies.

Although there is some justification in the argument that, being readily divisible, some modern farm inputs are 'scale neutral', this claim cannot be made with respect to other inputs associated with the first group. So, for example, improved (i.e. higher yielding) seeds and chemical fertilisers contrast with irrigation equipment and field machinery such as tractors and mechanical harvesters. Even if large- and small-scale farmers have equal access to 'modern' agricultural inputs (which is doubtful), yields on large farms can be expected to be at least as high as yields on small farms. But, due to the indivisibility of fixed capital inputs which are an integral part of the new technology, unit costs must tend to be lower, and profits higher on the larger farms, save where hiring the services of indivisible or 'lumpy' inputs such as field machinery or cooperative machinery ownership are feasible options for small-scale farmers. To the extent that a higher profit per unit of output reflects superior efficiency, large farms may well become more efficient than small as a consequence of the modernisation of traditional agriculture. Unfortunately, there is a dearth of empirical evidence on the relative efficiency of large and small farms in LDCs *in areas where modern agricultural technology has been adopted.* So, it is not possible to 'prove' that small farms have lost (or are losing) their traditional advantage in terms of 'efficiency' where the scarcest resource is land. But the *a priori* argument for expecting such a trend in all LDCs affected by the Green Revolution is compelling.

Returning to the issue of farm size policy in LDCs, a dynamic view anticipating technical change might lead to the rejection of a policy of uniform farm size in favour of one of size diversity. There are at least two arguments in favour of diversity (Bachman and Christensen, 1967, p. 255). First, since larger farmers generally lead in adopting new methods, it might encourage the more rapid adoption of technological changes in traditional agriculture. Second, because types of agricultural enterprises vary in the scope they afford for economies of size (e.g. contrast plantation type crops produced primarily for export with domestic food crops), an adequate degree of flexibility in farm size policy is desirable in any case.

Even though *production* activities in traditional agriculture may tend to show constant returns to scale, due to the absence of major capital inputs, the constant returns hypothesis lacks credibility with respect to the provision of external credit. Because a good many categories of marketing cost are fixed and indivisible (e.g. the depreciation of transport, storage and processing equipment), the relationship between total marketing costs per sales unit and total sales volume must be inverse, all other things being equal. Thus, larger farmers with more to sell must enjoy a marketing cost advantage compared with smaller producers with less output and low sales. The economies of large-scale marketing operations tend to be especially marked with respect to tropical export crops such as coffee, tea, sugar and bananas (Bachman and Christensen, 1967, p. 250). A similar argument applies to the purchase of farm requisites, such as seeds, fertilisers and pesticides. Suppliers are often prepared to reduce the price for larger-sized deliveries, quite apart from the possibility that larger farmers can exert more bargaining power. The argument can also be extended to the procurement of agricultural credit. Not only do larger-scale farmers tend to enjoy preferential access to credit, particularly from institutional sources, they are also in a better position to bargain for or otherwise obtain loans on preferential terms.

There are two obvious policy approaches to ameliorating or removing the small farmer's external services cost handicap. The first approach, favoured in socialist countries, is 'collectivisation', under which large numbers of formerly independent small farms are merged to form very large-scale production and marketing units with centralised administration and control. The rank-and-file members of the collective lose virtually all of their former independence with respect to both production and marketing. The second approach is 'co-operation' for the procurement of specific agricultural services such as marketing and credit. Under this approach farmers' resources are pooled and their independence compromised only in relation to the provision of specific common services which are expected to yield economies of scale, or even enable small producers to exert some group bargaining power. In this way, small farmers aim to procure services of the standard enjoyed by large farmers at a similar cost. Co-operative mambership may be either voluntary or compulsory, provided that government is prepared to legislate the necessary powers of enforcement. Whereas voluntary co-operatives tend to be more actively supported by their members, compulsory co-operatives or, as they are more usually termed, marketing boards, are generally able to exert great bargaining power in negotiating prices. A marketing board may even enjoy monopoly trading rights.

Collective farming, the principles of co-operative marketing and credit, and the activities of agricultural marketing boards are all discussed at more length in chapter 8. The Chinese system of collective farming is referred to in chapter 11.

2 Technological Change in Agriculture

We open this section by examining the character of agricultural technical change, with special reference to HYV technology. We then consider the generation of technological change in agriculture, including the theory of induced technical and institutional change. Next, we review the origins, spread and economic consequences of HYV technology in LDCs, including both direct and indirect effects together with relevant empirical evidence. This leads on to considering the distributional consequences of factor biases embodied in new agricultural technology, both in theory and specifically with respect to the effects of farm mechanisation on agricultural employment in practice. Finally, we consider policy implications relative to the discovery and selection of new agricultural technologies that are appropriate to the economic circumstances of LDCs, emphasising the role of government.

2.1 Nature of technological progress in agriculture

A technological improvement possesses two general properties. First, a new production function is created such that any given quantity of resources yields a larger product, at least above some threshold level of resource input. Second, the proportions in which resources are combined to produce a given output at least cost are generally changed. We now proceed to examine the first property in more detail, but defer further discussion of the second property to section 2.6 below.

Technological progress in commonly conceptualised as a 'shift variable' which, in graphical terms, shifts the production function positively or 'vertically'. In Figure 6.4(a) and (b), X represents any variable input (with all complementary inputs assumed fixed) and Y output. The functional relationship between X and Y depends upon the choice of technology, with t_2 representing an improvement on t_1.

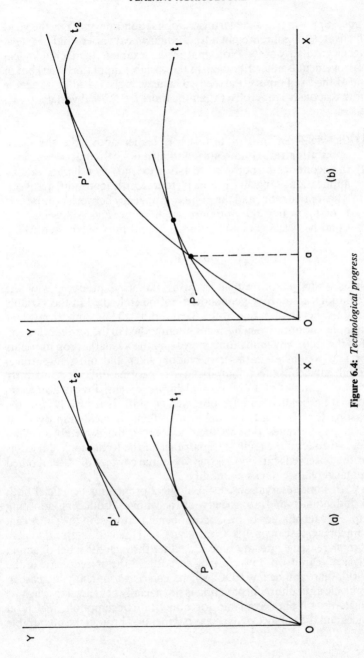

Figure 6.4: *Technological progress*

Thus, supposing X is the fertiliser application rate and Y is the yield of wheat, t_1 might represent a traditional wheat variety and t_2 a new, higher yielding variety. Note that in this example the technological improvement is not 'embodied' in the variable input (fertiliser) but in one of the fixed inputs (the seed). The new, higher yielding variety is more responsive to fertiliser than the more traditional variety in two senses:

(1) the yield per unit of fertiliser is higher above the threshold application rate (0 in 6.4(a), and a in 6.4(b)); and
(2) the economic response to fertiliser extends to a higher rate of application: tangents p and p′ represent a (constant) fertiliser: wheat price ratio, and the points of tangency between p(or p′) and t_1(or t_2) signify the optimum application rates (subject to the usual neo-classical assumptions), i.e. profit is maximised where

$$\delta y/\delta x = P_x/P_y.$$

There is no reason, either in principle or in practice, why the technological improvement should not be embodied in the variable input (rather than in one of the fixed inputs). Thus, to return to our example, the improvement might be embodied in the fertiliser instead of in the seed. Any innovation increasing the available crop nutrients per unit weight of fertiliser (including filler and other impurities) would have this effect. Empirically, it may be difficult to identify precisely sources of technological improvement. Production functions are estimated and are observed to shift dynamically. But the reasons for the shift are unknown. Technological improvement effectively becomes 'disembodied'. However, this is merely an expedient to conceal the analyst's ignorance of the technological reasons for the observed shift and, in this sense, the concept of 'disembodied technical change' lacks credibility.

With some reservations, technological progress is beneficial both to individuals and to society as a whole. Producers adopting improved techniques of production benefit, at least in the short run, from an increase in profits. Consumers, and the nation, stand to gain from increased aggregate supplies, either through the relief of actual physical scarcity, or lower prices, or both. In the present state of the world, there can be few LDCs where, *ceteris paribus*, any increase in domestic agricultural production is not socially beneficial through its effects on supplies available for domestic consumption, real food prices and the balance of overseas payments. Through the favourable effect on crop yields, even *subsistence* farmers stand to gain from adopting HYV varieties. They are able to produce more for their own consumption with the same effort (or the same amount with less effort). *Commercial* farmer benefits are largely confined to an increase in producer surplus, which may be eroded as time passes. If the rate of growth in aggregate food supplies should eventually excceed the rate of growth in aggregate food demand, agricultural product prices must decline in real terms to the detriment of commercial farmers and the benefit of consumers. Whereas commercial farmers as a group suffer a loss of producer surplus, there is a gain in the surplus accruing to consumers. As the sole or principal consumers of their own output, *subsistence* farmers also benefit from any gain in consumer surplus. But, selling little or nothing, they are unaffected by losses (or gains) in producer surplus (Hayami and Herdt, 1977).

But there tend to be losers as well as gainers from technological progress–hence the reservations. As far as the modernisation of agriculture is concerned, the possible losers include farmers who are unable or unwilling to adopt better methods. They may also include landless labourers, and even some classes of domestic food consumers. We discuss the contrasting effects of land-augmenting and labour-displacing agricultural technologies on output, employment and income distribution in section 4 of this chapter. Possible barriers to the adoption of improved agricultural technology in LDCs are fully discussed in chapter 2 section 1.3.2. As shown there, vulnerability to such barriers is particularly marked in the case of large numbers of peasant farmers operating on a very small scale, due to the combination of inadequate farm size, poverty, risk-aversion, imperfect knowledge and other market failures resulting in discrimination against small producers. Non-adopters are unable to share the benefits of increased physical production and productivity (especially of land) that are enjoyed by adopters. If they produce a marketable surplus, their lower physical productivity (e.g. in crop yields) is particularly costly in terms of income forgone if, due to a shift in the *overall* balance of aggregate supply and demand, there is a real decline in the market price. Although, *ceteris paribus*, adopters and non-adopters alike suffer an *absolute* loss of income because of the price decline, adopters remain *relatively* better off due to the buffer provided by their larger output and sales volume. The non-adopters lack this buffer.

It is also recognised that the benefits of technological progress are more certain for landowners than for tenant farmers. The latter, whether they are potential adopters of the improved technology, or not, risk dispossession by their landlord. The increased farm profits afforded by technological advances give landlords an added inducement to re-possess tenanted land in order to farm it themselves. Even without repossession, a major proportion of the gains from the adoption of new technology is likely to accrue to landlords through increased rents. Rent equilibriates the demand for rented land with the available supply. The demand for rented land function shifts to the right as the expected level of farm profit rises. But the supply is inelastic for two reasons. First, the area of land beyond the existing extensive margin of cultivation is extremely limited, especially in overpopulated countries. Second, landlords have a greater incentive to retain, or even repossess land for their own cultivation. Hence most of the excess demand for agricultural land resulting from technological progress in farming (and the consequent rise in farm profit expectations) is removed, not by any substantial increase in the supply of land suitable for cultivation, but by the escalation of rent to a higher equilibrium level. Tenant farmers must pay a higher rent to secure (or retain) tenancies. Landlords benefit from an increase in wealth as well as from a higher rental income. Since, theoretically, the value of land is merely its capitalised rent, a landowner's wealth must rise as a consequence of each rent increment. A principal argument for land reform, under which the ownership of agricultural land is transferred from large landlords to their former tenants, is to achieve a more equitable distribution of the benefits of technological progress in agriculture. However, even after an equalising land reform, technological progress still drives up the price of land to those wishing to enter agriculture by farm purchase, including landless labourers. With land wholly or mainly under private ownership, controlling the price or taxing windfall gains is fraught with many difficulties, including the undesirable consequences of discouraging landowners from offering land for sale.

Despite reservations concerning the distribution of its benefits, there can be little serious doubt that LDCs are considerably better-off with technical advances in agriculture than without them. In terms of the relief of hunger alone, the Green Revolution must have been beneficial in the countries affected by it (Lipton, 1978). Under favourable conditions, nearly all categories of farmers in LDCs, regardless of farm size, tenure, access to credit, attitude to risk, and other constraints, can benefit from adopting HYV technology. Despite a tendency for smaller farms to lag behind larger farms in the adoption process they eventually catch up (Feder et al., 1981). As discussed in chapter 2, the consequences of technological stagnation are continuing low labour productivity on farms and, for most farmers, very low and stagnant incomes. Because of the scarcity of certain agricultural resources – particularly land – the rate of aggregate agricultural growth and development in most LDCs is critically dependent on the rate of technological progress. Moreover, due to agriculture's critical role in fostering overall development in LDCs, as discussed in chapter 3, the importance of an adequate rate of technological progress in farming extends beyond agriculture itself to the national economy as a whole.

3 Generation of New Agricultural Technology

Technological innovation is a two-stage process of invention or discovery, and adoption of the improved input or method of production by producers. Although an innovation can have no economic impact unless and until it is adopted by producers, adoption is logically preceded by invention. Despite the recognition by economists that scientific discoveries and inventions can explain faster economic growth and development (or their absence explain economic stagnation) the traditional approach to modelling economic growth has been to treat invention as an exogenous shift variable. In other words, traditionally, there has been no economic theory of invention or technological improvement.

An alternative to the exogenous shift variable approach is to assume that a stock of unused inventions is always at the disposal of producers, i.e. adoption never catches up with invention, so that invention *per se* can never impede growth. Such an assumption underlies Boserup's contra-Malthusian theory that, even in poor countries, the response of food supplies to population growth is elastic (see chapter 9). Boserup believes that in primitive agriculture farmers in the aggregate do not actually *adopt* more productive technologies until forced to do so by population pressure. However, the credibility of this theory is dependent on what many critics regard as dubious assumptions.

As far as agriculture is concerned, there is a more credible alternative to the traditional 'manna from heaven' approach to the generation of new technological discoveries. This is the theory of induced technical and institutional change (Hayami and Ruttan, 1971, ch. 3; Ruttan, 1974). The crux of this theory is that the research and investment which necessarily precedes new discoveries leading to technical progress is induced by market forces. In agriculture, changes in the relative scarcities of resources (as expressed by price changes), especially land and labour, induce a derived demand for technological innovations to facilitate the substitution of relatively less scarce and cheap factors for more scarce and expensive ones. For example, in a labour-scarce economy there is a tendency for capital in the form of labour-saving machinery to be substituted for human labour. But, in a land-scarce economy, yield-increasing and land-saving inputs such as fertilisers, irrigation and HYVs are substituted for land.

3.1 Induced technical change

Hayami and Ruttan have evolved a meta-production function hypothesis to explain how induced technical change increases the elasticity of response to factor price changes. Consider first varying the amount of a single input factor, such as fertiliser, in response to a change in the factor: product price ratio. Although farmers can normally be expected to increase fertiliser inputs in response to a decline in the fertiliser: crop price ratio, the amount of the fertiliser increment and crop yield increment corresponding with it may both be only comparatively small *unless* new crop varieties are developed which are *more responsive* than traditional varieties to fertiliser application.

In Figure 6.5 the fertiliser application rate is measured along the horizontal axis and the crop yield along the vertical axis. All other factor inputs, including land, are assumed to be fixed: u_0 and u_1 are fertiliser response curves, respectively, representing traditional and improved, higher-yielding crop varieties. The tangents P_0 and P_1 represent different fertiliser: crop price ratios. With the farmer's choice of variety confined to u_0, a decline in the fertiliser: product price ratio from P_0 and P_1 would justify only a relatively small increase in the rate of fertiliser application, from Of_1 to Of_2, corresponding with a rise in crop yield from Oy_1 to Oy_2. Now suppose in response to the same price change, the higher-yielding crop variety u_1 is evolved. Farmers are now able to switch from u_0 to u_1 as well as varying the fertiliser rate. Compared with the original equilibrium at Of_1, Oy_1, the new profit-maximising equilibrium is at Of_3, Oy_3. The fertiliser increment has increased from $Of_2 - Of_1$ to $Of_3 - Of_1$, and the yield increment from $Oy_2 - Oy_1$ to $Oy_3 - Oy_1$. It is concluded that with

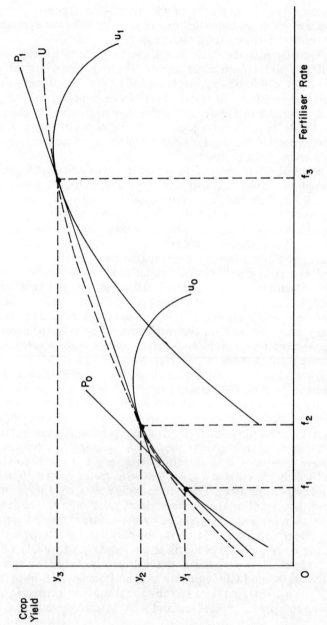

Figure 6.5: *Induced technical change*

the discovery or development of more fertiliser-responsive crop varieties, any given decline in the price of fertiliser will have a greater impact on both fertiliser usage and crop yield.

Conceptually, u_0 and u_1 are a pair of short-term fertiliser response curves drawn from a much larger 'family'. The long-term relationship between the rate of fertiliser application and crop yield, assuming the dynamic development of even more fertiliser-responsive crop varieties, is shown in Figure 6.5 by the envelope curve U. This is Hayami and Ruttan's meta-production function. In effect, this function represents the very long-run when all conceivable technical alternatives might be discovered.

The behavioural rationale of this model is that a decline in the price of a single variable factor (fertiliser in the example) provides producers (or their industry representatives) with an economic incentive to press the R & D 'industry' to discover and develop new technology possessing the property of making output more input responsive. That is, the economic rate of input usage at the new lower price level would be substantially increased compared with the rate justified by existing technology. Similarly, it can be argued that changes in factor prices give guidance to the discoverers of improved technology regarding types of technological advance with the best 'market prospects'. However, this version of the model oversimplifies reality because, in the real world of input factor substitution, the incentive for promoting technological change may equally well derive from changes in *factor/factor* price ratios.

In Figure 6.6 units of fertiliser are measured on the horizontal axis and units of land on the vertical axis above the origin. Below the origin the vertical axis measures research input directed to the discovery of more fertiliser-responsive crop varieties. In the upper quadrant the short run isoquants u_0 and u_1 represent the same level of output, but different crop varieties, with u_0 being less fertiliser-responsive than u_1. Given an initial fertiliser: land price ratio as represented by the tangent P_0, the least-cost factor combination is Ol_1 land and Of_1 fertiliser. The corresponding research input is Ot_1. Suppose that a fall in the fertiliser: land price ratio to the level represented by tangent P_1 is matched by an increase in research input to Ot_2. The research pay-off is shown by movement down the scale-line in the lower quadrant of the diagram from Ot_1, Of_1 to Ot_2, Of_2. Fertiliser level Of_2 corresponds with the discovery of crop variety u_1. With that discovery LCFC equilibrium shifts from Ol_1, Of_1 on u_0, to Ol_2, Of_2 on u_2. This shift entails a substantial absolute increase in the use of the cheaper input (fertiliser) and a large reduction in the use of

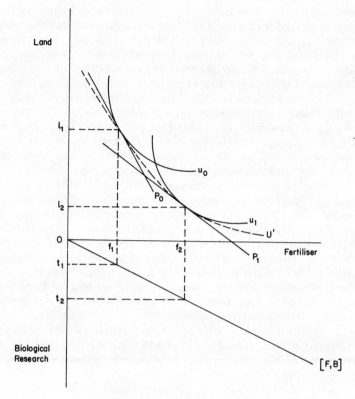

Figure 6.6: *Induced factor substitution*

the dearer input (land). This substitution of factors may be regarded as taking place along the envelope curve U', or the long-term isoquant corresponding with the short-term isoquants u_0 and u_1. In this model the meta-production function is the complete 'family' of long-term isoquants representing all possible levels of production.

This model emphasises the link between the technological or research input and the discovery of innovations which broaden the scope for factor substitution in response to price change. It can readily be extended to all factors and from biological to mechanical technology. Advances in mechanical technology are generally labour-saving. When the price of labour rises relative to the price of land, farmers are consequently induced to press agricultural engineers to develop new machinery with an enhanced capacity for substituting land for labour, i.e. bigger and more powerful machines with the capacity for enabling each worker to handle a larger land area are demanded. We shall return to capital – labour substitution, under the heading of the distributional consequences of adopting factor-biased technology, in a subsequent section of this chapter.

The meta-production function hypothesis has been shown to give a statistically acceptable explanation of contrasting past patterns of agricultural growth in the USA and Japan (Hayami and Ruttan, 1971, ch. 6; Ruttan, 1974). The historical record shows that both countries achieved similar rates of agricultural growth over the period 1880–1960, though with different technologies and factor mixes. Whereas growth in the output of US agriculture was primarily based on improvements in mechanical technology (reflecting labour scarcity), in Japan improvements in yield-increasing biological technology (reflecting land scarcity) were dominant. Variations in factor proportions – measured by land: labour, machine – power: labour and fertiliser: land ratios – were satisfactorily explained in both countries by changes in factor price ratios. Elasticities of factor substitution were estimated and used as indicators of the influence of induced technical innovations on the rate of growth. It was considered *a priori* that fixed technology would afford comparatively little scope for substitution. Thus, the fact that empirically the elasticity coefficients proved to be quite large was interpreted as evidence in favour of the induced technical innovation hypothesis. The study concluded that

> Development of a continuous stream of new technology, which altered the production surface to conform to long-term trends in factor prices, was the key to success in agricultural growth in the US and Japan. (Hayami and Ruttan, 1971, p. 135)

3.2 Induced institutional change

In an LDC context an important element of Hayami and Ruttan's theory is that induced technical change tends to be impeded by institutional barriers. In particular, LDCs generally lack adequate agricultural research institutions to foster the discovery and application of new scientific and technical knowledge. Institutional innovation is consequently, needed to break this bottleneck. In other words, technical innovation and institutional innovation are *complementary*.

Knowledge of how to create effective research institutions in LDCs is only fragmentary. Though true, the observation that it involves the recruitment of suitable personnel, training and finance amounts to little more than a statement of the obvious. However, the main responsibility for organising and financing research is likely to fall on government for two reasons. First, the prospective commercial pay-off from agricultural research in LDCs may be too small or uncertain to attract private investors. The manufacture and distribution of agricultural machinery, chemicals and other modern inputs is increasingly dominated by large international companies. Although such companies spend heavily on R & D they are primarily interested in developing new products for mass markets, principally in the developed countries. The local problems and research needs of small-scale peasant farmers in poor countries tend to be neglected. Because the private sector is unlikely to cater adequately for its research needs, peasant agriculture must perforce rely principally on government for a research programme tailored to its particular requirements.

The second reason for arguing that government must assume the main responsibility for agricultural research in LDCs is that, for reasons of self-interest, research conducted in the private sector is likely to be biased in favour of capital intensive and labour-saving *mechanical* innovations. These are generally not well-suited to the requirements of peasant farmers, for whom capital is scarce and dear whereas labour is plentiful and cheap. Rather their need is for labour-intensive, biological innovations such as HYVs. The bias towards mechanical technology stems from the partitioning of the benefits of innovations between their suppliers and their users, i.e. between manufacturers and farmers. The distribution of benefits is principally determined by the extent to which the discoverer of an improved product is subsequently able to monopolise its supply. With competitors free to copy and market his design, most of the benefits rapidly pass to them and ultimately to users who get the new product at a competitive price. Whereas machinery manufacturers can protect new designs by taking out patent rights, which may even be respected internationally, the fruits of biological research, (such as the breeding of improved crop varieties) are much harder to protect. Because they are inconspicuous, some biological materials – such as seeds – are easily stolen and smuggled across international frontiers. All such materials are readily producible in the hands of competitors. Whereas the research programmes of firms in the private sector are necessarily based on self-interest, research in the public sector is directed towards discoveries which promote social interests.

Government is responsible for national agricultural policy and the only way of ensuring that a country's agricultural research programme is compatible with its policy objectives is for government to oversee the research programme and provide many of the resources needed for its implementation. By this means the institutional barriers to induced technical change may be removed.

3.3 International agricultural research institutions

Much of the agricultural research which has benefited LDCs in recent years has been conducted by international research institutions, such as the International Maize and Wheat Improvement Centre (better known as CIMMYT, the initial letters of its Spanish title) located in Mexico, and the International Rice Research Institute (IRRI) situated in the Philippines. These institutions, originally financed by the Rockefeller Foundation and other large charitable foundations in the developed countries, but more recently by an international consortium of aid agencies and financial institutions – the Consultative Group on International Agricultural Research (CGIAR) – have conducted disinterested agricultural research on behalf of the LDCs as a contribution to their agricultural development. The main emphasis of this research has been the discovery of biological innovations to raise crop yields. The fruits of this research include the high yielding varieties (HYVs), particularly of wheat and rice, which have revolutionised agriculture in some LDCs since the mid-1960s. The work of the international research institutions is complementary with that of national institutions in the LDCs. An important function of national research organisations is to conduct local trials to select imported HYVs for adaptation to local conditions.

3.4 Evaluation

Although the theory of induced technical and institutional change represents an undoubted advance on treating technical progress as 'manna from heaven', it is vulnerable to the criticism that whereas its underlying behavioural assumptions are neo-classical, these do not fit the conditions of peasant agriculture in LDCs. Thus, it is assumed that farmers are profit-maximisers and that their attitude to technical change is strictly market oriented. Moreover, it is taken for granted that farmer possess full information on factor substitution rates and factor price ratios. Whist these assumptions may be acceptable as an approximation to reality as far as farmers in developed capitalist countries are concerned, their validity for LDCs is much more questionable. The meta-production function model explains the contrasting patterns of past agricultural development in the USA and Japan quite well, but these are both developed countries. For reasons discussed in chapter 2, as well as in section 1 above, the neo-classical paradigm may not adequately explain the behaviour of peasant farmers. The barriers to technical change in peasant agriculture may be virtually insurmountable in the short term, for cultural and social reasons, as well as because of economic uncertainty, risk-aversion and lack of knowledge. Thus, where agriculture is most backward (by western capitalist standards) the theory of induced technical change may have little immediate relevance. Under such conditions, it is not a question of agricultural research institutions responding to market forces initiated by profit-oriented farmers. On the contrary, assuming that technical progress in agriculture is a valid policy goal, the initiative must come from the research institutions and government to persuade farmers to adopt new technology in their own and the national interest. The rate at which farmers can be won over to accepting new technology is likely to depend in part on the success of measures to reduce the number and severity of economic uncertainties confronting them. It may be possible to reduce uncertainty relatively quickly by means such as minimum price guarantees, but to the extent that it is also necessary to break down social and cultural barriers to technical change, progress in achieving the goal of a rapid rate of technical advance in agriculture may be much slower.

We have discussed reasons for technological stagnation in LDC agriculture and the nature of barriers to technical change in chapter 2 section 1.3.2. We shall not repeat the points made in that discussion here but merely remark that they are also relevant to the subject-matter of this chapter.

4 Factor-Biased Technological Change and its Distributional Consequences

Economic theorists distinguish between three types of technological change based on factor input ratios as follows:

(1) Hicks-neutral technological change under which despite the change in output, the capital – labour (K/L) ratio remains constant.

(2) Harrod-neutral technological change under which the capital-output (K/O) ratio remains constant, so that the factor proportions are biased in favour of saving labour, i.e. the K/L ratio increases.
(3) Solow-neutral technological change under which the labour-output (L/O) ratio remains constant, so that factor proportions are biased in favour of saving capital, i.e. the K/L ratio diminishes.

Thus Hicks-neutrality marks the dividing line between labour-saving Harrod-neutrality, and capital-saving Solow-neutrality.

Although the concept of a constant K/L ratio possesses some theoretical interest as a reference datum, in practice technological change usually involves a change in factor proportions. Moreover, the change in factor proportions must imply some change in the total levels at which factors are employed. However, the adoption of factor-biased new technology does not necessarily imply that, in total, less of the 'disfavoured' factor will be employed. So, for example, far from diminishing the total demand for labour, the adoption of labour-saving Harrod-neutral technology may initiate a process of economic adjustment which actually creates more jobs or lengthens working hours. A 2-factor isoquant diagram will serve to clarify the analytical argument (Donaldson and McInerney, 1973). In Figure 6.7, isoquants P_0 and P_t represent identical levels of output but different techniques of production, i.e. they exist on different production surfaces. But the technique underlying P_0 is more labour-intensive (or less capital-intensive) than the technique underlying P_t. The higher level isoquant $\bar{P}_t$ belongs to the same family as P_t. The slope of the isocost line O_1 represents the K/L price ratio, and similarly with the isocost lines O_2 and O_3, representing higher levels of total outlay. Suppose that initially a producer is in profit maximising equilibrium where P_0 is tangent to O_2, giving employment to Ol_1 labour and OK_1 capital. Factor proportions are signified by the slope of the ray L/K. Suppose now that the producer switches to a more capital intensive technique of production *without adjusting the level of output*. At the new equilibrium where P_t is tangent to O_1, a lower labour input of Ol_2 is combined with a higher capital input of Ok_2. The difference in slope between the rays L/K and L'/K' signifies the capital-using bias of the shift to the new technique.

But, assuming constant prices, P_t is not the *profit-maximising* level of output. Compared with the initial equilibrium on P_0, total costs

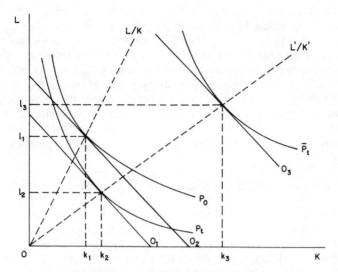

Figure 6.7: *Factor-biased technological change*

have been reduced from level O_2 to level O_1. Thus, assuming that factor proportions are now fixed at L'/K', the new profit-maximising level of output must be further from the origin than Ol_2, Ok_2. Expanding output along any scale-line, such as L'/K', necessarily entails employing more of *both* factors. Suppose that, following the switch to the more capital-intensive production technique, the profit-maximising level of output increases to $\bar{P}_t$. Then, the factor input levels are correspondingly increased to Ol_3 labour and Ok_3 capital. At Ol_3 the employment of labour is in fact at a higher level than the original profit-maximising equilibrium level of Ol_1 on P_0.

Depending upon the direction of the change, a simultaneous change in the factor-price ratio may either reinforce or offset the effect of a change in the choice of production technology on total factor employment levels. Thus, the effect of any decline in the K/L price ratio will be to 'flatten' the isocost lines shown in Figure 6.7. This gives producers the inducement to adopt an even more capital-intensive alternative technology than before, i.e. the ray L'/K' is also flattened, signifying an even larger change in factor proportions compared with the initial choice of technology. Contrarily, if the emergence of the more efficient capital-using technology happened to coincide with a permanent *increase* in the K/L price ratio (implying cheaper labour and dearer capital) a 'steeper' L'/K' would *reduce* the change in factor proportions induced by the new technology. But whatever the change in the factor price ratio, and its effect upon the choice of technology and associated factor proportions, the conclusion regarding the effect of technological choice on factor employment levels remains unchanged. This is that changes in factor input requirements are *jointly* determined by the change in factor proportions *and* the change in the profit maximising level of output. For this reason, it is impossible to determine *a priori* whether, for example, the adoption of labour saving technology will really displace labour or actually create more jobs.

4.1 Land-augmenting and labour-displacing technical change

Although the adoption of new technology with a labour-saving bias does not *necessarily* displace labour, it may do so, depending on the relative strengths of the 'changed factor proportions' and 'output adjustment' effects. From the standpoint of policy, the risk of labour displacement may be reduced by discouraging technological change with a capital bias. Indeed, if increasing agricultural employment is a prime policy objective, encouraging producers to adopt *labour-using* technologies is an appropriate means of achieving that goal. In this context, and particularly in the literature on farm mechanisation policy in LDCs, a distinction may be made between land-augmenting and labour-displacing technical change (e.g. Yudelman *et al.*, 1973, chapter III.)

Land-augmenting technical change induces an increase in total output, either through enlarging the cultivated area or through higher crop yields – *ceteris paribus*, higher output will tend to increase labour requirements. Labour-displacing technical change saves labour but does not direcly affect output. Its economic rationale is to substitute some other input for labour in order to reduce unit costs. But, as readers of this chapter have already been reminded, cost reduction usually has a disequilibriating effect upon the level of output which maximises profits. Thus, in theory at least, technical change which is ostensibly labour-displacing must also exert some countervailing pressure on the demand for labour through the indirect effect on output. Whether this occurs in practice is an empirical question.

The most obvious examples of LATC are biological and chemical innovations which increase crop yields. But mechanical innovations cannot be excluded, *ex hypothesi*, because they also may be output increasing, either through a crop area or a crop yield effect. If a farmer has unlimited land at his disposal the maximum area which he can feasibly cultivate during any time period is constrained by the available supplies of human labour and mechanical power, particularly during critical periods of the year such as sowing and harvesting seasons. Thus, if a mechanical innovation is used to break a seasonal bottleneck in the supply of farm labour its effect can be to increase output by bringing more land into cultivation.

Employment remains unchanged, or may even increase during other periods of the year, due to the enlargement of the cultivated area. A mechanical innovation can still be land-augmenting, despite an *inelastic* land supply, if it is the means of increasing the intensity of cropping. So, for example, given a fixed land area, the feasibility of growing two crops a year rather than only one may hinge on the use of mechanical power to accomplish essential field operations within a critical time period. The conditions which are least conducive to LATC are the combination of a plentiful labour supply with very inelastic land and a cropping system which is already highly intensive. Under such conditions, it is difficult to see how farm mechanisation could be land-augmenting, unless mechanisation possessed some *yield increasing* property. This is possible provided that the application of mechanical power enables some yield-augmenting task to be accomplished which is beyond the capacity of unaided human labour. So, for example, if there is a direct cause- and-effect relationship between depth of cultivation and crop yield, and the optimum depth is attainable only with the assistance of machinery, this condition is satisfied. Seminal work on the economic forces underlying agricultural mechanisation in LDCs, with special reference to the adoption of tractors in Brazil, is reported in Binswanger and Ruttan (1978), chapter 10.

Our discussion of land-augmenting and labour-displacing technical change points to the conclusion that they are ambiguous concepts. LATC is supposed to increase output whilst maintaining, or even increasing, the demand for labour. LDTC is supposed to reduce costs by saving labour without varying output. Although it is tempting to think of LATC and LDTC as policy alternatives, they may in fact occur at sequential stages of technological progress. Thus, in the first stage, output and the demand for labour both increase due to the adoption of LATC. At first, the supply of labour is elastic, but becomes increasingly inelastic as agricultural output and the demand for labour both continue to grow. Eventually, seasonal bottlenecks in labour supply emerge and, at this stage, LATC leads to LDTC.

Although the ostensible motivation for LDTC may be cost saving, the relief of an acute seasonal labour scarcity may be a more fundamental but hidden reason. Once LDTC has been adopted for one purpose, it tends to be extended to other tasks at the same or a different period of the year. One reason for this is that, other things being equal, it pays to intensify the use of a capital asset in order to spread its fixed costs over a larger output.

A further reason for the ambiguity of the supposed distinction between LATC and LDTC is that some new agricultural inputs embodying technical change can either be land-augmenting or labour-displacing, depending on how they are applied. Thus, for example, a farmer may in principle acquire a tractor either to save labour (by substitution) or to increase output (by extending his cultivated area). But these opposite motivations are difficult to observe in practice and, in any case, having acquired the tractor for one purpose, the farmer may change his mind and use it for the other one, or even for both. The near impossibility of identifying inputs which are *invariably* labour-displacing has important policy implications, especially for mechanisation policy. But before proceeding to policy conclusions in the final section of the chapter, we briefly consider some empirical evidence on how agricultural technical change, including farm mechanisation, has affected employment in agriculture in certain LDCs.

5 Agricultural Technical Change and Agricultural Employment: Empirical Evidence

We remarked in chapter 3 that because of employment linkages between agriculture and other industries, agricultural growth creates additional jobs outside agriculture, as well as in farming itself. In the same way, agricultural technical change may affect employment, not only in agriculture itself (as discussed in the previous section) but also in all industries linked with agriculture in the input–output matrix. The first of these may be termed the *direct* employment effects, the second the *indirect* effects.

5.1 Direct employment effects

An empirical study of the direct employment effects of agricultural technical change in the Indian Punjab from 1968–9 to 1973–4

(Krishna, 1975) was based on a theoretical model attributing the technologically-induced change in agricultural employment between two periods to the following three causal variables:

(1) a pure 'technology effect', reflecting factor bias;
(2) a 'crop-mix effect', reflecting the change in the allocation of land from a given area amongst different crops (with different labour requirements) associated with the change in technology; and
(3) an 'intensity effect', reflecting change in the index of cropping intensity (100 = one crop per annum).

The principal crops were wheat and rice. The empirical findings showed an overall decline of 64 man hours per hectare per annum in the total labour input for both crops over five years. But the crop mix and intensity effects were both positive (totalling a gain of approximately 45 man hours per hectare). Thus the net loss in direct employment was entirely due to a large negative technology effect (approximately 109 man hours per hectare).

The results of a study of introducing tractors to farms, also in the Punjab, confirmed Krishna's finding that more intensive cropping is favourable to on-farm employment, but not the finding of an overall net loss of employment due to technical change (Roy and Blase, 1978). In order to find out how the introduction of tractors affects cropping intensity, crop yields, and on-farm employment, a cross-sectional study of farms with tractors (TFs) and farms without tractors (NTFs) was undertaken. The results indicated that the principal effects of tractorisation had been more intensive cropping and higher crop yields. These gains were attributed in part to the better 'quality' of tractor ploughing. But despite some differences between TFs and NTFs in the proportions of hired labour and family labour, as well as in the relative numbers of permanent and casual workers, there was no significant difference between them in *total* labour use per unit of net crop area.

An on-farm sample survey of farm tractors supplied to India under the British overseas aid programme reached a similar conclusion, namely, that although tractors had affected cropping and increased farm profits, they had not displaced human labour (ODM, 1976). But there is also contrary evidence on the effects of mechanisation supporting Krishna's more general finding that technical change in agriculture tends to be labour-displacing. So, for example, the results of a survey of cross-sectional studies similar to the one conducted by Roy and Blase, mainly in India, indicated that large-scale mechanisation *is* generally labour-displacing (Yudelman, *et al.*, 1971, ch. IV). A review of studies in East Africa concluded that although tractor mechanisation is often profitable, especially on large farms, it is usually labour-displacing. But, exceptionally, a labour supply constraint may be binding on farm output and profit, especially on farms in the vicinity of large towns where farmers must compete with urban employers for labour: under such conditions tractorisation may not be labour-displacing (Clayton, 1972).

In summary, Krishna's study of farm labour displacement due to agricultural technical change in the Punjab is outstanding for its generality and analytical rigour. Its somewhat 'pessimistic' findings therefore command close attention. Apart from this, most empirical evidence on how farm technical change affects the employment of labour relates specifically to mechanisation, particularly with tractors. The evidence on whether tractors displace labour is mixed. This is not surprising since, as we have agreed *a priori*, the direct substitution of machine power for the power of human labour is not the sole motive farmers have for adopting machinery. Moreover, mechanisation tends to be a gradual and piecemeal process taking a long time to complete. At any given time the process is likely to be more advanced in some places, be they individual farms, regions or whole countries, than in others. Thus, the evidence provided by contemporaneous but geographically separate studies may cover many different stages of the mechanisation process. It would not be surprising if the results of 'earlier stage' studies indicated little or no labour displacement, whereas those of 'later stage' studies indicated the opposite.

The hypothesis that the labour-displacing effect of farm mechanisation in LDCs is progressive needs rigorous empirical verification. However, positive verification would strengthen the case for government intervention to control the rate of mechanisation. In the final section of this chapter we consider whether 'selective mechanisation' is a feasible policy instrument.

5.2 Indirect employment effects

Krishna describes a comparative static input–output model which he used to measure the aggregate output and employment effects of technological innovation in the agricultural sector. Readers are reminded that relative to a given change in the final demand for the product of a particular sector, input–output analysis measures not only the consequential changes in that sector's output but also the additional output which is induced in other sectors. The size of the total induced demand is determined by the intensity of inter-industry transactions and the size of the relevant demand multipliers.

Empirically, a 77 × 77 input–output table of the Indian economy was condensed into a 2-sector table by grouping sectors into a composite farm sector (FS) and a composite non-farm sector (NFS). Appropriate assumptions were made concerning the values of the parameters of the model's exogenous variables. These included the aggregate labour force growth rate, the agricultural growth rate, technologically-induced rates of change in agricultural input coefficients, and rates of change in marginal propensities to spend on FS goods, and NFS goods. Alternative parameter values were used in some cases (e.g. agricultural growth rate) to suit different study objectives.

The model was used to project the effects of an assumed rate of technical progress in the FS on aggregate (i.e. FS + NFS) output and employment. Three basic variants of the model were used to separate the effects of technical change *per se* from technical change with growth. The variants were (i) technical change without growth in the FS; (ii) 5 per cent per annum growth *without* technical change in the FS; and (iii) 5 per cent per annum growth *with* technical change in the FS. For each of these variants the projected rates of growth in output and employment are shown in Table 6.1.

As expected, variant 1 actually displaces labour from the FS and, despite positive employment growth in the NFS (induced by farm technical change) aggregate employment also declines. An interesting

Table 6.1: *Projected growth and employment effects of farm technological change in India (Base year: 1964/65)*

Variant	FS		NFS		FS + NFS	
	X	L	X	L	X	L
	Growth rate (per cent per annum)					
1	0.0	−4.0	1.0	1.0	0.7	−2.6
2	5.0	5.0	4.0	4.0	0.4	4.7
3	5.0	0.7	5.1	5.1	0.5	2.0

Note: X = sectoral output; L = sectoral employment.
Source: Adapted from Krishna (1975), Table 11.7.

feature of variant 2 results is that, without farm technical change, employment grows more slowly in the NFS than in the FS. As its inherent assumptions are probably closer to reality, variant 3 results are the most interesting of all. The combination of growth and technical progress in the FS boosts employment and growth in the NFS to the high level of more than 5 per cent. But technologically-induced labour displacement reduced FS employment growth to less than 1 per cent per annum. The growth rate of aggregate (FS + NFS) employment is consequently reduced to only 2 per cent per annum. In terms of absolute values, and the context in which we discuss it, the results of this study are of no more than passing interest. Krishna himself warns against attaching too much credence to projections based on a particular set of assumptions. However, this reservation in no way detracts from the value of the study in demonstrating the difference between viewing the consequences of farm technical change for factor employment in the FS alone, and the economy as a whole. The choice of approach must obviously depend on the investigator's objectives. But it is quite clear that in many contexts the partial or single-sector approach can be unduly myopic and therefore very misleading.

6 Agricultural Resources and Technical Change in LDCs: Policy Conclusions

Our policy conclusions fall principally under the headings 'research and information' and 'mechanisation'.

6.1 Research and information

Agricultural research is needed to hasten socially beneficial farm technical change in LDCs. For institutional reasons, the initiation and finance of agricultural research in LDCs is primarily the responsibility of government. If government neglects this responsibility the likely consequence is either a research vacuum or a research programme financed by the private sector and biased in favour of private vested interests. In many LDCs agricultural progress is currently retarded by an unfavourable labour:land ratio. This points to a research programme emphasising biological innovations to promote land-augmenting technical change. The search for mechanical innovations to save labour will often be better postponed until agricultural labour becomes scarcer at a later stage of development (this point is elaborated in chapter 3 under labour surplus development models).

Substantial scope exists for international co-operation in the promotion of agricultural research to benefit LDCs. The fruits of the existing CIMMYT and IRRI research programmes, as well as those of many less publicised bilateral arrangements between DCs and LDCs (often under technical aid programmes), show the benefits of such co-operation. A sensible division of labour is for DCs to concentrate on solving more basic research problems with a broad spread of possible applications. LDCs can then concentrate on adopting what has been discovered in the DCs to suit their particular natural environment and economic conditions.

Advancing agricultural research is not synonymous with agricultural technical change. The actual adoption of new technology in agriculture depends on the spread of information, and the removal of barriers to adoption. Responsibility for bridging the gap between the research worker and the farmer rests primarily with the agricultural extension service, particularly with respect to the dissemination of information. The removal of barriers to adoption is, for reasons discussed in chapter 2, a much harder nut to crack, but there is frequently a case for local research specifically to identify the major barriers to adoption.

Local farm trials and demonstration plots are important means of disseminating knowledge of new discoveries to farmers, including their net economic benefits. But the extension service function is not confined to the dissemination of *new* knowledge it also includes showing imperfectly informed farmers how to manage their existing resources more efficiently to yield a larger profit or greater utility. Economic efficiency combines allocative with technical efficiency. As far as peasant agriculture is concerned, the evidence that farmers are allocatively efficient is more convincing than the evidence that they are technically efficient. Thus, with better knowledge of existing farm practices, many farmers may be able to move on to a higher production function without waiting for new discoveries.

In chapter 2 we referred to the argument that, because of their different resource endowments and technical skills, western-style technology is 'inappropriate' for adoption by the LDCs. This argument has the obvious policy implication that part of the LDC agricultural research effort must be directed to the development of more appropriate technologies to accelerate the economic betterment of peasant farmers. A central aim of such research must be to raise agricultural labour productivity at a low capital cost and without undue labour displacement.

6.2 Mechanisation

Here we consider whether it is desirable and feasible for LDCs to pursue 'selective' mechanisation as a policy objective. We also consider how the private costs and social consequences of farm mechanisation in LDCs may be affected by non-specific government policies.

Selective mechanisation means 'restricting mechanisation to where it contributes to increasing employment or is necessary to break a seasonal bottleneck' (FAO, 1973, ch. 3). This is a very general definition and little attempt has been made to make it sufficiently specific for use as a working principle or rule. We do not know of anywhere where selective mechanisation has been adopted as a specific policy goal, although a few examples of its 'feasibility' have appeared in the literature. But, before considering the feasibility of selective mechanisation, we first discuss the conditions of its 'desirability' as a policy objective. There are two aspects of this. First, there is the familiar trade-off between short-term gains and long-run losses (or vice versa). Whereas the *long-term* social interest requires capital to be substitued for labour in order to raise agricultural labour productivity and farm incomes, the *short-term* interest requires that jobs be preserved and living standards maintained for the many. In practice, there has to be a compromise between the pursuit of short- and long-term policy goals. Selective mechanisation is biased towards the short-term, and its desirability as a policy goal has to be judged within a broader policy framework reflecting government's views on the relative importance of long- and short-run policy objectives.

The second reason for doubting the desirability of selective mechanisation is that the favourable concealed or indirect effects of farm mechanisation on employment may out-weigh the direct effects. We cited earlier Krishna's projection based on Indian data, in which the additional work-time created outside agriculture by farm technical change exceeded the worker time lost in agriculture itself. Thus, taking a wider view than that of employment in agriculture *per se*, selective mechanisation could actually be inimical to the maximisation of aggregate employment, even in the short term.

It might be argued that the indirect employment benefits of farm technical change can be discounted if the relevant inputs are imported, so that the extra non-farm employment generated by their purchase leaks abroad. However, this argument is open to the criticism that, because of trade reciprocity, it may be necessary to admit more imports of machinery (as well as imports of other goods) in order to expand exports. Thus, indirectly, curtailing machinery imports may also limit employment in export industries.

Taken together, these two aspects point strongly to the conclusion that, even if selective mechanisation's objective of preventing displacement of farm labour is feasible, it is not *necessarily* a desirable policy goal.

Turning from the desirability to the feasibility of selective mechanisation, most published examples have merely stipulated the technical *conditions* of feasibility, rather than demonstrating that it is economically feasible in practice (e.g. Yudelman, *et al.*, 1971, ch. IV). In one example, involving the mechanisation of rice production in Thailand, the displacement of labour could be avoided *only* if the rice was transplanted (rather than the seed being broadcast), the use of a tractor was restricted to the first of two ploughings, and the rice acreage was expanded to absorb the consequent 'saving' of labour. In another example, the feasibility of double cropping wheat with cotton in Pakistan depended on introducing a stationary threshing machine (to replace threshing by hand) to break a labour bottleneck created by an overlap of the wheat harvesting and cotton planting seasons. If mechanisation had been extended to non-threshing operations labour would have been displaced; but this could be the more profitable option for the farmer.

These two examples illustrate the very restrictive conditions governing the successful attainment of the objectives of selective mechanisation. It is also apparent that selective mechanisation implies a substantial degree of external control over the types of machinery that are permitted to be used on farms, and even more important, the operations that the machines admitted are allowed to be used for. Even if it were possible to identify systems of selective mechanisation which, if widely adopted by farmers, would be socially beneficial in a particular place at a particular time, the *administration* of such a scheme would scarcely be feasible under *private* machine ownership. Even if private machine ownership is banned in favour of a government monopoly, with farmers being restricted to hiring machine services from the government, the administrative burden of control, to ensure that machines are restricted to authorised uses, is still considerable. There is also the danger of monopoly breeding inefficiency due to the absence of competition. Even in LDCs where private agricultural contractors have been allowed to compete with machinery services supplied by the government, the latter have often operated at a substantial financial loss. Thus, even if the objective of selective mechanisation – to prevent or retard the displacement of labour from agriculture – is desirable in itself, there are serious doubts concerning its feasibility, particularly in market-type economies based on decentralised decision-making and producers' freedom of choice. Binswanger and Ruttan (1978), chapter 10, contains a good discussion of farm mechanisation policy, including the feasibility of selective mechanisation, with specific reference to Brazil.

Despite the question mark against the feasibility of selective mechanisation, there is no doubt that government can influence the rate of mechanisation *unselectively*, by controlling the distribution of agricultural machinery and the terms on which it is acquired by farmers. In many LDCs virtually all modern agricultural machinery is imported due to the lack of domestic manufacturing capacity. The variety of instruments government can use to regulate imports is considerable. Imports can be reduced or curtailed by means of import duties and quotas. Alternatively they can be encouraged by removing duties or increasing quotas, or by granting positive import subsidies. Inducements to import may or may not be specific to a particular category of import, depending on whether government operates an import licensing system. The currencies of LDCs frequently trade at a discount in the open market. *Ceteris paribus*, over-valuation increases the volume of imports (including machinery imports) because of its effect on the terms of trade. A dual exchange rate system may operate to regulate the balance of payments. Importers who are licensed to import at the market rate of exchange, rather than at the higher 'official' rate, enjoy a concealed import subsidy. Countries pursuing a policy of economic self-sufficiency (or autarky) erect high barriers against virtually all imports. But, apart from these, governments of LDCs have probably tended to encourage rather than discourage imports of agricultural machinery. In some cases the encouragement is explicit. Import licences are liberally granted to would-be importers, with or without an import subsidy, in pursuance of a policy of speeding the 'modernisation' of agriculture. Many countries have also admitted substantial quantities of agricultural machinery from abroad under overseas aid programmes. But elsewhere the encouragement has only been implicit – or even unintentional – notably in the case of countries with chronically over-valued currencies.

Regardless of whether agricultural machinery is imported or domestically produced, there are numerous ways in which government may indirectly encourage the substitution of machinery for human labour. For example, advancing credit to farmers at a subsidised interest rate stimulates all forms of investment, including machinery. Similarly, a fuel subsidy encourages the acquisition of machines such as tractors and self-propelled combine harvesters. The encouragement to mechanise is not necessarily *intentional*. It may well be the unintentional effect of policies designed to achieve other objectives. So, for example, subsidised credit may be intended to encourage farmers to purchase non-capital inputs, such as fertilisers and pesticides; the stimulus to machinery investment is unforeseen.

This review of farm mechanisation policy in LDCs points to four major conclusions. First, although the substitution of capital for labour is inevitable in the long run, the optimum rate of mechanisation is for public decision. Secondly, government's ability to control the rate of farm labour displacement, by pursuing a policy of selective mechanisation, is strictly limited. Thirdly, government can choose to encourage or discourage mechanisation *unselectively*, by applying specific taxes or subsidies. Fourthly, because numerous non-specific government policies, such as the exchange rate, influence the purchase of farm machinery, such policies need frequent review and possible revision to avoid unintentional effects on the rate of mechanisation.

References

Bachman, K.L. and Christensen, R.P. (1967), 'The Economics of Farm Size', in Southworth, H.M. and Johnston, B.F. (eds), *Agricultural Development and Economic Growth*, Cornell University Press. Ithaca.

Berry, D.A. and Cline, W.R. (1979), *Agrarian Structure and Productivity in Developing Countries*, Johns Hopkins University Press, Baltimore.

Binswanger, H.P. and Ruttan, V.W. (1978), *Induced Innovation, Technology, Institutions and Development*, Johns Hopkins University Press, Baltimore.

Clayton, E.S. (1972), 'Mechanisation and employment in East African agriculture', *Int. Labour Rev.*, **105**(4).

Donaldson, G.F. and McInerney, J.P. (1973), 'Changing machinery, technology and agricultural adjustment', *American J. Agric. Economics*, **55**(5).

Eckart, J.B. (1977), 'Farmer response to high-yielding wheat in Pakistan's Punjab', in Stevens, R.D. (ed.) *Tradition and Dynamics in Small-Farm Agriculture*, Iowa State Univ. Press, Ames, Iowa.

FAO (1973), *The State of Food & Agriculture*, Food and Agriculture Organisation of the United Nations, Rome.

Farrell, M.J. (1957), 'The measurement of productive efficiency', *J. Royal Stat. Soc.*, Series A, vol. 120, Part pp. 254–6.

Feder, G., Just, R. and Silberman, D. (1981), *Adoption of Agricultural Innovations in Developing Countries: A Survey*, World Bank Staff Working Paper no. 444, The World Bank, Washington DC.

Hayami, Y. and Herdt, R.W. (1977), 'Market price effects of technological change on income distribution in semi-subsistence agriculture, *American J. Agric. Econ.* **59**(2).

Hayami, Y. and Ruttan, V.W. (1971), *Agricultural Development: An International Perspective*, Johns Hopkins University Press, Baltimore.

Hopper, W.D. (1965), 'Allocation efficiency in a traditional Indian Agriculture', *J. Farm Economics*, **47**(3).

Jones, William O. (1977–8), 'Turnips, the Seventh Day Adventist principle and management bias', *Food Res. Inst. Studies*, **XVI**(3).

Junankar, P.N. (1980), 'Do Indian farmers maximise profits?' *J. Dev. Studies* **17**(1).

Krishna, R. (1975), 'Measurement of the direct and indirect employment effects of agricultural growth with technical change', in Reynolds, L.G. (ed.), *Agriculture in Development Theory*, Yale University Press, New Haven.

Lipton, M. (1968), 'The theory of the optimising peasant', *J. Devel. Studies*, **4**(3).

Lipton, M. (1978), 'Inter-farm, inter-regional and farm-non-farm income distribution: the impact of the new cereal varieties', *World Development*, **6**(3).

Newbery, D.M.G. and Stiglitz, J.E. (1981), *The Theory of Commodity Price Stabilization*, Oxford University Press.

ODM (1976), *British Aid Tractors in India: An Ex-Post Evaluation*, Ministry of Overseas Development, London.

Rask, N. (1977), 'Factors limiting change on traditional small farms in southern Brazil, in Stevens, R.D (ed.) *Tradition and Dynamics in Small-Farm Agriculture*, Iowa State Univ. Press, Ames, Iowa.

Roumasset, J. (1976), *Rice and Risk*, North-Holland, Amsterdam.

Roy, S. and Blase, M.G. (1978), 'Farm tractorisation, productivity and labour employment: a case study of Indian Punjab', *J. Devel. Studies*, **14**(2).

Ruttan, V.W. (1974), 'Indeed technical and institutional change and the future of agriculture' in Hunt, K.E. (ed.), *The Future of Agriculture*, Institute of Agricultural Economics, Oxford.

Schulter, M. and Mount, T. (1974), *Management Objectives of the Peasant Farmer: An Analysis of Risk Aversion in the Choice of Cropping Pattern, Surat District, India*, Occasional Paper no. 78, Cornell University, Department of Agricultural Economics.

Schultz, T.W. (1964), *Transforming Traditional Agriculture*, Yale U.P., New Haven.

Shapiro, K. (1977), 'Efficiency differentials in peasant agriculture and their implications for development policies', in *Contributed Papers Read at the 16th International Conference of Agricultural Economists*, Institute of Agricultural Economics, Oxford.

Wolgin, J.M. (1975), 'Resource allocation and risk: a case study of smallholder agriculture in Kenya', *Amer. J. Agric. Econ.*, **54** (4).

World Bank (1978), *World Development Report*, World Bank, Washington DC.

World Bank (1981), *Accelerated Development in Sub-Saharan Africa*, World Bank, Washington DC.

Yotopoulos, P.A. and Nugent, J.B. (1976), *Economics of Development*, Harper & Row, New York.

Yudelman, M., Butler, G. and Banerji, R. (1971), *Technological Change in Agriculture and Employment in Developing Countries*, Organisation of Economic Co-operation and Development, Paris.

Notes

1. Specifically, the inputs were land area, bullock time (hours), human labour (hours), and irrigation water (volume): the production alternatives were barley, wheat, peas and gram, all measured by weight.
2. Technically, the individual producer will choose the point on the production frontier which is tangent to his indifference curve (see Newbery and Stiglitz, 1981, pp. 170–1).
3. Factor productivity has physical attributes affected by resource endowments. A poor farmer with very little land and an excess of labour cannot be efficient in terms of *total* factor productivity without access to additional land. Berry and Cline (1979) overlook this aspect of the relationship between factor productivity and farm size.

7 Supply Response

1 Introduction

In economics 'supply response' in underdeveloped agriculture generally means the variation of agricultural output and acreage mainly due to a variation in price. Let Q_a be the volume of agricultural output and P indicate the price level, W_t weather condition (e.g. rainfall), A acreage, and t any time period. A typical simple supply response function can then be written as follows:

$$Q_a = f(P_{t-1}, A_t, W_t, U_t) \cdots \qquad (1)$$

where P_{t-1} is really a proxy for the expected price and U_t a statistical error term. Such a supply response functon indicates that the volume of agricultural output will depend upon past price level of such output, the acreage under cultivation at present and the level of rainfall in the current year plus a catch-all variable which is not specifically known, i.e. Y_t – the yield variation – is implicit in U_t. The exact nature of such a function can vary due to difference in the nature of the economies. But, at the outset, the reason behind formulating such a function is quite clear. It is to be expected that:

$$\frac{\partial f}{\partial P_{t-1}} > 0; \quad \frac{\partial f}{\partial A_t} > 0; \quad \frac{\partial f}{\partial W_t} \gtrless 0$$

This means that output is expected to vary positively with past price and land under cultivation but it could either rise or fall with changes in rainfall depending upon whether or not we have a normal rainfall or flood and drought.

Sometimes, acreage under cultivation is taken as a proxy for output due to a very close and direct relationship between acreage and output, i.e.

$$Q_{at} = a_1 + a_2 A_t \quad a_2 > 0 \cdots \qquad (2)$$

SUPPLY RESPONSE

Hence in many cases, supply response has been assumed to be equivalent to response of acreage under cultivation to changes in economic and non-economic factors. Thus, we have:

$$A_t = f(P_{t-1}, W_t, \cdots, U_t) \cdots \quad (3)$$

Sometimes, equation (3) has been modified to take into account the impact of *relative* rather than absolute prices. Assume that P_{at} stands

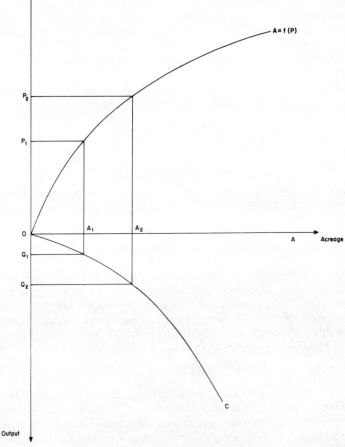

Figure 7.1: *Output and acreage response to price*

for the price of alternative crop that can be raised from the same plot of land, whereas P_{ot} is the own price of the crop. Hence, we obtain

$$A_t = f(P_{ot-1}/P_{at-1}, W_t, \cdots, U_t) \cdots \quad (4)$$

The mechanism of supply response can be illustrated with the help of a simple figure. In Figure 7.1, we measure acreage ($= A$) on the horizontal axis and price (relative or absolute) on the vertical axis. The curve OA represents the supply response of acreage to changes in the level of prices. Such a curve is clearly upward sloping. In the bottom section of the figure we show the relationship between output and acreage. As it has been already suggested, such a relationship is positive and this has been approximated by the curve OC. It can be easily shown from Figure 7.1 that when the price is OP_1, the amount of acreage under cultivation is OA_1 and output is OQ_1. If the price level rises to OP_2, acreage under cultivation rises to OA_2 and output increases to OQ_2.

However, it is possible to postulate a direct relationship between price of agricultural output and its supply. Here we can write:

$$Q_{at} = f(P_{at-1}) \cdots \quad (5)$$

or

$$Q_{at} = f(P_{ot-1}/P_{at-1}) \cdots \quad (6)$$

Such a supply response model is generally known as the Cobweb model. Note that P_a = own Price and P_0 = Price of other crops.

2 The Cobweb Model: An Illustration

The Cobweb model is based on the theory of lagged adjustment. Given the annual cycle of agricultural crops, markets in agriculture operate period by period and such a phenomenon is also known as a 'hog cycle' in the USA because production there is cyclical. If this is true, then the continuous adjustment principle should be replaced by a discrete one in the following way:

$$P_t - P_{t-1} = f[D_p(P_{t-1}) - S(P_{t-1})] \cdots \quad (7)$$

If the value of f were found to be very high, then the price amplitude of fluctuations, after an initial disturbance, could be ever-increasing around the equilibrium price and this could render the model 'explosive' (i.e. once the equilibrium price has been disturbed, there will be a tendency to move away from the point of equilibrium). Let us define an equilibrium as the point where demand ($=D_t$) is equal to supply ($=S_t$). Hence we have:

$$S_t = D_t \cdots \qquad (8)$$

S_t is *not* instantaneous supply curve but *lagged output* function. Let the demand function be written as:

$$D_t = a_0 - a_1 P_t \cdots \qquad (9)$$

This simply means that demand is inversely related to price, and as price goes up, demand goes down (and vice versa). The form of the demand curve is assumed to be linear to facilitate exposition.

The supply curve can be written as a function of the price in the last period (P_{t-1}) since it is assumed that current supply decisions are taken solely on the basis of past prices. Thus we have on the basis of our rather naive assumption:

$$S_t = b_0 + b_2 P_{t-1} \cdots \qquad (10)$$

Such a model is 'recursive' as it is argued that current year's supply is given by previous year's price and such a supply determines current price, given the current demand and market-clearing conditions. Once we obtain current year's price, we can determine supply in the next year. Such a situation is shown in Figure 7.2.

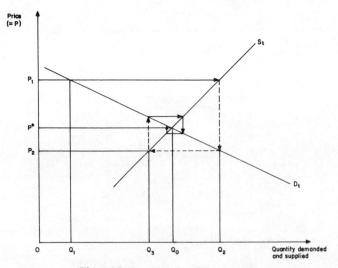

Figure 7.2: *A 'convergent' Cobweb model*

It is obvious from the diagram that the equilibrium price will be given at P^e price and output will be OQ_0. Let us assume that a drought leads to a crop failure and we have OQ_1 amount of output. Given an excess demand the price level rises to OP_1. Such an increase in price level encourages farmers to produce OQ_2. But then we have an excess supply, price level falls to OP_2 which, in its turn, leads to a reduction of supply to OQ_3. At this output we have an excess demand situation which results in a rise in price level. The oscillation process ends when the equilibrium price and output are obtained at OP^e and OQ_0.

It is necessary to point out that both price and output can overshoot (*above* the equilibrium values) or undershoot (*below* the equilibrium values) due to the inherent mechanism of a Cobweb theorem. Notice further that in Figure 7.2, the amplitude of fluctuations in price and output gradually falls and eventually convergence (i.e. movement towards the equilibrium) occurs at the equilibrium. This is due to the elasticities of the demand and supply curves. There is no reason to assume that a Cobweb cycle will always converge. Indeed, it can diverge, in which case there will be a movement away from the equilibrium values. This is demonstrated in Figure 7.3 simply by changing the slopes of the D and S curves. Readers can check for themselves that a Cobweb cycle may also be continuous, which implies no movement towards equilibrium or away from the initial cycle (see Figure. 7.4).

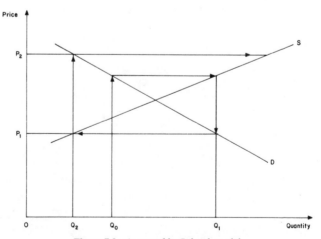

Figure 7.3: *An unstable Cobweb model*

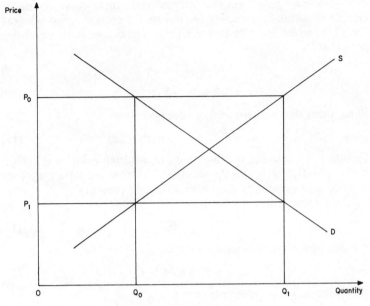

Figure 7.4: *A 'neutral' Cobweb model*

If the Cobweb movements are divergent, then the market is *unstable*. If it is convergent, then the market is *stable*.

2.1 Stability of the equilibrium in a Cobweb cycle

A close look at Figures 7.2 and 7.3 suggests why some markets for agricultural products are stable whereas others are unstable. The degree of stability in Figure 7.3 depends critically upon the elasticity of the supply curve. The more elastic the supply curve, the less stable the market. In other words, if the response of the suppliers to a small change in the expected price is drastic and large, then the market will tend to be unstable. Check also that the greater the *inelasticity* of demand, the more a sure change in supply will alter prices. As a general rule, it is possible to formulate the following proposition:

2.1.1 The Proposition
If the absolute elasticity of supply is greater than the absolute elasticity of demand, then the equilibrium is unstable. This is easily verified by drawing diagrams.

An algebraic proof. Let the demand and supply curves be linear and let demand be inversely related to the current price whereas supply be given by past period's price. Hence, we have equations (9) and (10):

$$D_t = a_0 - a_1 P_t \cdots \tag{9}$$

$$S_t = b_0 - b_1 P_{t-1} \cdots \tag{10}$$

Thus, from the market-clearing equation (8):

$$a_0 - a_1 P_t = b_0 + b_1 P_{t-1} \cdots \tag{11}$$

Equation (11) is useful to evaluate the equilibrium price ($= P^e$). Then if the initial price $(P_0) \neq P_e$, we want to know the time trajectory of price that eventually goes back to equilibrium price.

In the equilibrium, we have:

$$P^e = P_t = P_{t-1} \cdots \tag{12}$$

We can now rewrite equation (11) as:

$$a_0 - a_1 P^e = b_0 + b_1 P^e \cdots \tag{13}$$

The solution for P^e is then given by:

$$a_0 - b_0 = b_1 P^e + a_1 P^e \cdots \tag{14}$$

or

$$a_0 - b_0 = (a_1 + b_1) P^e \cdots \tag{15}$$

or

$$P^e = \frac{a_0 - b_0}{a_1 + b_1} \cdots \tag{16}$$

To find out the line trajectory of prices, beginning with $P = P_0$, we must find a solution for the first-order difference equation given by equation (11). Here, we consider price in t period as a function of price in the initial period, P_0. This yields equation (17) for P_t:

$$P_t = \frac{a_0 - b_0}{a_1 + b_1} + \left(-\frac{b_1}{a_1}\right)^t \left(P_0 - \frac{a_0 - b_0}{a_1 + b_1}\right) \cdots \tag{17}$$

$$= P^e + \left(-\frac{b_1}{a_1}\right)^t (P_0 - P^e) \cdots \tag{18}$$

In Figure 7.2, given the slopes of the supply and demand curves (e.g. b_1 is positive as the supply curve slopes up and so on) and $P_0 > P^e$, the change in the market in the first period will be given as follows:

$$P_0 = P^e + \left(-\frac{b_1}{a_1}\right)^0 (P_0 - P^e) \cdots \quad (19)$$

$$= P^e + P_0 - P^e = P_0 \cdots \quad (20)$$

For the next period, we can write:

$$P_1 = P^e - \left(\frac{b_1}{a_1}\right)(P_0 - P^e) \cdots \quad (21)$$

Price in this period is $P_1 < P^e$, whereas in the preceding period it was $P_0 > P^e$, and the market will attain stability if $(-b_1/a_1)^t$ tends to fall to 0 as t approaches infinity. This simply means that the market will be stable only if the elasticity of demand is greater than the elasticity of supply. Fortunately, and perhaps a bit surprisingly, markets for agricultural goods are usually stable (see any good textbook, e.g. Layard and Watters, 1978).

3 'Perverse Supply Response' in Backward Agriculture

In the previous sections we have discussed the problem of supply response in underdeveloped agriculture under the implicit assumption that farmers in LDCs behave in a *normal way*. This means that the higher the price level, the greater will be the level of output per acre. In other words, the slope of the supply curve is positive.

Under such a condition, the elasticity of supply with respect to price change is greater than zero. Suppose P denotes the price level and S denotes supply and ∂P and ∂S denote change in price and change in supply respectively. Therefore, the elasticity of supply η_s is:

$$\eta_s = \frac{\partial S/S}{\partial P/P} = \frac{\partial S}{S} \cdot \frac{P}{\partial P} = \frac{\partial S}{\partial P} \cdot \frac{P}{S} \cdots \quad (22)$$

The *normal* supply response occurs when $\eta_s > 0$.

However, two other behavioural postulates are possible. These are that the response to a price increase is either zero or negative. In the first case, supply of output may not respond to price changes at all. For instance, where farmers produce only for subsistence and not for the market, positive price changes may not induce them to produce more. In such cases, the supply elasticity will be equal to zero, i.e. $\eta_s = 0$. On the other hand, a price rise may actually induce the farmers to supply less output. This is generally known as a *'perverse* supply' response. Here the supply curve slopes *downwards* whereas in the 'normal' case the curve will slope upwards. In such a case, the elasticity of supply with respect to price will be negative, i.e. $\eta_s < 0$. Although perverse supply responses appears to be economically irrational it has merited some attention in the literature (see Mathur and Ezekiel, 1961; Khatkhate, 1962).

The case of the 'perverse supply' response crucially depend upon the assumption of a unit elasticity of the demand for money income so that when price falls by a certain proportion, farmers *raise* the *marketed* output by the same proportion to retain the same amount of money income. Hence, as price falls, marketed output *rises*, marketed output being defined as the amount of production net of consumption by the farmers. Others have obtained the same result by assuming that the elasticity of the money demand function of the subsistence farmers is less than unity (see Mathur and Ezekiel, 1961). Demand for money of the farmers is generally given by the requirements to make payments in cash (e.g. taxes, rent, demand for non-agricultural, say, industrial goods, etc). Although it is obvious that such assumptions will generate a perverse supply response curve (i.e. as price falls, supply rises), it is doubtful whether these assumptions are realistic enough to produce a perverse behaviour in practice. For one thing, the assumption that a fall in price of agricultural goods wll have no substitution effect is not very realistic. The substitution effect usually increases the demand for a product due to a fall in its relative price. For another, it is assumed that the income elasticity of the demand for non-agricultural goods is zero in these models as farmers actually reduce the sale of crops when price rises due to a fixed demand for money. This could only happen when farmers do not have any extra need for non-agricultural goods given an additional increase in their money income. Once again, such a situation seems implausible.

A perverse supply response may also be observed due to the impact of the surplus foreign food disposal programme. Under the U.S. Public Law 480, food export from the U.S.A. at times of acute food scarcity in LDCs has no doubt prevented a mass starvation and death. However, some have argued that when foreign supply competes directly with domestic production the market price at home will be depressed. It has been pointed out that although India succeeded in avoiding mass starvation during the severe drought-years of 1965–67 and subsequent crop-failures, nevertheless, the domestic supply of foodgrains failed to increase as prices of foodgrains were kept at an artificially low level due to the American export policy. Notice that had America been exporting to another poor country to prevent acute scarcity and famine, indirect competition (via arbitrage) with foreign supplies would have hindered prices to attain their 'normal' equilibrium values. The actual shortfall of domestic production will depend upon the values of elasticities. When foreign food is imported in the Indian market, then with a given Indian food supply, the degree of a fall in domestic price will be given by the value of the demand elasticity. As we have seen in the Cobweb model, the increase in supply in the next period will be determined by the value of the elasticity of supply with respect to prices. If we assume that the supply elasticity is normal i.e. positive, the import of food in a LDC will reduce increases in domestic supply (Schultz, 1960). Indeed, the higher the values of elasticity, the greater will be the reduction in supply. However, if we assume, along with Dantwala, that farmers do not respond at all to changes in agricultural prices, then the analysis of the effects of imports of foodgrains on domestic production is not applicable (Dantwala, 1963).

Food imports from abroad can help to set up buffer-stocks in developing countries. Such buffer-stocks are supposed to iron out sharp fluctuations in prices. If it is assumed that producers of agricultural goods are usually risk averse, then reduction of price fluctuations via imports may actually raise the production of agricultural goods at home. This will be the case even when the buffer-stock price is lower than the average of the oscillating price that would prevail in the absence of stabilization programmes (Fisher, 1963; see also chapter 10 of this volume).

4 A Simple Supply Response Model

It is possible to develop a simple supply response model to analyse the impact of price on output. Such a study usually involves the specification of output equation with reference to only two variables in the first instance. Other variables may be introduced at a later stage of the analysis (see e.g. Yotopoulos and Nugent, 1976). Let us assume that total output (= Q of any crop) is given by the product of acreage (= A) and yield (= Y). Thus we have

$$Q = A \cdot Y \cdots \qquad (23)$$

Let us also assume that both acreage under cultivation and yield respond to price (= P) changes. Totally differentiating, we obtain,

$$\frac{\partial Q}{\partial P} = Y\frac{\partial A}{\partial P} + A\frac{\partial Y}{\partial P} \ldots \quad (24)$$

Assuming constant returns to scale (i.e. an increase in inputs will raise output by the same proportion) and dividing the above equation by Q/P, we get,

$$\frac{dQ/dP}{Q/P} = \frac{Q/A \cdot \partial A/\partial P}{Q/P} + \frac{Q/Y \cdot \partial Y/\partial P}{Q/P} \ldots \quad (25)$$

or

$$\ell_{qp} = \frac{\partial A/\partial P}{A/P} + \frac{\partial Y/\partial P}{Y/P} \ldots \quad (26)$$

or

$$\ell_{qp} = \ell_{ap} + \ell_{yp} \ldots \quad (27)$$

where, ℓ_{ap} = the elasticity of acreage with respect to price, ℓ_{yp} = the elasticity of yield with respect to price, and ℓ_{qp} = the elasticity of output with respect to price.

It is generally assumed that ℓ_{yp} is non-negative. To obtain a lower bound estimate of ℓ_{qp}, it is only necessary to regress acreage on price. In empirical studies some have assumed that the price elasticity of yield is negligible (e.g. Falcon, 1964; Behrman, 1968). Others have emphasised the possibility of important yield responses to prices in countries like the USA where a substantial increase in yields have substituted for a limit on acreage (Nerlove, 1958).

The idea of regressing acreage rather than output on price is not very difficult to understand. Acreage is really a *proxy* variable since *planned* output (the variable to be explained) cannot be directly observed. It is, however, possible to use realised output as a proxy for planned output. But realised output could differ significantly from planned output due to the impact of random factors (e.g. weather) on agricultural output. However, acreage is generally under the direct control of the cultivator and hence it is usually accepted as the proxy variable for output.

As regards yield response and technical change, it should be mentioned that the substitution of yield for crop area in the USA has been substantially due to adoption of higher yielding crop varieties (as well as to heavier fertiliser application) and the same could be said of some LDCs during 1960s and 1970s due to the Green Revolution.

But, in the short run, *without technical change*, yield response is likely to be confined to variation in the intensity of harvesting, etc.

For example, if the price at harvest is very low (i.e. below the marginal cost of harvesting) farmers may not bother to harvest at all. But such varieties in short-term yield response may *not* be picked up empirically by using only annual data.

4.1 Role of expected price in supply response

It is important to bear in mind that acreage planted could vary with *expected* price of the output. Hence it may be necessary to develop a model which includes expected price as one of the explanatory variables. The price expectation models have been developed by Cagan (1956) and Nerlove (1958), *inter alia*, in the economic literature. Following Cagan, elasticity of expectation (ℓ_{ex}) can be defined as the percentage change in expected future price divided by percentage change in present prices.

$$\ell_{ex} = \frac{P_{t+1} - P_t}{P_t} \bigg| \frac{P_t - P_{t-1}}{P_{t-1}} \ldots \quad (28)$$

Hence a unit elasticity of expectation will imply that the proportion of expected price change to present price change will be the same. However, in Cagan's model, expectations are adaptive, which means people change their expectations in proportion to the error related with previous levels of expectation. Expected price (P_t^*) will thus be the sum of past expected price ($= P_{t-1}^*$) plus a proportion ($= \beta$) of the difference between actual past price ($= P_{t-1}$) and the past expected price. Hence, we can write:

$$P_t^* = P_{t-1}^* + \beta[P_{t-1} - P_{t-1}^*] \\ 0 < \beta \leq 1 \ldots \quad (29)$$

where β = coefficient of price expectation (associated with price uncertainty). It can be shown that:

$$P_t^* = \sum_{t=0}^{T} \beta(1-\beta)^i P_{t-1-i} \ldots \quad (30)$$

Since:

$$P_t^* = (1-\beta)P_{t-1}^* + \beta P_{t-1} \ldots \quad (31)$$
$$= \beta P_{t-1} + (1-\beta)[(1-\beta)P_{t-2}^* + \beta P_{t-2}] \ldots \quad (32)$$

and so on.

The above equation shows that anticipated prices form a geometrically falling lag structure as a function of all past prices (but cannot be estimated empirically because we are unable to observe P_{t-1}^*, P_{t-2}^*, etc).

When output is a function of anticipated prices, we have:

$$Q_t = a + bP_t^* + u_t \ldots \tag{33}$$

It is clear from the last three equations that we have:

$$Q_t = a + b\Sigma\beta(1-\beta)^i P_{t-1-i} + u_t \ldots \tag{34}$$

In the Cobweb model Equation (34) reduces to $Q_t = a + bP_{t-1}$.

To estimate the above equation, it is necessary to *follow an interactive procedure* to iterate β in order to obtain the highest proportion of explained variation (by the independent variables) to total variation.

In Nerlove's model, anticipated prices determine equilibrium output ($=Q_t^*$) and, in each production period, output is partially altered in proportion to the difference between last periods actual output and the long-run equilibrium output. Hence, we have the following basic equations:

$$Q_t^* = a + bP_t^* \ldots \tag{35}$$

$$P_t^* = \beta P_{t-1} + (1-\beta)P_{t-1}^* \ldots \tag{36}$$

$$Q_t = \gamma Q_t^* + (1-\gamma)Q_{t-1} \ldots \tag{37}$$

where $\gamma =$ the rate of adjustment associated with technical and institutional rigidity.

Manipulation to eliminate non-observable variables gives:

$$Q_t = \beta\gamma a + \beta\gamma b P_{t-1} + [(1-\beta)+(1-\gamma)]Q_{t-1} \\ - (1-\beta)(1-\gamma)Q_{t-2} \ldots \tag{38}$$

Equation (38) can be rewritten as:

$$Q_t = d + eP_{t-1} + fQ_{t-1} - gQ_{t-2} \ldots \tag{39}$$

where e and $e/1-f$ are the short and long-run supply elasticity parameters.

(Note: Since β and γ cannot be separately identified, Equation (39) can be solved empirically only by assuming either that $\beta = 1$ or that $\gamma = 1$ (which is obviously restrictive)). This assumption enables the lag 2 taken in Q_t to be ignored.

The Nerlove's supply response model is obviously more general than the simple supply models which we have discussed at the outset (see equation (10) above):

$$Q_t = b_2 + b_3 P_{t-1} \ldots \tag{10'}$$

A close look at equations (10) and (10′) reveal that the dependent variable (i.e. the variable to be explained) in (10′) is Q_t rather than S_t. The reason for such a change lies in the distinction between supply response in general and *marketed* supply or 'surplus' response in particular. This distinction is important and deserves special consideration. But before we discuss the problem of *marketed* surplus, it is necessary to clarify some issues related to the supply response in the agricultural labour market in LDCs.

5 Supply Response in the Underdeveloped Agricultural Labour Market

The term supply response in the agricultural labour market in the LDCs implies changes in the supply of labour in the agricultural sector in LDCs with respect to changes in the level of wages or income. It is sometimes assumed that the supply curve of labour in underdeveloped agriculture may be backward bending. The point can be further illustrated with the aid of figures. Consider a 'normal' supply curve of labour which is expected to rise upwards when wages/income rise (see Figure 7.5 and the curve S′).

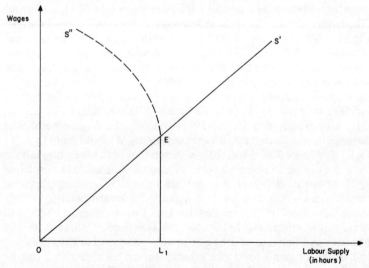

Figure 7.5: *Supply curve of labour*

A 'backward' bending supply curve is shown as OS″. Notice that up to the point E, the supply curve is 'normal' whereas after that point the curve bends backwards as less labour hours will be forthcoming despite a rise in the level of wages. Such a phenomenon could occur due to high leisure preferences when income rises. The perverse or backward bending supply curve of labour can also be illustrated with the help of a figure in which income/wages are measured on the vertical axis, and *leisure* (rather than labour) is measured in the horizontal axis. In Figure 7.6 the maximum amount of leisure is shown by the line O_1L_1. It is possible to measure *labour* hours if we move from right to left with O_1 as the point of origin. Thus, with O_1 as the point of start, a movement from right to left will indicate more labour and less leisure. Assume that there is a family of indifference curves between leisure and income/wages, and these are shown by curves like $i_0, i_1, i_2, i_3, \ldots$ in figure 7.6(a). Farmers have the choice not to do anything and enjoy leisure, in which case they consume O_1L_1 of leisure but do not earn much income. If they decide to work all the time they can initially get OW_1 level of wages. In other words, the line O_1W_1 is the 'budget' line of a farmer who has the opportunity to trade his leisure against a maximum income of OW_1. The leisure–income choice is given by the indifference curve which shows various combinations of leisure and income to which farmers are indifferent. Given the principles of indifference curve analysis, the initial equilibrium income and *leisure* are given by OW'_1 income and OL_1 leisure at point E on the indifference curve i_0 in Figure 7.6(a). In Figure 7.6(b), such an equilibrium combination is given by the level of wage at OW'_1 with OL_1 amount of *labour* supplied by a farmer.

Let us now assume that the farmer faces a new budget line as the level of income goes up from O_1W_1 to O_1W_2. The new equilibrium is, once again, attained where the budget line is tangent to the highest indifference curve i.e. E_1. Consumption of leisure falls from OL_1 to OL_2 and wages rise from OW'_1 to OW'_2. In Figure 7.6(b) this movement is indicated by an increased supply of labour from O_1L_1 to O_1L_2. In Figure 7.6(a) and (b) it is shown that both wages and labour supply will rise in the next point of equilibrium as the budget line shifts upwards. However, if wages rise above OW'_3, the supply curve 'bends' backwards. It is evident that the backward bending supply curve of labour is determined by two forces: income effect and substitution effect. We know that the substitution effect is always negative. An increase in the price of any 'good' will always lead to a fall in the demand for these goods. But an income effect can be negative or positive. An increase in income will normally raise the

SUPPLY RESPONSE

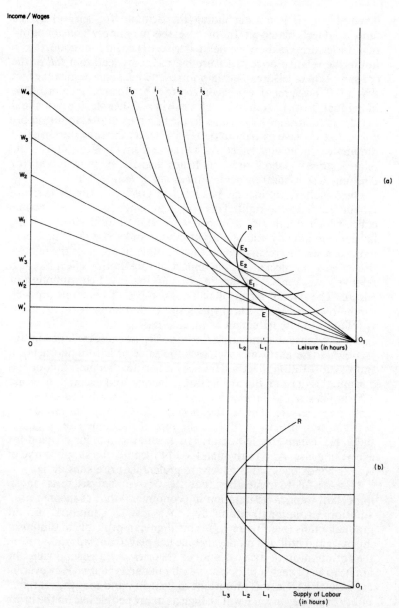

Figure 7.6: *Different cases of labour supply response in developed agriculture*

demand for all goods (including the demand for leisure) as the individual feels better-off. In this case, this may imply 'consumption' of more leisure. On the other hand, if leisure is an inferior good, then a fall in the relative price of leisure may actually lead to a *fall* in the consumption of leisure (consumption of *more labour*) despite the fact that a fall in the relative price of leisure raises the *real* income of the individual. Notwithstanding such a rise in real income, the individual decides to 'consume' less leisure. In the first part of the labour supply curve that we have traced out (i.e. O_1R) clearly the substitution effect dominates the income effect. As we move from O_1W_1 to O_1W_3, the relative price of *labour* rises and hence farmers sell more labour (or consume less leisure), hence *consumption of leisure* falls.

Once the level of income has reached OW'_3, the supply curve of labour bends to the right. This is due to a strong 'normal' income effect which dominates a rather weak substitution effect. At E_2, farmers are already selling a lot of labour and a rise in its price will indicate a strong rise in income. On the other hand, since the relative rise in price has decreased, the impact of substitution effect has also relatively weakened. The net result is an increased consumption of leisure. The overall result of these two types of reactions are well summarised in Figure 7.6(b) which shows that an increase in labour is followed by its reduction as income rises.

It is now easy to summarise the basic arguments developed in this section. A rise in income increases the price of leisure and hence a simple substitution effect will induce farmers to offer more labour and consume less leisure. But an increase in income and the price of leisure also implies a *fall* in *real* income of the farmer because he also 'consumes' leisure. This is the *income effect* which will reduce the consumption of leisure. But as the farmer's income rises substantially, his *demand* for leisure also rises with a rise in his demand for everything else. At a fairly high level of income, the supply curve of labour bends backwards as leisure preferences rise substantially.

It is useful to remember that the above analysis rests on an important implicit assumption in economics – that as income rises, additional (marginal) utility due to a rise in additional income gradually falls (see Figure 7.7). This Pigovian principle of diminishing marginal utility of money income has played an important role in the formulation of many economic theories. As a general rule, the application of such a principle is easily understandable. However, in the case of LDCs, it is important to bear in mind that the marginal utility of money income will be high as many people live on the brink of subsistence. As such, even if the level of farm income rises (from a

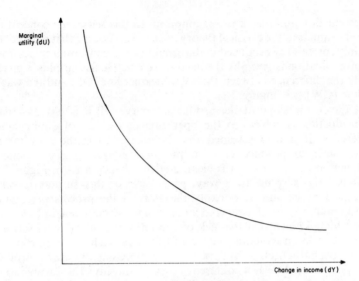

Figure 7.7: *Marginal utility of income*

very low level) it is very unlikely that the income effect will dominate the substitution effect. On the contrary, it will be more plausible to argue that the substitution effect will dominate the income effect and as such leisure will fall. Hence, in theory, it is not easy to make a case for a backward bending supply curve of labour in underdeveloped agriculture. Also, the empirical evidence available so far does not lend much support to the theory of a backward bending supply curve in the agricultural sector of LDCs. Not so much that available evidence fails to support the backward bending hypothesis, as that relevant evidence of any kind is extremely scarce. More research is needed in an LDC farming context.

6 The Concept of 'Marketed Surplus': Some Methods of Estimation

The concept of 'marketed' surplus has been regarded as important in the literature of economic development. 'Marketed' surplus refers to the amount of output which is sold in the market net of farmers' own consumption. So, in simplest terms, it is the difference between production and on-farm consumption. In this sense, the concept is very similar to the classical theory of surplus. Assume that corn is the only commodity produced in the economy and consumption per unit of population is given by the fixed wage line. Hence, surplus is given by the difference between the total product line (OP) and the wage line (OW) (see Figure 7.8).

Given the slope of the production curve OP it is obvious that production is subject to the operation of the law of diminishing returns. It is also assumed that labour ($=L$) is the only input available to produce corn output in the aggregate production function i.e. $Q = f(L)$. It is clear that the shaded area in Figure 7.8 shows the surplus. In a way, generation of this surplus is very important because it is really the start of the process of capital accumulation which is so vital for economic development in LDCs. It is widely known that the lack of capital is one of the most serious bottlenecks on economic growth of LDCs. As such, it is important to ease the bottleneck as far as possible by increasing the size of surplus.

There are, however, different types of surplus. The surplus that we have discussed so far is really food surplus. Other types include labour surplus in the Lewis and Fei-Ranis-type models. Here the concept of labour surplus implies that the marginal productivity of some labourers in agriculture is equal to zero so that their transfer

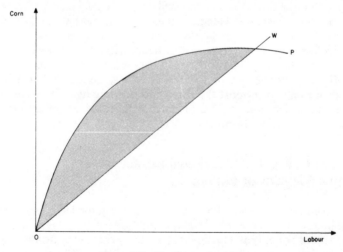

Figure 7.8: *The origin of surplus*

from the agricultural to the non-agricultural sector will not reduce the level of output within the agricultural sector. Such surplus labour may then be used for the industrial sector (see chapter 4).

It is also necessary to indicate that the agricultural sector can generate financial surplus. This can be measured by the difference between monetary receipts obtained by the agricultural sector and the monetary payments made by such sectors. Problems of capital formation should ease considerably if the agricultural sector can generate substantial financial surplus or profit which could then be reinvested for growth of both agriculture and industry as it has been emphasized in the Jorgenson type model. Public policies like taxation and borrowing could play a very vital role in the mobilisation of such surplus from agriculture. It is also possible to use a positive monetary policy whereby a *high* interest rate will be offered to attract more savings, sometimes held in the form of gold, land and estates.

As regards the estimation of marketed surplus we shall first set out the method used by Krishna (1964) in his seminal contribution. Later we shall introduce some modifications introduced by Behrman (1968) and others.

Let M be the marketed surplus, Q be the agricultural output produced by the farmer, and C be on-farm consumption. Hence we have the following definitional relationship:

$$M \equiv Q - C \cdots \qquad (38)$$

If we differentiate the above equation with respect to the market price of crop, we have:

$$\frac{\partial M}{\partial P} = \frac{\partial Q}{\partial P} - \frac{\partial C}{\partial P} - \frac{\partial C}{\partial I} \cdot \frac{\partial I}{\partial P} \quad \because C = C[P, I(P)] \qquad (39)$$

where I = the total net income of farmers. Krishna assumes that the farmer produces only *one* crop. A rise in price of this crop will raise his income. If the farmer were only a consumer of the crop, a rise in the price of the crop would reduce his total income in relation to the quantity of crop actually consumed. In Krishna's model, the farmer is both a consumer and a producer. Hence the net effect of a rise in price on income is the sum of these two conflicting effects (because, due to a price rise, as a producer the farmer gains whereas as a consumer he loses). Thus:

$$\frac{\partial I}{\partial P} = Q - C - M \cdots \qquad (40)$$

If we substitute equation (40) into (39), we have:

$$\frac{\partial M}{\partial P} = \frac{\partial Q}{\partial P} - \frac{\partial C}{\partial P} - M\frac{\partial C}{\partial I} \ldots \quad (41)$$

If we multiply the above equation through by P/M, and then arrange them to obtain results in terms of elasticities, we have:

$$\frac{P}{M} \times \frac{\partial M}{\partial P} = \frac{Q}{M} \cdot \frac{P}{Q} \cdot \frac{\partial Q}{\partial P} - \left[\frac{Q}{M} - 1\right]$$

$$\cdot \left[\frac{C}{P} \times \frac{\partial C}{\partial P} \times \frac{M}{Q} \cdot \frac{PQ}{I} \times \frac{I}{C} \times \frac{\partial C}{\partial I}\right] \ldots \quad (42)$$

$$\therefore e = rb - (r-1)(g + mkh) \ldots \quad (43)$$

The last equation has been obtained by using the following notations:

e ≡ the price elasticity of marketed surplus
r ≡ the ratio of total crop to marketed crop ≡ Q/M
b ≡ the price elasticity of cash crop ≡ P/Q × ∂Q/∂P
g ≡ the consumption elasticity of cash crop on the farm with respect to its price
k ≡ the ratio of the total value of cash crop produced to the total net income of the cultivator ≡ PQ/I
h ≡ the consumption elasticity of the cash crop with respect to total income of the former ≡ I/C · ∂C/∂I

By using the common statistical techniques it is possible to estimate the parameters like r, b, etc. to evaluate the overall value of e in equation (43).

7 Some Criticisms of Krishna's Method and the Alternative Approach of Behrman

Krishna's method of estimating the price elasticity of marketed surplus has been criticised on the following grounds:

(a) Krishna has used one crop and one price only. Such an assumption implies that the income of the cultivator is obtained from the sale of one crop only. Clearly, such an assumption is questionable. As Behrman points out, the relevant income consideration in the demand function for on-farm consumption of the crop of concern is the *total* net income of the farmer.

(b) Krishna defines P as a relative price, but his use of it sometimes suggests an absolute price. In his estimation, Krishna tacitly assumes that the first partial derivative of income with respect to the *relative* price is the same as the first partial derivative of such income with regard to the *absolute* price. The ambiguity in the use of the price variable may bias the model specification and its estimates.

(c) Krishna tacitly assumes complete adjustment which implies the elasticity that he estimates is relevant only for a sizeable number of production periods after a change in price. If the number of production periods which are necessary for complete adjustment is high, the partial adjustment after one or two or three periods may be of much more interest to policy makers than the complete adjustment. Krishna does not mention what this elasticity would be if only enough time for partial adjustment has passed.

7.1 Behrman's estimation

Behrman follows Krishna in his definition of marketed surplus:

$$M = Q - C$$

Since it is not adequate to allow only the absolute price of cash crop to alter (i.e. P_1) Behrman proceed to consider the total price from all sources of income of a producer of the crop other than that from the particular cash crop, (P_2), and the aggregate price for all goods other than this particular crop which is consumed by the producer of the cash crop, (P_3), Hence we have:

$$\frac{\partial M}{\partial P_1} = \frac{\partial(\partial Q)}{\partial(P_1/P_2)} \cdot \frac{\partial(P_1/P_2)}{\partial P_1} - \frac{\partial C}{\partial(P_1/P_3)} \cdot \frac{\partial(P_1/P_3)}{\partial P_1} - \frac{\partial C}{\partial I} \cdot \frac{\partial I}{\partial P_1} \ldots \tag{44}$$

If we use approximations, then the first partial derivative of *gross* income with respect to P_1 can be substituted for the first partial derivative of *net* income with reference to P_1. Writing in terms of elasticities, we have:

$$\frac{P_1}{M_1} \cdot \frac{\partial M_1}{\partial P_1} \simeq \frac{Q^I}{M_1} \left[\frac{P_1/P_2}{Q} \cdot \frac{\partial \partial Q}{\partial(P_1/P_2)} \right] - \left[\frac{Q^I}{M_1} - 1 \right]$$
$$\cdot \left[\frac{P_1/P_3}{C} \cdot \frac{\partial C}{\partial(P_1/P_3)} + \left\{ \frac{I}{C} \cdot \frac{\partial C}{\partial I} \right\} \left\{ \frac{P_1 Q^I}{I} \right\} \right]$$

$$\cdot\left\{1+\frac{P_1/P_2}{Q}\cdot\frac{\partial Q^I}{\partial(P_1/P_2)}\right\}\right]-\left[\frac{Q^I}{M_1}-1\right]$$
$$\cdot\left[\frac{I}{C}\cdot\frac{\partial C_3}{\partial I}\cdot\frac{P_1/P_2}{Q_A^I}\cdot\frac{\partial Q_A^I}{\partial(P_1/P_2)}\right]\left[1-\frac{P_1Q^I}{I}\right]\cdots \quad (45)$$

$$e \simeq rb_1 - (r-1)[g + hk(1+b_1)] - (r-1)hb_2(1-k)\cdots \quad (46)$$

where Q^I = planned production of all other goods, b_1 = the price elasticity of the cash crop with reference to relative price (i.e. P_1/P_2), and b_2 = the price elasticity of the planned goods (Q^I) *other* than Q, the cash crop produced with reference to P_1/P_2.

Behrman then re-estimates positive elasticities obtained by Krishna and comes up with negative elasticities. Needless to say that the implications of positive price elasticities are completely the opposite to those which are negative. In the first case, 'policy implication for mobilising surplus will be to raise the price of crops whereas in the latter case, the implication is to reduce price to raise surplus.' It is obvious why so much empirical research has been done in the LDCs to obtain a fairly accurate estimate of price elasticities of marketed surplus. However, when Behrman uses his own equation to obtain the elasticities of marketed surplus for Thai cultivators with respect to changes in both the absolute and the relative price level, he has obtained *positive* elasticities since total production response has been positive and the offsetting income effects on consumption are absent. In the absence of income effects on consumption, the estimated price elasticity of the marketed surplus of Thai rice is found to be higher than the elasticity of price of total supply.

7.2 Some estimation methods

Behrman uses the following per capital demand equation in order to derive income and price elasticities for on-farm rice consumption in Thai agriculture:

$$C_t = a_0 + a_1(P_1/P_3)_t + a_2I_t^n + u_t \cdots \quad (47)$$

where P_1/P_3 = the ratio of price to the alternative crop price, and I_t^n = the normal income of the consumer in period t. It is given by the following equation:

$$I_t^n = b_0 + I_{t-1}^n + b_1(I_{t-1} - I_{t-1}^n) + V_t$$
$$= \sum_{i=0}^{\infty} (1-b_1)^i (b_1 I_{t-1-i} + V_t) \cdots \quad (48)$$

If we substitute equation (48) into equation (47), the *reduced* form can be written in the following way:

$$C_t = a_2 b_0 + a_0 b_1 + (1 - b_1)C_{t-1} + a_1(P_1/P_3)_t \\ - a_1(1 - b_1)(P_1/P_3)_{t-1} + a_2 b_1 I_{t-1} + W_t \ldots \quad (49)$$

where W_t = error term.

Readers familiar with econometric methods will recognise that the structural parameters of the model of Behrman are overidentified. Given some assumptions about the error term, Behrman uses a non-linear maximum likelihood estimation method to estimate the parameters. Interestingly enough, Behrman does not find any substantial change in demand due to changes in income or price. Behrman himself admits that such a surprising result may be due to the following reasons:

(a) Data aggregation: the income or price effect on *total* rice consumption may not be significant since a higher price of a certain quality of rice may induce farmers to switch over to the consumption of lower quality rice without changing the quantity consumed.
(b) The quality of data that have been used might not have been good enough to yield a more reliable set of estimates.

Notwithstanding these limitations, Behrman uses zero estimates for price and income elasticities (i.e. g and h, respectively) in his final equation for the elasticity of the marketed surplus with respect to price (i.e. e in equation (46)), and with an estimated sales ratio, $r = 2.375$, *positive* values for elasticities of marketed surplus have been generally found for two different formulations for aggregate supply.

Behrman concludes that for both types of formulations, estimates of price elasticity of marketed surplus are positive as the income effect on consumption was absent while the price effect on production has been positive. Next, the partial adjustment may be of more interest than total adjustment as the latter may need a long period. Also, in the absence of an income effect, the measured elasticity of *marketed* surplus with respect to price is higher than the elasticity of *total* supply with respect to price. One of the major implications of Behrman's findings is to follow a positive price policy in Thai agriculture.

The estimation method of Behrman has not been universally followed. Nowshirvani, for instance, used a modified Nerlove model to obtain the desired level of acreage under cultivation (A_t^D) with the following set of equations:

$$A_t^D = b_1 + b_2 P_t^e + b_3 R_t + b_4 T + U_t \ldots \quad (50)$$
$$P_t^e - P_{t-1}^e = \beta(P_{t-1} - P_{t-1}^e) + \partial(Y_{t-1} - Y_{t-1}^e) \ldots \quad (51)$$
$$A_t - A_{t-1} = \gamma(A_t^D - A_{t-1}) \ldots \quad (52)$$

where T = time trend, R = rainfall, Y = yield of a crop, β = price expectation coefficient, and y = area adjustment coefficient. It is interesting to note that equation (51) suggests that peasants form their price expectations on the basis of both past price *and yield*-changes. Farmers may also discount that part of price alteration which is due to random supply changes. In the case of rice (though not in the case of other crops) it has been observed that such a hypothesis is valid when first differences of prices have been regressed on the first differences of yields.

On the basis of his study for two states in India (UP and Bihar), Nowshirvani concludes that *generally* the estimated price coefficients for subsistence crops have been insignificant but significant for cash crops. However, a simple Nerlovian model which uses only a weighted average of past prices to account for price expectation may not be adequate. The degree of commercialisation of a crop may have a crucial effect on expectation formation.

Other forms of the Nerlovian model which have been used can be stated as follows:

$$\log A_t = \log b_0 + \log b_1 \log A_{t-1} + b_1 \log P_{t-1} + b_3 \log T \ldots \quad (53)$$

where A = acreage under cultivation, P = average farm price/wholesale price, and T = the trend variable. In incorporating the effect of rainfall ($= W_t$), and allowing for the effects of relative prices and relative yields, we can write:

$$A_t = b_0 + b_1 P_{t-1} + b_2 Y_{t-1} + b_3 W_t + b_4 A_{t-1} + U_t \ldots \quad (54)$$

where P_{t-1} = one period lagged own crop price to alternative crop price and Y_{t-1} = one period lagged crop yield relative to alternative crop yield. If lagged relative gross income per acre is GR_{t-1} then the equation (54) may be rewritten in the following way:

$$A_t = C_0 + C_1 GR_{t-1} + C_3 A_{t-1} + C_4 W_t + U_t \ldots \quad (55)$$

However, care must be taken to define gross income ($= GR_t$) to avoid the problem of multicollinearity (interrelationships between any two variables in a multiple regression equation which are supposed to be independent of one another) by using both gross revenue and acreage together.

A number of other distributed lag models have been reported by Parikh (1971) in his econometric analysis of supply response models for Indian cereals. Parikh has obtained mixed results (i.e. both positive and negative price coefficients for wheat and rice) for difference Indian states – a conclusion which has been confirmed by the findings of Cummings (1974) for rice crops in India.

It must, however, be mentioned that most evidence from India tends to suggest a greater positive influence of price on the production of rice and wheat than other food crops. The most consistent pattern of positive estimates on the district level was found in Punjab (India and Pakistan), Andhra and Tamil Nadu (previously Madras province). But in Assam, Maharashtra and Karnataka (in India), mixed results have been found. In a large majority of districts, both rainfall and trend coefficient estimates are found to be positive and statistically significant to account for variations in both rice and wheat acreage. In arid and semi-arid regions, the impact of rainfall has been found to be striking.

Not surprisingly perhaps, in arid and semi-arid regions of India and Pakistan, the impact of rainfall has been particularly significant. Cummings claims that his supply model has performed satisfactorily in explaining rice and wheat acreage variations in most parts of India and Pakistan. It is useful to note that he has rejected the use of yield variable in his model because of its unreliability, though his supply model is basically Nerlovian, and includes both area adjustments and price expectations. His basic model can be written in the following simple way:

$$A_t = f(P^h_{t-1}, W_t, A_{t-1}, T) \ldots \qquad (56)$$

where A_t = crop acreage in period t, P^h_{t-1} = farm harvest price divided by working class cost of living index, W_t = an index for rainfall which shows deviation from normal rainfall during the period just preceding and during sowing, and T = the trend variable. Notice that multicollinearity may exist between T and A_{t-1}. Cummings uses OLS regression analysis to estimate equation (56). Due to the difficulties in the identification of the parameters which influenced both the adjustment and the expectation coefficients, Cummings, in his subsequent formulation of the model has imposed restrictions on the expectation coefficient. A different regression has been run for a specific value of the restriction which varied from zero to two and on the basis of the minimum sum of squares, the best one was chosen.

The index of weather that Cummings and Askari have discussed is also of special interest as it really stands for an index of *variation* in rainfall during the critical period of sowing and growing crops when soil moisture is badly needed. However, other types of weather indices may also be used. But the use of the working-class cost of living index as a deflator of farm harvest price may raise some questions about the assumed similarity of taste (demand) of the working class and the peasants. Although yield has not entered as an argument in the postulated function of equation (56), in one subsequent study Askari and Cummings have used yield expectation as one of the explanatory variables of acreage.

As regards the 'testing' of the price responsiveness hypothesis for other countries, Table 7.1 at the end of this chapter shows results for different crops in difference countries. In the case of the Philippines, although the price mechanism played a reasonably efficient role in resource allocation (i.e. inter-crop cultivation), its positive role in raising agricultural output has not always been confirmed due to weak response of yield and output growth to a positive price change (Managhas *et al.*, 1966). The supply response model that has been used in Philippines is basically similar to the one formulated by Krishna. On the other hand, in the case of Indonesia, positive price elasticities have been observed in many, but not all regions (Mubyarto, 1965).

The basic Nerlovian model, in a slightly modified form, has also been used to test the impact of land reform on supply responsiveness of peasants to price changes. Using the data for some selective Middle East countries (Egypt, Syria, Iraq, Jordan and Lebanon), the authors have hypothesised that if land reform had a favourable impact on the responsiveness of the farmers, supply elasticities should tend to have higher positive (less negative) values after land reform. Despite the lack of adequate data on time series in the case of Iraq and Syria, Askari and Cummings claim that land reform had a favourable effect in raising output and surplus. However, in the case of Iraq and Syria, land reform did not have much impact on output.

It must be borne in mind that farmers react differently to different types of crops, and a high supply elasticity for one crop does not necessarily mean that peasants will also react in the same way for all agricultural crops in general – a point that has been confirmed by an empirical study by Swift (1969) for Chile. Positive price elasticity for Peruvian rice cultivators has been observed by Merrill who used a Nerlovian price expectation model to obtain his estimates of elasticities. Merrill argues such 'economic rationality' is due to the strong market orientation of a few large landowners in Peru (Merrill, 1967). It has already been indicated that in a large part of South Asia farmers have shown a fair degree of price sensitivity. Evidence from

African countries also tends to confirm that farmers do respond to price movements. In the light of such evidence, the case for a positive price policy for agricultural development is strong indeed.

8 Perennial Crops

In the case of perennial and cash crops, similar supply response models have been used and tested (see Bateman, 1968; Behrman, 1968; Olayide, 1972, Tomek, 1972; Wickens and Greenfield, 1973; Saylor, 1974; and Ghosal, 1975). Price responsiveness of crops like cocoa, coffee, cotton and rubber has been estimated by the use of econometric models which tried to explain, say, supply of actual stock of cocoa trees ($= S_c$) by the expected real producer prices for cocoa ($= P_c^e$) and expected real producer prices for alternative cash crops ($= P_a^e$). The adjustment in the true stock is basically Nerlovian as it is a proportion to the difference between actual and planned stock of tree. Hence:

$$S_t - S_{t-1} = \delta(S_t^p - S_{t-1})\ldots \quad (57)$$

where $\delta =$ the stock adjustment coefficient.

It is assumed that S_t^p is a function of expected prices. Thus:

$$S_t = C_0 + C_1 P_{ct}^e + C_2 P_a^e + U_t \ldots \quad (58)$$

Equations (57) and (58) have then been used with a price expectation equation (discussed earlier) to obtain a relationship between stock adjustment and price movements. The estimates of short- and long-run elasticities for four different regions in Ghana in this sort of stock adjustment model have been found as positive (Bateman, 1968).

Behrman points out that it is more useful to consider *planned area* under cocoa (A_t^p) as a function of expected prices of both cocoa and the major substitute crop, which is coffee. Thus, Behrman obtains the following equation:

$$A_t^p = a_0 + a_1 P_{ct}^e + P_{at}^e + U_t \ldots \quad (59)$$

Due to the lack of information about area under cocoa, Behrman writes equation (59) in the following way:

$$Q_t = b_0 + \Sigma Y_i A_{t-i} + b_2 P_t^c + b_3 P_{t-1}^c + V_t \ldots \quad (60)$$

where $P_t^c =$ the actual cocoa price at time t, $Y_i =$ the average yield per unit area i years after planting, and $Q_t =$ production of cocoa. The problem of infinite series has been solved by a first difference formulation that could be written as follows:

$$Q_t = C_0 + C_1 \Delta Q_{t-1} + C_2 \Delta Q_{t-2} + C_3 \Delta P_t \\ + C_4 \Delta P_{t-1} + C_5 \Delta P_{t-2} + C_6 \Delta P_{t-3} + C_7 \Delta P_{t-n_1} \\ + C_8 \Delta P_{t-n_2} + C_9 \Delta P^a_{t-n_1} + C_{10} \Delta P^a_{t-n_2} + W_t \ldots \quad (61)$$

where n_1 = the age at which trees first bear fruit and n_2 = the age at which second major rise in yield takes place.

Although Behrman obtains positive short- and long-run price elasticities in eight African and Latin American countries, he points out that in six of the countries there is little evidence of significant short-run price responsiveness. Despite this finding, Behrman cautions against a policy of raising high prices by supply restrictions unless such a policy is pursued along with stockpiling. It is important to stress that, except for Venezuela, Ecuador and the Dominican Republic, Behrman obtains high values of long-run supply elasticities. As such, the policy of maintaining a high and stable price may eventually produce 'cocoa mountains'. If this is true then it follows that, in practice, it is necessary to follow the following three policies: (a) a strict restriction on cocoa planting; (b) producers should operate independent of the movements of world prices; and (c) cocoa marketing boards should work as efficiently as possible.

In the case of coffee, attempts have been made to use the basic Nerlovian model to test the price responsiveness of coffee producers in a number of LDCs (see e.g. Frederick, 1965, 1969; Saylor, 1974). Interestingly enough, prices did not turn out to be significant variable in many cases in explaining variation in coffee acreage in Uganda. Clearly Frederick's study suggests that other factors (e.g. weather) should also be taken into consideration in explaining coffee acreage variation – a point which has been emphasised by Arak (1967) in her study on the Brazilian coffee supply. Taxes and quotas could influence price expectations considerably. Planting may be regarded as a function of the proportion of existing trees (which are at least 10-years-old) and the 'optimal' age distribution of coffee trees. The stock adjustment model can be regarded as a function, albeit simplistically, of expected coffee prices, existing stock of trees and the cumulative planting. In a different formulation of the same type of model, readers may wish to explain the *ratio* of new planting to the existing stock, of coffee trees. The change in the age induced planting at $t(=\Delta A^a_t)$ can be considered as a function of the *proportion* of existing trees older than ten years ($=T_t$). Hence:

$$\Delta A_t^a = f(T_t) = \lambda(T_t - T_t^*) \qquad \lambda > 0 \ldots \qquad (62)$$

where λ = a function of expected coffee prices, T_t^* = the 'optimal' tree age distribution, and $\delta \Delta A_t^a / \delta T_t > 0$.

Expected coffee prices have been generated by Arak with the help of the exchange rate and the New York market price of coffee. Thus we have:

$$P^e = P_m X m \ldots \qquad (63)$$

where P^e = the expected price of coffee, P_m = the market price of coffee, and m = the fraction of quota. The overall supply equation, following Arak, can then be written as follows:

$$\Delta A_t^a = u_1 \left[(a_0 + a_1 P_t^e) - T_{1920} - \sum_{i=1921}^{t-1} \Delta A_i \right] \\ + \lambda_1 [T_t - (b_0 + b_1 P_t^e)] \ldots \qquad (64)$$

If we wish to explain the proportion of new planting to the existing stock of trees in t, the above equation can then be rewritten as follows:

$$\frac{\Delta A_t^a}{S_t} = u_2 \left[(a_0 + a_1 P_t^e) - T_{1920} - \sum_{i=1921}^{t-1} \Delta A_i \right] \\ + \lambda_2 [T_t - (b_0 + b_1 P_t^e)] \ldots \qquad (65)$$

where ΔA_i = planting of trees in time i, T_{1920} = stock of coffee trees in 1920, S_t = the stock of coffee trees in time t, and T_t = the proportion of existing trees other than ten years. With the aid of equations (64) and (65), Arak obtains the reduced form of her model with which she estimates that the short-run elasticity of São Paulo coffee planting (measured annually) with respect to expected coffee price is $+2.02$. However, when she assumes that the age distribution of the tree-stock can change the difference between planned and actual stock (i.e. μ_1 and μ_2 shown as a proportion of trees which are at least 10-year-old, i.e. T_t), then the estimated short-run elasticity rises slightly to $+2.28$. Limitations on data have prevented Arak from reaching conclusive results in other areas, though the estimated elasticity (planting) with respect to expected price has been positive in all areas. She concludes that high export taxes have significant impact on price expectations though public policies designed to control the planting of new coffee trees have been virtually impotent.

Arak's results are, however, questioned by Bacha (1968) who has introduced production cycles in his model of the coffee economy. Bacha uses the following equations to show the relationship between output of coffee (Q_t) and area (A_t) with the following equations:

$$Q_t = aA_t \ldots \quad (66)$$

if t is even

$$Q_t = bA_t \ldots \quad (67)$$

if t is odd

where a = the output to area ratio in off years and b = output/area ratio in peak years.

Bacha's supply equation is then expressed in terms of lagged output ($= Q_{t-1}$) and past prices. It is assumed that price expectations of Brazilian producers change slowly. We then have:

$$Q_t = a_0 + a_1 D_t + a_2 P_{t-4} + a_3 D_t P_{t-4} + a_4 Q_{t-2} \ldots \quad (68)$$

where D_t = the dummy variable and equals zero if t is even and 1 if t is odd. Bacha's estimates of short and long-run supply elasticities for coffee have turned out to be $+0.23$ and $+1.00$, respectively. For Latin America as a whole (1943–60), such short- and long-run elasticities have been observed as $+0.276$ and $+0.518$, respectively, while for Africa (1943–60) they are estimated as $+0.239$ and $+0.374$, respectively.

In the case of other perennial crops like tea, evidence from India suggests that although the short-run acreage elasticity with respect to price has been positive, its value has not been high and usually varies between $+0.024$ to $+0.157$, whereas output elasticity has been slightly higher (varying between $+0.142$ and $+0.351$). However, all price parameters have been found to be statistically significant at the 5 per cent level (see Rajagopalan and Meenakshisundaram, 1969).

As regards rubber, Behrman's estimates of elasticity of supply in Malaysian and other producing countries may be mentioned. Using the simple supply response model in terms of current price and introducting *expected* yield, rainfall and area as explanatory variables, Behrman obtains significant price responsiveness and relatively important estimates for small producers. But for large estates, the estimates have proved to be insignificant. Similarly, other studies by Ghosal suggest a short-run elasticity for the Nerlovian model as $+0.22$. When Ghosal (1975) uses a simple model with the help of current prices and a trend variable, he obtains an estimate of elasticity of price which is $+0.12$. Such a low value of elasticity has been explained partly by technical reasons where output could not have

Table 7.1: Supply elasticities, by crop and region

Crop	Region	Period	Author	Short-Run Elasticity	Long-Run Elasticity
RICE	Punjab	1914–46	Raj Krishna	+0.31	+0.59
	Punjab	1960–69	Kaul and Sidhu	+0.19 to 0.24	+0.64 to 0.68
	Punjab	1955–66	Askari and Cummings	+0.18	+0.42
	Haryana	1950–70	Singh, Singh, and Rai	+0.83	—
	Uttar Pradesh (14 districts)	1953–63	Nowshirvani	—	−0.11 to +0.27
	Bihar (12 districts)	1953–63	Nowshirvani	—	+0.01 to 0.022
	Bihar-Orissa	1900–39	Parikh	+0.16 to 0.24	—
	Orissa	1938–51	NCAER	+0.05	—
	Madras (Tamil Nadu)	1937–66	Subramanian	negative	—
	Madras (Tamil Nadu)	1946–67	Cummings	+0.08	+0.08
	Madras (Tamil Nadu)	1952–65	Askari and Cummings	−0.26	+0.76
	Andhra Pradesh	1952–67	Askari and Cummings	+0.46	+0.66
	Himachal Pradesh	1949–66	Cummings	−0.07	−0.06
	Assam	1950–67	Cummings	+0.07	+0.07
	Gujarat	1954–67	Cummings	−0.07	−0.07
	Kerala	1955–67	Askari and Cummings	−0.14	−0.10
	Maharashtra	1955–67	Cummings	−0.12	−0.14
	Mysore	1951–67	Cummings	+0.06	+0.07
	Bengal	1911–38	S. Krishna	+0.06	+0.19
	West Bengal (autumn rice)	1949–66	Cummings	+0.37	+0.38
	West Bengal (winter rice)	1919–66	Cummings	−0.05	−0.08
	India	1938–57	NCAER	+0.22	—
	Pakistan (west)	1949–68	Cummings	+0.12	+0.17
	Pakistan (west)	1950–68	Askari and Cummings	−0.03	−0.07
	Bangladesh	1948–63	Hussain	+0.03 to 0.09	—
	Bangladesh	1950–68	Askari and Cummings	+0.23	+1.28
	Egypt	1920–40	Askari, Cummings, and Harik	−0.21	−0.24

Table 7.1 (continued)

Crop	Region	Period	Author	Short-Run Elasticity	Long-Run Elasticity
RICE (continued)	Egypt	1953–72	Askari, Cummings, and Harik	+0.08	+0.08
	Iraq	1961–71	Askari, Cummings, and Harik	+0.66	+1.57
	Thailand (50 provinces)	1937–63	Behrman	+0.19	+0.28
	Thailand (50 provinces)	1937–63	Behrman	+0.08 to 0.33	+0.16 to 0.45
	West Malaysia	1951–65	Aromdee	+0.23 to 0.25	+1.35
	Japan	1951–65	Aromdee	0	0
	Philippines Ilocos	1953–64	Mangahas, Recto, and Ruttan	+0.04 to 0.28	+0.51
	Philippines Central Luzon	1953–64	Mangahas, Recto, and Ruttan	neg. to +0.55	neg. to +2.15
	Philippines South Tagalog	1953–64	Mangahas, Recto, and Ruttan	+0.19 to 1.95	+0.42 to 2.06
WHEAT	Gujarat	1954–67	Cummings	+0.93	+1.00
	Maharashtra	1955–67	Cummings	+0.24	+0.33
	Punjab (irrigated)	1914–46	Raj Krishna	+0.08	+0.14
	Punjab	1900–39	Parikh	+0.06	+0.10
	Punjab (irrigated)	1951–64	Kaul	+0.08	+0.09
	Punjab	1950–67	Cummings	+0.10	+0.13
	Punjab	1948–65	Maji, Jha, and Venkataraman	+0.11 to 0.67	+0.51 to 1.02
	Madhya Pradesh-Berar	1900–39	Parikh	neg.	neg.
	Haryana	1950–70	Singh, Singh, and Rai	+0.60	—
	Delhi	1953–67	Askari and Cummings	+0.25	+0.28
	Himachal Pradesh	1953–66	Askari and Cummings	+0.04	+0.04

Crop	Region	Period	Author	Short-Run Elasticity	Long-Run Elasticity
	Uttar Pradesh	1953–62	Nowshirvani	—	−0.13 to +0.76
	Uttar Pradesh	1950–62	Rao and Krishna	+0.03 to 0.21	+0.09 to 0.64
	Bihar (3 districts)	1952–64	Nowshirvani	—	+0.41
	Mysore	1955–67	Askari and Cummings	+0.48	+0.52
	Rajasthan	1954–68	Askari and Cummings	+0.13	+0.29
	West Bengal	1946–67	Cummings	+0.23	+0.20
	Pakistan (West) (dry farming)	1933–59	Falcon	0	—
	Pakistan (West) (irrigated)	1933–59	Falcon	+0.10 to 0.20	—
	Pakistan (West)	1950–68	Askari and Cummings	+0.07	+0.21
	Egypt	1953–72	Askari, Cummings, and Harik	+0.91	+0.44
	Syria	1947–60	Askari, Cummings, and Harik	−0.02	−0.03
	Syria	1961–72	Askari, Cummings, and Harik	+0.64	+3.23
	Iraq	1951–60	Askari, Cummings, and Harik	+0.40	−4.42
	Iraq	1962–71	Askari, Cummings, and Harik	−0.85	−0.34
	Jordan	1955–67	Askari, Cummings, and Harik	+0.20	+0.23
BARLEY	Punjab	1914–46	Raj Krishna	+0.39	+0.50
	Punjab	1950–67	Cummings	+0.22	+0.27
	Punjab	1955–66	Askari and Cummings	−0.63	+0.94

Table 7.1 (continued)

Crop	Region	Period	Author	Short-Run Elasticity	Long-Run Elasticity
BARLEY (continued)	Haryana	1950–70	Singh, Singh, and Rai	+0.58	—
	Uttar Pradesh (4 divisions)	1953–63	Nowshirvani	—	+0.04 to 0.50
	Bihar	1953–63	Nowshirvani	—	+0.17 to 0.40
	Rajasthan	1954–68	Askari and Cummings	+0.31	+1.19
	Delhi	1953–67	Askari and Cummings	+0.53	−3.60
	Himachal Pradesh	1949–66	Cummings	−0.10	−0.26
	India	1938–57	NCAER	+0.10	—
	Pakistan (West)	1951–68	Cummings	+0.03	+0.02
	Pakistan (West)	1954–68	Askari and Cummings	+0.01	+0.01
	Syria	1953–72	Askari, Cummings, and Harik	−0.57	−0.57
	Iraq	1951–60	Askari, Cummings, and Harik	+0.16	−0.28
	Iraq	1962–71	Askari, Cummings, and Harik	−0.05	−0.19
MAIZE	Punjab	1914–46	Raj Krishna	+0.23	+0.56
	Punjab	1960–69	Kaul and Sidhu	+0.11 to 0.13	+0.14 to 0.16
	Haryana	1950–70	Singh, Singh, and Rai	+0.33	—
	Egypt	1953–72	Askari, Cummings, and Harik	+0.04	+0.09
	Sudan	1951–65	Medani	+0.23	+0.56
	Syria	1961–72	Askari, Cummings, and Harik	+2.27	+2.16
	Jordan	1955–66	Askar, Cummings, and Harik	−0.21	−0.25
	Lebanon	1953–72	Askari, Cummings, and Harik	+0.13	+0.29

Crop	Region	Period	Author	Short-Run Elasticity	Long-Run Elasticity
	Philippines	1946–64	Mangahas, Recto, and Ruttan	neg. to +0.23	+0.42 to 1.14
	Thailand (4 provinces)	1949–63	Behrman	+0.27 to 4.47	+0.41 to 14.17
	Thailand (1 province)	1955–63	Behrman	+1.09	+1.09
CASSAVA	Punjab	1914–46	Raj Krishna	+0.09	+0.36
BAJRA	Punjab	1951–64	Kaul	−0.05	−0.06
	Haryana	1950–70	Singh, Singh, and Rai	pos.	
	Madhya Pradesh	1951–64	Kaul	−0.08	−0.16
	Madras (Tamil Nadu)	1947–65	Madhavan	−0.22 to +0.03	−2.50 to +0.15
CUMBU	Madras (Tamil Nadu)	1951–65	Rajagopalan	+0.83 to 0.90	
RAGI	Madras (Tamil Nadu)	1947–65	Madhavan	+0.09 to 0.15	+0.16 to 0.31
	Madras (Tamil Nadu)	1951–65	Rajagopalan	pos.	
	Punjab	1914–46	Raj Krishna	0	
JOWAR	Sholapur (Maharashta)	1938–57	NCAER	+0.50	−0.58
	Madhya Pradesh	1951–64	Kaul	−0.04	−0.06
	Madras (Tamil Nadu)	1947–65	Madhavan	+0.02 to 0.20	+0.03 to 0.28
SORGHUM	Sudan	1951–65	Medani	+0.31	+0.59
	Sudan (traditional farms)	1966–69	Medani	+0.10 to 0.21	+0.23 to 0.31
	Sudan (modern farms)	1966–69	Medani	+0.50	+0.63 to 0.70
	Punjab (3 districts)	1951–64	Kaul	−1.00 to +0.49	−1.52 to +1.38
GRAM	Haryana	1950–70	Singh, Singh, and Rai	pos.	
	Andhra Pradesh	1957–67	Bhadur and Haridasan	+0.06 to 0.67	
GREEN GRAM	Andhra Pradesh	1957–67	Bhadur and Haridasan	+0.18 to 0.41	
	Iraq	1961–70	Askari, Cummings, and Harik	−0.24	−0.32
VETCH	Syria	1961–72	Askari, Cummings, and Harik	−0.50	−0.53
	Jordan	1955–67	Askari, Cummings, and Harik	−0.37	−0.62

Table 7.1 (continued)

Crop	Region	Period	Author	Short-Run Elasticity	Long-Run Elasticity
MILLET	Syria	1961–72	Askari, Cummings, and Harik	+1.21	+1.60
	Iraq	1961–71	Askari, Cummings, and Harik	−0.84	−3.30
GIAN MILLET	Sudan	1951–65	Medani	+0.09	+0.36
	Iraq	1961–71	Askari, Cummings, and Harik	+0.88	+1.85
BROAD BEANS	Egypt	1953–72	Askari, Cummings, and Harik	+0.19	+0.14
SESAMUM	Andhra Pradesh	1955–68	Cummings	+0.29	+0.23
	Gujarat	1955–68	Cummings	+0.08	+0.10
	Maharashta	1955–68	Cummings	+0.23	+0.30
	Mysore	1955–68	Cummings	+0.03	+0.04
	Punjab	1949–67	Cummings	−0.93	−2.33
	Rajasthan	1951–68	Cummings	+0.37	+0.34
	Madras (Tamil Nadu)	1947–65	Madhavan	+0.42 to 0.48	+0.31 to 0.32
	Assam	1949–67	Cummings	−0.42	−0.98
	Tripura	1954–67	Cummings	+0.40	+0.56
	Pakistan (West)	1951–67	Cummings	−0.09	−0.09
	Bangladesh (winter)	1953–64	Cummings	+0.21	+0.60
	Bangladesh (summer)	1953–64	Cummings	−0.28	−0.20
RAPE AND MUSTARD	Pakistan (West)	1950–62	Cummings	+0.38	+0.48
	Bangladesh	1950–62	Cummings	+0.23	+0.42
	Punjab	1951–67	Cummings	+0.89	+4.05
	Punjab	1960–69	Kaul and Sidhu	+0.51 to 0.78	+3.05 to 3.25
	Uttar Pradesh	1953–64	Nowshirvani		+0.89
	Bombay	1953–68	Boon-raung et al.	+0.24	—
GROUNDNUTS	Gujarat	1955–67	Cummings	−0.11	−0.11
	Maharashtra	1956–67	Askari and Cummings	+0.10	+0.33

208 AGRICULTURE AND ECONOMIC DEVELOPMENT

Crop	Region	Period	Author	Short-Run Elasticity	Long-Run Elasticity
	Mysore	1955–66	Askari and Cummings	−0.17	−0.15
	Madras (Tamil Nadu)	1900–39	Parikh	+0.80	+3.71
	Madras (Tamil Nadu)	1947–65	Madhavan	+0.03 to 0.34	+0.04 to 0.65
	Pondicherry	1958–68	Cummings	+0.16	+0.14
	Andhara Pradesh	1951–67	Cummings	+0.69	+0.52
	Rajasthan	1954–68	Askari and Cummings	−0.31	−1.24
	India	1938–57	NCAER	+0.22	—
	India	1953–68	Boom-raung et al.	+0.22	—
	Sudan	1951–65	Medani	+0.72	+1.62
	Nigeria	1948–67	Olayide	+0.24 to 0.79	—
LINSEED	Iraq	1950–60	Askari, Cummings, and harik	+2.33	+2.44
	Iraq	1961–71	Askari, Cummings, and Harik	−2.85	−14.23
COTTON	Punjab (Desi variety)	1922–43	Raj Krishna	+0.59	+1.08
	Punjab (American variety)	1922–43	Raj Krishna	+0.78	+1.62
	Punjab (American)	1951–64	Kaul	+0.34	+2.84
	Punjab (Desi)	1960–69	Kaul and Sidhu	+0.45 to 0.68	+0.79 to 1.17
	Bombay	1904–39	S. Krishna	+0.15	+0.25
	Gujarat	1954–68	Cummings	+0.05	+0.08
	Madras (Tamil Nadu)	1947–65	Madhavan	+0.01 to 0.31	+0.02 to 0.54
	Andhra Pradesh	1953–68	Askari and Cummings	+0.40	+0.64
	Kerala	1957–68	Cummings	−0.39	−0.41
	Mysore	1955–67	Askari and Cummings	+0.33	+0.54
	Assam	1954–68	Askari and Cummings	+0.12	+0.25
	Tripura	1951–69	Cummings	+0.20	+0.29
	India	1938–57	NCAER	+0.75	—

Table 7.1 (continued)

Crop	Region	Period	Author	Short-Run Elasticity	Long-Run Elasticity
COTTON (continued)	India	1948–61	Raj Krishna	+0.64	+1.33
	Pakistan (West)	1933–59	Falcon	+0.41	—
	Pakistan (Desi)	1950–67	Cummings	+0.41	+0.28
	Pakistan (American)	1950–67	Cummings	+0.40	+0.47
	Egypt	1899–1937	Stern	+0.38	—
	Egypt	1914–37	Stern	+0.52	—
	Egypt	1953–72	Askari, Cummings, and Harik	−0.09	−0.08
	Syria	1961–72	Askari, Cummings, and Harik	+1.49	+1.09
	Iraq	1961–71	Askari, Cummings, and Harik	−0.85	−1.44
	Sudan	1951–65	Medani	+0.39	+0.50
	Nigeria	1948–67	Oni	+0.38	+0.28
	Uganda Buganda	1945–66	Alibaruho	+0.50	+0.63
	United States	1909–32	Nerlove	+0.20 to 0.67	—
KENAF	Thailand (8 provinces)	1954–63	Behrman	+0.88 to 7.71	+1.19 to 42.60
JUTE	Bengal (undivided)	1911–39	Stern	+0.68	+1.03
	West Bengal	1900–39	Parikh	+0.01	+0.12
	West Bengal	1951–62	Rabbani	+0.70	+0.71 to 0.74
	West Bengal	1954–69	Askari and Cummings	+0.58	+0.89
	Assam	1949–69	Cummings	+0.07	+0.05
	Bihar	1949–62	Rabbani	+0.78 to 0.80	+0.88 to 0.97
	Bihar	1955–69	Askari and Cummings	+0.57	+0.65
	Orissa	1950–62	Rabbani	+0.75 to 0.79	+0.77 to 0.88
	Uttar Pradesh	1957–68	Cummings	+0.14	+0.14
	Tripura	1949–69	Cummings	+0.80	+1.60
	India (undivided)	1911–39	Rabbani	+0.38 to 0.47	+0.65 to 0.80

Crop	Region	Period	Author	Short-Run Elasticity	Long-Run Elasticity
	India	1951–62	Rabbani	+0.74 to 0.76	+0.96 to 0.99
	Bangladesh	1931–54	Clark	+0.60	—
	Bangladesh	1949–63	Rabbani	+0.39 to 0.40	+0.65 to 0.66
	Bangladesh	1948–66	Huq	+0.35	+0.83
	Bangladesh	1949–68	Cummings	+0.40	+0.48
SISAL	Tanzania	1945–67	Gwyer	+0.42 to 0.50	+0.24 to 0.42
TOBACCO	Andhra Pradesh	1940–67	Rao and Singh	+0.25	+0.42
	Assam	1956–67	Askari and Cummings	−0.30	−0.26
	Bihar	1954–67	Askari and Cummings	+0.05	+0.06
	Gujarat	1954–67	Cummings	+0.16	+1.00
	Maharashtra	1954–68	Cummings	−0.08	−0.12
	Mysore	1955–67	Askari and Cummings	+0.06	+0.07
	Madras (Tamil Nadu)	1949–67	Cummings	+0.22	+0.25
	India	1938–57	NCAER	+0.71	—
	Pakistan (West)	1951–67	Askari and Cummings	−0.08	−0.11
	Bangladesh	1950–66	Cummings	+0.51	+0.53
	Nigeria	1945–64	Adesimi	+0.60	+0.82
	Malawi	1926–60	Dean	+0.48	—
SUGAR	Uttar Pradesh (6 divisions)	1909–42	Nowshirvani	+0.19 to 1.47	+0.24 to 2.46
	Uttar Pradesh	1950–68	Rathod	+0.25	—
	Punjab	1915–43	Raj Krishna	+0.34	+0.60
	Bihar	1951–64	Nowshirvani	—	+1.38
	Bihar Tirkut division	1950–64	Jha	+0.65	+0.79
	Andhra Pradesh	1952–64	Subbarao	+0.50	—
	Madras (Tamil Nadu)	1947–65	Madhavan	+0.52 to 0.63	+0.66 to 1.21
	Philippines	1914–64	Askari	+0.08 to 0.13	+0.13 to 0.16

Table 7.1 (continued)

Crop	Region	Period	Author	Short-Run Elasticity	Long-Run Elasticity
COCOA	Ghana (medium areas)^f	1949–62	Bateman	+0.42 to 0.51	+1.28
	Ghana (new areas)^f	1949–62	Bateman	+0.61 to 0.87	+1.06
	Ghana	1947–63	Behrman	—	+0.71
	Nigeria	1947–63	Behrman	—	+0.45
	Nigeria	1948–67	Olayide	+0.15 to 0.20	—
	Ivory Coast	1947–63	Behrman		+0.80
	Cameroon	1947–63	Behrman	+0.68	+1.81
	Brazil	1947–63	Behrman	+0.53	+0.95
	Ecuador	1947–63	Behrman		+0.28
	Dominican Republic	1947–63	Behrman	+0.03	+0.15
	Venezuela	1947–63	Behrman	+0.12	+0.38
COFFEE	Uganda	1926–38	Frederick	+0.63	—
	Uganda Buganda	1926–38	Frederick	+0.42	—
	Kenya	1946–64	Maitha	+0.15	+0.38
	Kenya (estates)	1946–64	Maitha	+0.16	+0.40
	Kenya (smallholders)	1946–64	Maitha	+0.20	+0.51
	Africa	1943–60	Bacha	+0.14 to 0.24	+0.37 to 0.60
	Brazil	1948–64	Behrman and Klein	+0.10	+0.11
	Brazil Sao Paulo	1930–55	Arak	+2.02 to 2.28	—
	Brazil Sao Paulo	1925–33 and 1951–61	Bacha	+0.23	+1.00
	Brazil Sao Paulo	1948–70	Saylor	+0.10 to 0.16	+0.51 to 0.64
	Colombia	1952–65	Bateman	—	+0.84
	Latin America (excluding Brazil and Colombia)	1943–60	Bacha	+0.28	+0.52
	Jamaica	1953–68	Williams	+0.70 to 0.82	—

Crop	Region	Period	Author	Short-Run Elasticity	Long-Run Elasticity
TEA	India	1921–61	Rajagopalan	+0.02 to 0.06	+0.09 to 0.16
RUBBER	Malaysia (smallholders)	1948–61	Chan	+0.12 to 0.34	—
	Malaysia (smallholders)	1953–60	Stern	+0.20	—
	Malaysia (estates)	1949–63	Behrman	−0.09 to +0.09	+0.15
	Thailand	1947–65	Behrman	+0.04 to 0.41	+0.19
	Indonesia (estates)	1949–64	Behrman	0 to +0.05	+0.40
	Nigeria	1948–67	Olayide	+0.17 to 0.24	+0.21 to 0.94
	Liberia	1950–72	Ghoshal	+0.14	+0.22
PALM OIL	Nigeria	1948–67	Olayide	+0.22 to 0.26	—
PALM	Nigeria	1948–67	Olayide	+0.05 to 0.10	—
KERNELS	Nigeria	1949–64	Oni	+0.22 to 0.28	—
	Nigeria, Eastern Nigeria	1949–64	Oni	+0.28 to 0.39	—

Source: Askari and Cummings, (1977)

been varied much in the short-run and partly by the argument that producers are not profit maximisers.

9 Conclusion

It is interesting to observe on the basis of the evidence available so far that producers of primary products in developing countries tend to behave 'rationally' in general, though there are exceptions, particularly in highly subsistence economies. Although it has been observed that in many cases, farmers are 'price responsive', it should not necessarily imply that other variables do not matter. Indeed, in many cases other variables, like weather condition (rainfall and temperature), soil moisture, family size, education, irrigation, farm size, wealth, income and resource endowments, could have significant effects. A number of these variables have not always been tested vigorously. Future research may well be directed in testing the significance of both price and non-price variables in affecting area and output of agricultural crops in LDCs.

References

Ady, P. (1968), 'Supply functions in tropical agriculture', *Bull. Oxford Inst. Econs and Stats*, vol. 30.
Arak, M. (1968), 'The Price Responsiveness of São Paulo coffee growers', *Food Res. Inst. Studies*, vol. 8, no. 3.
Askari H. and Cummings, J.T. (1976), *Agricultural Supply Response: A Survey of the Econometric Evidence*, Praeger.
Bacha, E.L. (1968), 'An Economic Model for the World Coffee Market: The Impact of Brazilian Price Policy', PhD thesis, Yale University.
Bardhan, P.K. and Bardhan, K. (1970), 'Price Response of Marketed Surplus of foodgrains', *Oxford Economic Papers*, July.
Bateman, M. (1968), *Cocoa in the Ghanaian Economy: An Econometric Model*, North-Holland, Amsterdam.
Behrman, J. (1968), *Supply Response in Underdeveloped Agriculture: A Case Study of Four Major Annual Crops in Thailand: 1937–1963*, North-Holland, Amsterdam.
Cagan, P. (1956), 'The monetary dynamics of hyper inflation' in Friedman, M. (ed.), *Studies in the Quantity Theory of Money*, University of Chicago Press, Chicago.

Cummings, T.J. (1974), *Supply Response in Peasant Agriculture. Price and non-price factors*, PhD thesis, Tuft University.

Dantwala, M.L. (ed.) (1970) 'Symposium on farmers' response to prices', *J. Indian Soc. of Agr. Stat.*, vol. 22, June.

Falcon, W.P. (1964), 'Farmers' response to price in a subsistence economy: the case of West Pakistan', *Amer. Econ. Revi.*, Papers and Proceedings, May.

Fisher, F. (1963), 'A theoretical analysis of the impact of food surplus disposal on agricultural production in recipient countries', *J. Farm Econo.*, vol. 45, November.

Frederick, K.D. (1965), 'Coffee Production in Uganda', PhD thesis, MIT.

Frederick, K.D. (1969), 'The role of market forces and planning in Uganda's economic development 1900–1938', *Eastern African Economic Review*, vol. 1.

Freebairn, D.K. (1969), 'The Dichotomy of prosperity and poverty in Mexican Agriculture', Land Economics, 31–42, February.

Ghoshal, A. (1975), 'The price responsiveness of primary producers: a relative supply approach', *Amer. J. Agr. Econ.*, vol. 52, February.

Khatkhate, D.R. (1962), Some Notes on the Real Effect of Foreign Surplus Disposal in Underdeveloped Economies, *Quarterly Journal of Economics*, vol. 76, May.

Krishna, R. (1963), 'Farm supply response in India and Pakistan: a case study of the Punjab regions', *Economic Journal*, vol. 73.

Lipton, M. (1967), 'Should reasonable farmers respond to price changes?', *Modern Asian Studies*, vol. 1.

Managhas, M., Recto, A.E. and Ruttan, V.W. (1966), 'Price and market relationships for rice and corn in the Philippines', *J. Farm Econ.*, vol. 48, August.

Mathur, P.N. and Ezekiel, H. (1961), 'Marketed surplus of food and price fluctuations in a developing economy', *Kyklos*, vol. 14.

Merrill, W.C. (1967), 'Setting the price of Peruvian rice', *J. of Farm Econ.* vol. 49, February.

Mubyarto, S. (1965), 'The elasticity of marketable surplus of rice in Indonesia: a study of Java-Madura', PhD thesis, Iowa.

Nerlove, M. (1958), *The Dynamics of Supply Estimation of Farmers' Response to Price*, Johns Hopkins University Press, Baltimore.

Nowshirvani, V. (1968), 'Agricultural supply in India. Some theoretical and empirical studies', PhD thesis, MIT.

Olayide, S.O. (1968), 'Some estimates of supply and demand elasticity for selected commodities in Nigeria's foreign trade', *J. Bus. and Soc. Studies.* vol. 1, September.

Owen, W.F. (1966), 'The Double developmental squeeze on agriculture', *The American Review*, vol. 56, No. 1, 43–70, March.

Parikh A. (1971), 'Farm supply response: a distributed log analysis', *Oxford Inst. Stats. Bull.*, vol. 33.

Rajagopalan, V. and Meenakshisundaram, V. (1969), 'Travails of the tea industry – An economic appraisal', *Indian J. Agr. Econ.* vol. 24.

Saylor, R.G. (1974), 'Alternative measures of supply elasticities: the case of São Paulo coffee', *Amer. J. Agr. Econ.*, vol. 56, February.

Schultz, T. (1960), 'Value of US farm surpluses to underdeveloped countries', *J. Farm Econ.*, vol. 42.

Swift, J. (1969), 'An economic stwdy of the Chilean agrarian reform', PhD thesis, MIT.

Tomek, W. (1972), 'Distributed log models of cotton average response: a further result', *Amer. J. Agr. Econ.*, vo. 64, February.

Wickens M. and Greenfield, J. (1973), 'The econometrics of Agricultural supply: an application to the world coffee market', *Rev. Econ. and Stats.*, vol. 55, November.

Yotopoulos, P.A. and Nugent, J.B. (1976), *Economics of Development*, Harper & Row, New York.

8 Institutional Constraints on Agricultural Development and Remedial Policies

In this chapter we consider the constraints imposed on agricultural development by defects in the institutions controlling the distribution of land, access to agricultural capital and credit, and the competitive structure of agricultural markets. We also consider the motives for and possible effects of reforming policies in these areas.

1 Inequitable Landownership and Land Reform

In this section we examine systems of agricultural landownership and land tenure; the meaning of and motives for land reform; the possible benefits and costs of land reform, including its effects on farm output and the marketed surplus; the limitations of land reform; and cooperative farming and tenancy reform as policy alternatives to individual farm ownership.

1.1 Landownership and tenure systems

Taking a world view, systems of agricultural landownership and tenure are both diverse and complex. To simplify, we make the broad distinction between 'traditional' and 'modern' systems. Under *traditional* systems the ownership of land may be either communal or private. Under communal or tribal ownership, farmers have individual rights of cultivation, but not necessarily exclusive use of the land. For example, grazing rights are often held in common. Under private ownership, the rights of ownership and cultivation may either be exercised by the same person (the owner-farmer) or separately by landlords and tenants. In the traditional mode the landlord–tenant system has generally been feudal in character with tenants paying and landlords receiving rent in the form of either a share of the crop or labour services. Traditional systems of commercial ownership still persist, in much of Africa, for example. Feudal landlord–tenant systems still prevail in much of Latin America, the Middle East and Southern Asia.

Under *modern* systems of land ownership the dichotomy is capitalism versus socialism. In the capitalist model landownership is private, with landowners exercising the option of either farming the land themselves or letting it to a tenant usually paying a fixed cash rent. In the socialist model the land is owned by the state, although the responsibility for cultivation is often given to co-operative groups or 'collectives'. The collective farmers are required to meet production norms and delivery quotas set by the state.

This system of classification which is summarised in Figure 8.1, oversimplifies reality. In the LDCs, many gradations exist between traditional and modern modes of ownership and tenure, as well as between pure capitalism and pure socialism.

The use of the term 'land reform' most commonly refers to the redistribution of landownership from traditional and feudal-type landlords to their previous tenants or wage labourers. But, in principle, the meaning of land reform can be extended to cover any *socially beneficial* change in a country's system of agricultural landownership or tenure arrangements. So, for example, under some circumstances the change from a primitive system of communal or tribal landownership to advanced communal or even private ownership might yield a substantial social gain in terms of more intensive use of the land and a larger marketed food surplus. However, the remainder of our remarks about land reform relate mainly to the redistribution of privately owned land from the few to the many.

1.2 Meaning of land reform: redistribution of land and agrarian reform

Even when land reform is restricted to the transfer of private ownership rights, the term is ambiguous in that it fails to distinguish between the redistribution of land ownership alone, comprehensive 'agrarian reform', and tenancy reform.

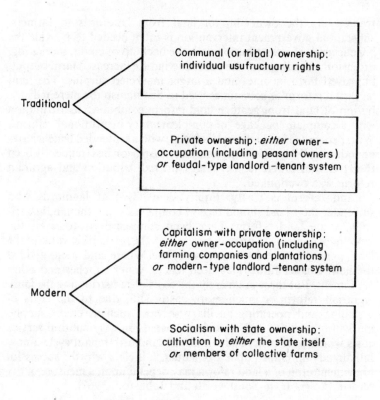

Figure 8.1: *Systems of land ownership and tenure*

Because even feudal-type landlords commonly supplement the cultivation rights granted to their tenants with services such as the provision of credit or a marketing outlet for farm products, the mere redistribution of landownership from landlords to tenants is unlikely to advance agricultural development. Even though a landlord may have 'exploited' his tenants, the freedom conferred on them by owning the land is a doubtful gain if they are unaccustomed to independence and therefore unable to fend for themselves as farmers. With the landlord gone, his former tenants are likely to need alternative sources of credit and marketing services, and even advice on how best to care for crops and livestock. Thus, in addition to transferring the ownership of land from landlords to farmers, substantial government intervention is often needed to provide the beneficiaries of redistribution with the back-up of credit, marketing, extension and other services needed to induce increased farm output, improved farm incomes and a larger marketed surplus. The term 'agrarian reform' is sometimes used to distinguish the mere redistribution of land from genuine land reform combining redistribution with a complete 'package' of complementary institutional reforms. Where so-called land reform has failed, with its intended beneficiaries abandoning the land in large numbers, the reason has frequently been that the vital difference between land redistribution and agrarian reform was overlooked.

Land reform is costly. Equity demands that landlords who surrender their land should receive compensation – though, historically, some land reforms have been confiscatory. However, the amount of compensation may be less than the full market value of the land, particularly under a socialist regime. Where land is sub-divided the new, smaller farms may need fences, water supplies and other equipment. But these are merely the costs of redistributing the land. Agrarian reform is much more costly still, due to the costs of installing and operating ancillary services such as credit supply, marketing facilities and agricultural extension. Agricultural service costs which were concealed within the landlord–tenant system now fall directly upon the public Exchequer. It is clear why the successful implementation of a land reform may depend upon a measure of *tax* reform (Lewis Jr, in Southworth and Johnston, 1967).

1.3 Motives for land reform

Although the demand for land reform frequently comes from the 'grass roots', its underlying goals are primarily *social* and *political*, rather than economic. Thus, although development economists tend to see land reform as a possible means of accelerating LDC development through the economic betterment of peasant farmers, the farmers themselves may be unaware of this possibility. However, in societies where the mere ownership of land gives social status to the holder, the desire to own land can be intense. The possible use of land as a factor of production is a secondary consideration. A landowner's wealth derives from the value of the land, which in turn reflects the capitalised value of the rent. But, in the context of land reform, the source of rent payment is the *psychic* income the owner derives from merely possessing the land, rather than income from farming. Thus grass roots pressure for land reform comes from the popular demand for a fairer distribution of wealth rather than for a better farm income. But the motives for land reform also include the redistribution of *political* power.

In feudal societies wealth is concentrated in the hands of large landlords who also dominate politically. In true traditional societies the peasants may accept their politically subordinate role as the price of their landlord's protection, particularly in times of economic hardship. But, in the course of the transition from traditional to modern society, landlords may seek to exert their economic power. It is then that the landlords' role may come to be seen as exploitative rather than paternalistic, leading to popular pressure to remove their political and economic power simultaneously (Dorner, 1972, ch. 3).

Evidence that the distribution of agricultural land in some LDCs is in fact heavily skewed in favour of the rich and powerful has been presented in chepter 2, section 2. The evidence is accompanied by a discussion of how unequal access to land exacerbates rural poverty. Despite our argument that the conscious motive for land reform amongst the people is primarily social and political, a large part of the objective economic case for it must be income redistribution in favour of the rural poor. But the actual success of land reform in this respect is uncertain depending on factors discussed in the next section.

1.4 Benefits and costs of land reform

An idealised view of the benefits of land reform may be taken. The argument runs that as a consequence of transferring landownership rights from landlords to their former tenants, newly-independent farmers have both the resources and the incentive to accumulate capital. The resources derive from the rent they no longer have to pay; complete security of tenure provides the incentive. Apart from taxation, the whole agricultural surplus (output minus personal consumption) now accrues to the cultivator who cannot be dispossessed. Aggregating from the individual farm to the level of the agricultural sector as a whole, the benefits of land reform are supposed to include both an enlarged domestic agricultural output and a bigger marketed surplus of food and other agricultural products. Thus food shortage is averted, food prices remain relatively low to the benefit of nascent industry and urban consumers, and pressure on the balance of the overseas payments is relieved by a declining food import bill. It is also claimed that land reform, which breaks down large farms or estates into smaller-sized agricultural holdings, promotes greater efficiency in the utilisation of agricultural resources, particularly land. We referred in chapter 6, section 1.2 to the evidence of Berry and Cline (1979) in support of the hypothesis that an inverse relationship exists between farm size and crop yields (or the productivity of land), at least in traditional agriculture. The same authors infer from this evidence that, assuming that the productivity of new small farms (created by land reform) is the same as existing farms of the same size, the twin goals of greater equity and higher productivity can be reached simultaneously by sub-dividing large farms into smaller ownership units virtually without limit (Berry and Cline, 1979, ch. 2). The only exceptions to this rather startling conclusion are differences in land quality and 'economies of scale in a few products'. Our own view is that in drawing this inference Berry and Cline overlook some rather obvious diseconomies of 'minifundia' which are too small to yield an adequate livelihood to an occupier.

These are some of the claimed benefits of land reform, but how do they compare with reality? Whether the immediate beneficiaries of land redistribution will in fact save and invest in agriculture is a moot point depending upon numerous economic and social factors. We have stressed how in peasant agriculture technical innovation is obstructed by economic uncertainty and lack of knowledge, as well as by ingrained social attitudes that are inimical to change (ch. 2, section 1.3). In this chapter we have emphasised that those to whom land is transferred do not necessarily see the reform as opening the way to augmenting their income from the land: rather they may view mere possession of the land as the ultimate goal. Moreover, newly-independent farmers may lack the experience to farm successfully on their own, particularly if ancillary services previously supplied by the landlords are not available from another source.

Without higher output at the level of the individual farm there can be no spread of land reform benefits from agriculture itself to the larger economy. But even if farmers do increase their output, they may increase their own consumption by an equal amount. Although such increased consumption may count as a gain in welfare, there is no increase in the marketable surplus. Indeed, as explained more fully in chapter 3, section 4, where we summarised Griffin's analysis of marketable surplus transfer mechanisms, the abolition of rent payments under land reform could even *reduce* the size of the marketable surplus. Were this to occur, the additional social costs of land reform, in terms of higher urban food prices, higher manufacturing costs and the balance of payments cost of larger food imports, could be considerable.

In breaking with the past so abruptly land reform is economically and socially disruptive. Time is needed to establish a new structure of agricultural production with supporting ancillary services such as credit, marketing and agricultural extension. Thus, in the *short term*, there is almost bound to be some decline in agricultural production as an aftermath to land reform. This decline may be especially marked where reform involves switching from feudal-type, large-scale agriculture to small-scale family farming. So, for example, the break-up of Latin American *hacienda*, with the land distributed amongst the former *peons*, contrasts with merely transferring the ownership of existing small-scale farms in South-East Asia from landlords to their former share-cropper tenants. *Ceteris paribus*, the expected disruption of production would tend to be much greater in the former situation than in the latter (Dorner, 1972, ch. 5).

The empirical evidence of the net benefits of land reform is mixed. The record indicates a variety of results ranging from successful to unsuccessful reforms with a number in between for which the verdict is indecisive. The group of countries with successful reforms includes Egypt, Taiwan, South Korea and, probably, Iran. In all these countries an effective redistribution of landownership was combined with improved agricultural productivity and higher output. The unsuccessful group may be considered to include Mexico, where the *ejidos* have lagged behind private farms in productivity and output growth, and Peru, where the economic viability of small-scale farms has been difficult to establish, despite the formation of co-operatives. This group also includes Boliva and Iraq, where the output decline following reform was both dramatic and persistent. The indecisive group includes Japan, where a successful first reform was followed by a second reform which worsened the problem of under-sized farms, and India, where the objectives of land reform legislation have tended to be defeated by incomplete implementation and evasion (World Bank, 1975, pp. 252–62: Southworth and Johnston, 1967, pp. 285–93).

The case of Mexico, where land reform was only partial, so that reformed and unreformed sub-sectors of agriculture continued to operate side by side, is especially interesting. The results of a recent comparative study of land reform (*ejidal*) and private farms across 32 Mexican states revealed a substantial discrepancy in average productivity of labour (and farm income) in favour of the non-reformed farms. However, this result was attributed not to less efficient resource utilisation on the land reform farms but to their unequal access to land and capital. A larger share of these scarce resources for the land reform sector was the policy implication (Nguyen and Martinez-Saldivar, 1979).

Even with relatively successful land reforms the success has been qualified by unresolved problems. One such problem is that the larger and more successful of pre-reform farmers have tended to gain a disproportionate share of the post-reform benefits. Another is that landless labourers, with no access to land before reform, have frequently gained little or nothing.

1.5 Limitations of land reform

If it is true that agricultural resource efficiency is inversely correlated with farm size, then the redistribution of land appears to achieve the twin goals of greater equity and improved efficiency. However, in some countries the density of the rural population is so high that the ideal of providing *every* rural household with enough land to yield an adequate livelihood from agriculture alone is unattainable. There is simply not enough land to go round amongst all those with a claim based on equity.

Some countries have sought to impose farm size 'ceilings' in an endeavour to release more land for redistribution to families with only very small areas of land or none. But in India it was found that even with the ceiling as low as 20 acres, the amount of land released would be sufficient only to bring the *minimum* size of holding amongst all existing farm occupiers up to 5 acres. Landless households, representing more than 20 per cent of all rural households (see chapter 2, section 2) would remain landless. The same dilemma has been observed in other LDCs, especially in South East Asia (World Bank, 1975, p. 219). Under such conditions land reform alone cannot cure rural poverty or eliminate unemployment. Indeed, it is clear that in large part a solution to these problems must come from the acceleration of employment opportunities created *outside* the agricultural sector.

1.6 Alternatives to 'land to the tiller'

Although the use of the term land reform is sometimes restricted to the granting of individual ownership rights to former agricultural tenants, there are alternative 'reformist' approaches to raising agricultural output and incomes. We refer briefly to two of these.

1.6.1 Co-operative Farming

As an alternative to breaking up large farms or estates into family-sized farms, large farms may be transferred intact to co-operative groups of producers. Compared with small-scale individual ownership, co-operative farming spreads the scarce factors of skilled management and capital over a larger output and facilitates group decisions on indivisible factors such as irrigation. It also reduces the risk that, following an equalising land reform under private ownership, marked disparities of income and wealth within the agricultural sector will eventually re-emerge as, due to greater enterprise or good fortune, some farmers succeed in bidding resources away from other farmers. Readers are referred to Putterman (1983) for a more extended discussion of the issue of 'restratification'. A possible drawback of co-operation is a loss of individual incentive to greater enterprise and effort, but this depends on how individual rewards within the group are organised.

The choice between individual and co-operative farm ownership may be influenced by ideology, as well as by economics, but even if eastern-style collective farms are excluded, there have been co-operative farming experiments in many parts of the world. The record of achievement has been varied with some successes and many failures. Some of the most successful examples have been concerned with 'land settlement', rather than with the reform of an existing farm structure. Settlement schemes in Israel (*kibbutzim*) are notable in this respect. These have been widely used as a model in other countries, but only rarely with anything like the same degree of success as in Israel (Southworth and Johnston, 1967, pp. 302–3; Dorner, 1972, pp. 54–62).

The case for the co-operative farming alternative may be strengthening with time due to changing agricultural technology. In discussing the relationship between farm size and efficiency in an LDC context, we earlier expressed reservations concerning the claim that HYV technology is neutral to scale (chapter 6, section 1.2). Much of the economic rationale of co-operative farming derives from the opportunities afforded to members to exploit economies of large-scale production. Thus, *ceteris paribus*, co-operative farming is favoured by the emergence of new agricultural technology with a potential for yielding economies of scale, such as field mechanisation for example.

1.6.2 Tenancy Reform

Some advantages of land reform may be attainable without actually changing the ownership of the land. It may be possible to improve the economic status and productivity of farm tenants through land *tenure* reform. An efficient landlord–tenant system precludes gross inequality of bargaining power between the parties. Yet, in practice, farm tenants in LDCs frequently lack legal rights. Thus, a first step in tenancy reform may be the enactment of legislation to codify and limit the powers of landlords. Such a code might limit the unilateral raising of rent, improve security of tenure and provide for independent arbitration to settle disputes. A second and complementary step would be to oblige landlords and tenants to enter into legally binding tenancy agreements consistent with the code of rights. If properly enforced, such measures would revolutionise feudal tenure systems by compelling landlords to let land on fair and equitable terms. But there must be enough political will to ensure both the enactment of legislation and the oversight of its enforcement.

A good landlord–tenant system possesses the virtue of functional specialisation. Landowners concentrate on estate management, maintenance and administration, leaving tenants free to concentrate on actually farming the land. As previously mentioned, landowners may also provide ancillary services, such as production credit, a marketing outlet, and even farming advice. Share-tenancy is a device for spreading the risks of production between tenant farmers and their landlords: without such a system farmers would have to bear the whole of the risk themselves. Contrary to much popular supposition, the principle of share-tenancy is not socially inefficient; but it may be socially oppressive in practice due to the abuse of monopoly power by landlords (Griffin, 1976, pp. 122–3).

Although tenancy reform avoids the destruction of landlords' ancillary services, complementary institutional reforms may be needed to supplement or improve those services. By providing farmers with another source of credit or an alternative market outlet, government can help to ensure that landlords offer their services at competitive prices.

A complete land reform can be very expensive due to the high cost of compensating landowners for the loss of their property rights. The comparative cheapness of tenancy reform is one of its major advantages. The major drawback of tenancy reform is political rather than economic, namely, that it does not alter the structure of political power (Mellor, 1966, pp. 260–1: World Bank, 1975, pp. 222–3).

1.7 Summary and policy conclusion

(1) The motives for land reform, as seen by its beneficiaries, derive primarily from the social and political aspirations of the rural population. Farmers may have to be educated to see land reform as a means of improving their productivity and incomes.
(2) The agricultural benefits of land reform depend on much more than the mere redistribution of land ownership. A complete agrarian reform is needed entailing redistribution plus a complete package of supporting ancillary services.
(3) Although the possible economic benefits of land reform include a larger agricultural output and higher farm income, together with a great marketed surplus of agricultural products, the realisation of these benefits in practice depends upon numerous factors, including the form and content of government policies. Due to its disruptive character, land reform is virtually bound to result in some short-term loss of agricultural output.
(4) The empirical evidence on the agricultural benefits of land reform reveals a mixture of successes and failures. A general failing of virtually all reforms is that very few of the benefits have reached very small-scale farmers and landless labourers.
(5) In very densely populated countries the agricultural area released by land reform may be insufficient to provide adequate-sized farms for all rural families; that is, rural unemployment and poverty cannot be eliminated by land reform alone.
(6) Co-operative farming and tenancy reform offer alternative and possibly cheaper reformist approaches to raising agricultural output and incomes than conventional land reform. But these alternatives fail to satisfy other goals such as the redistribution of private wealth and the re-alignment of political power.
(7) Strong government direction and leadership are needed to enact land reform legislation, ensure its implementation, and provide its beneficiaries with adequate incentives and ancillary services to induce higher investment, output and farm income as well as a larger marketed surplus of agricultural products.

2 Capital and Finance in Underdeveloped Agriculture

The problem of capital scarcity in underdeveloped agriculture is generally well known. The role of financial and credit institutions in promoting capital accumulation in this context has received considerable attention in many countries in recent times. It is important to understand the nature and composition of rural monetary institutions in LDCs to formulate appropriate monetary and credit policies to act as catalysts to promote the level of economic development. In this section we shall first discuss the nature of the credit problem in underdeveloped agriculture. Next, an attempt will be made to analyse the demand and the supply side of rural money markets. Then, we shall describe the structure of rural interest rates in some LDCs. It is generally believed that such interest rates are very high and sometimes approach the level of usury. Merchants-cum-moneylenders are sometimes considered as the main agents of exploitation through usury. In extreme cases, it has been argued that an 'exploitative' mode of production created through usury, prevents the growth of the agricultural sector as a whole in LDCs. (See, Bhaduri, 1973, 1977.) Later, we shall discuss such views and their criticisms. At this stage, it may be useful to note some general features of rural money markets.

2.1 Functions of rural money markets in LDCs

A *money* market (rather than a capital market) usually caters for the demand for an supply of *short-run* loanable funds. We may note the following major functions of rural money markets:

(a) A major function of a money market is to allocate savings into investment and promote a more rational allocation of resources.
(b) It is generally well acknowledged that the growth of an agrarian economy could be significantly retarded due to a lack of savings and investment. An efficient money market raises savings and investment by promoting liquidity and ensuring the safety of financial assets. This function is of crucial importance in LDCs where savings and investment behaviour leave considerable room for improvement. It is well known that in the rural economies of most LDCs savings too often take place in the form of gold-hoarding and land-holding rather than in the holding of financial assets. Hence, notwithstanding the ability to save, the lack of an efficient and developed money market deprives the society of financial assets which could lead savings into fruitful investment.
(c) A money market generally promotes financial mobility as funds could be transferred from one sector to the other in a manner which economises transaction costs. Such economies of scale are

considerable when many people are involved in such transactions with an overall growth in the flow of total funds. It is noteworthy that an efficient money market is crucial for providing elasticity in the flow of funds.
(d) An efficient rural money market is essential for implementing the monetary policies of the central bank. Clearly, monetary policies are unlikely to be very successful if a large part of a predominantly agrarian economy works independently of the monetary transactions. This may sometimes happen in a barter economy where goods are usually exchanged against goods, and money is no longer a medium of exchange. Under such conditions, changes in money supply are unlikely to alter substantially the aggregate demand in the society and monetary policies will be largely ineffective.
(e) Rural money markets in LDCs are seldom homogeneous. Broadly, they can be divided into two major parts: organised, and unorganised. The organised section usually comprises the central bank, the commercial banks, co-operative banks, credit societies. They usually operate within the provisions of the Banking Companies Act of different LDCs. The organised sector generally maintains accounts which are open to audit and periodic inspection.

On the other hand, the unorganised sector operates outside the legal framework. They maintain sorts of accounts which are not always very sophisticated. These accounts are not open to inspection. Indeed, a great deal of secrecy covers the financial operations of the unorganised financial sector. It generally consists of moneylenders, landlords, merchants, traders, indigenous bankers, pawnbrokers, and even friends and relatives. Blending of moneylending with other types of economic activities is a special feature of such an unorganised sector. The other features include informality in dealings with customers, personal contact with borrowers, simple systems of maintaing accounts, flexibility of loan operations and secrecy about financial transactions.

The *moneylenders* are important sources of funds in the agricultural sectors of many LDCs and it is important to discuss their operations at some length in a village economy. Usually they have a very good knowledge of the character and repaying capacity of the borrower. They are quite flexible in their operations, but grant loans largely against personal security. Sometimes they also offer consumption loans. In case where moneylenders are also the landlords, there is a tendency to 'over price' the loan at the time of lending and 'underprice' them at the time of repayment. This complex mechanism could be understood more clearly in the light of what has been called the 'characteristics of a semi-feudal agriculture' (Bhaduri, 1973, 1977). Such characteristics can be summarised as follows:

Share cropping. This is a system under which the landowner leases out his land for at least one full production cycle and the net harvest (i.e. gross harvest – seed required for the next harvest) is shared between the landlord and the tenant on some mutually agreed, legal basis. However, in many cases, tenancy rights are not legally very well protected and terms of contract are quite complicated.

Perpetual indebtedness. Quite often it has been witnessed that tenants are heavily indebted, particularly to their landlords. Repayment of debt along with interest payments by tenants generally signifies a large fall in the available harvest and the tenant is compelled to borrow for consumption for his survival. Because of the presence of such consumption loans, tenants remain perpetually in debt.

Landlords as the moneylander. The problem is tenants are aggravated by another characteristic of a semi-feudal agriculture – the absence of a complete specialisation. Landlords frequently act as creditors to their tenants and consumption loans are usually advanced by the landlords at high rates of interest. Hence the tenant leases his land from the same person to whom he is perpetually indebted and this reduces him to the abject status of a traditional serf. Hence, the semi-feudal landlord exploits the tenant both through usury and through his property rights on land.

Inaccessibility to the market. It has sometimes been claimed that the tenant has no access to the organised money market. Under such circumstances, it follows that the only creditor of the tenant is his landlord who assumes the power of a pure monopolist. He lends against a future harvest at a high rate of interest and the tenant has very little option open to him but to accept such rates. It is also argued that the tenant has little access to the commodity market, and as such he cannot sell at the highest price. Generally, he has to sell when harvest prices are at their lowest just after harvesting to meet the scheduled date of repayment. On the other hand, when prices are at their highest, tenants could be left with very little food for themselves and they frequently need consumption loans just to survive.

Hence, lack of access to money and commodity markets is supposed to be the reason for usurious interest rates in LDCs. In the extreme, the 'mode of production' can be totally exploitative as the landlord-moneylender as the single buyer of the product and as a single lender in the money market squeezes out the whole of monopolistic profit.

It must be admitted that although such a view of the determination of the rural interest rate and the mode of exploitation is interesting, the empirical evidence available so far has not substantiated the 'interlocking of factors' theory (see Bardhan and Rudra, 1978; Bardhan, 1980). More specifically, the hypothesis that tenants and agriculturists have no other option but to sell to a single landlord who is also the only moneylender has not always been upheld in practice. It is possible to note cases of usurious rural interest rates in an exploitative situation where the monopolist-landlord-moneylender does successfully extract monopoly gains. On the other hand, as a general rule, some find it difficult to accept, particularly in view of the evidende available so far. Hence, it may be useful to explain the *economic*, as opposed to the *institutionalist*, argument to account for high rural interest rates.

2.2 An economic theory of rural interest rate determination

From the supply side, the rural interest rate ($= r$) may be regarded as the sum of (a) administrative cost ($= \alpha$), (b) risk premium ($= \beta$) (due to the probability of default in repayment), (c) opportunity cost of lending ($= \gamma$), and (d) the monopoly profit ($= \pi$) (see Bottomley, 1971). Hence, we can write:

$$r = \alpha + \beta + \gamma + \pi$$

The administrative cost ($= \alpha$) is not supposed to be very high as the moneylenders, as explained above, do not incur high administrative costs to run their moneylending activities. The opportunity cost of lending is also very small as the organised financial institutions offer a very low rate of interest on savings – a system which has been described by some as one of financial repression (see McKinnon, 1973; and Shaw, 1973). Furthermore, interest rates advanced by the rural financial intermediaries on deposits (e.g. Post Office savings banks, primary credit co-operatives, rual banks) hardly change. All the interest, then, centres on the two other variables, β and π. For the purpose of simplification, we can then write:

$$r = f(\beta, \pi)$$

It is generally argued that the higher the risk premium (which can be approximated by the probability of default), the higher will be the rural interest rate. The relationship between the monopoly profit and r is also positive. However, if the real income and output in the agricultural sector rises, then this will reduce the probability of default and hence the risk premium will also be reduced; which will also account for a fall in r. A fall in rural interest rate will weaken the 'monopoly power' of moneylenders in the rural economy too. In sum, an increase in the level of output and real income in the agricultural sector will increase the repayment capacity of the farmers. This would clearly increase their credit worthiness by reducing risk-premium which in its turn will lower the high level of rural interest rates that prevail in LDCs.

Institutional factors, like the growth of the credit co-operatives and the rural banks, also pay an important role in influencing the rural rates. Usually, these credit co-operatives and the rural banks charge interest rates which are much lower than those charged by the moneylenders. Despite such a differential in the lending rates, it is interesting to point out that the credit co-operatives and the rural banks failed to make a substantial progress in lending and saving activities of the village economies of most LDCs. Such a failure may suggest that the demand for credit is not very sensitive to a low rate of interest. It may also suggest that the pattern of loan administration that has usually been followed by the organised societies in the rural areas is fairly complicated for rural people. Banks, in many instances, follow the orthodox principles of lending and, given the general impoverishment of agriculture in LDCs and the low level of average real income, it is obvious that farmers are not generally considered as very credit worthy customers. Formal business practices by the organised lending institutions, lack of an adequate number of rural banks and the amount of transport costs are some of the other factors which could explain the lack of progress of organised finance. Sometimes farmers in LDCs need consumption rather than production loans, so it may be useful for the organised lending agencies to open up the possibility of financing such consumption loans. Indeed, in many instances, it is difficult to distinguish between consumption and production loans in under developed countries. The traditional idea that all forms of consumption loans are unproductive may not be always valid. Mirrlees (1975) has shown how an increase in consumption in LDCs could add to the efficiency and productivity of labour which, in its turn, has a salutory effect on economic growth and development.

Some have argued (e.g. Bhaduri, 1973) that farmers in LDCs are prevented by their landlords from adopting modern farming methods which might have increased their income since such a rise will increase repayments of borrowing by farmers and release them from the clutches of the landlords-moneylenders. To maximise the income from usury, it is in the landlords' interests to keep farmers-tenants heavily indebted for as long as possible. As a result, in the agricultural sector of most LDCs, technical progress has not been introduced; and even when it has been introduced its scale has been very limited. A semi-feudal agriculture is once again perpetuated with usury and indebtedness as the chief instruments of exploitation by the land-owning class.

On the other hand, given the complete dominance of the landlords over the poor farmers and given the principle of profit and income maximisation, it may be argued that the landlord can maximise his income by 'expropriating' the gains from increased productivity flowing from the adoption (rather than non-adoption) of new technology without resorting to usury. It may be argued that investment in the land will be prevented by the landlords since gains through usury are larger than those from increased gains in productivity. But such an argument is quite unsatisfactory because if landlords have enough control to prevent innovation, they should also have sufficient power to obtain from the peasants additional gains from increasing productivity via technical progress. In any case, usury as a mode of expropriating the 'surplus' could be unneccessary in labour surplus economies where the bargaining strength of the poor peasants and tenants is very weak. It should also be mentioned that the empirical evidence available so far does not support the strong inter-linkage between the moneylenders, merchants and landlords (i.e. between the product and the credit market) as suggested by the institutional school (see Rahman, 1979, Bliss and Stern, 1981, Ghatak, 1983). On balance the evidence, albeit sketchy, seems to suggest that rural interest rates are likely to be lower the higher the growth rates of agricultural output and repayment, the lower the risk premium and the probability of default in loan repayment, and the greater the spread of the co-operatives. High rates of rural interest rates are also significantly correlated with low per capita rural income. It is certainly true to say that a near-subsistence agrarian economy only helps the moneylenders or loan-sharks to strengthen their strange-hold over the rural economy. It could also enable them to obtain usury. However, the real 'explanation' of the rural interest rates is quite a complex issue since, income apart, institutional factors like the *social* and *legal* relationships between debtors and lenders and the nature of the *contract* between them could play a major role. But, on the whole, it seems fairly likely that a rise in agricultural output and rural real income should reduce the risk premium and uncertainty in lending, reduce the rural interest rates and contribute substantially towards weakening the monopoly power of the moneylenders.

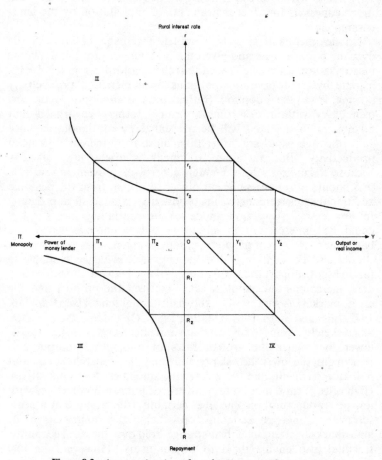

Figure 8.2: *A composite view of rural interest rate determination*

2.3 A simple geometric representation of the composite view of rural interest rate determination

It may be useful to describe a synthetic view of the rural interest rate determination by using a simple diagram (see Figure 8.3). On the vertical axis, we measure the interest rate and the level of income on the horizontal axis in the first quadrant. The relationship between r and Y is assumed to be inverse for the reasons already stated. In the second quadrant, the *positive* relationship between the monopoly power of the moneylenders ($=\pi$) and the rural interest rate is described. In the third quadrant, the association between repayments and the power of the moneylender is regarded as inverse because the higher the repayment, the lower will be the power of the moneylender to compel the farmers to pay high interest rates on grounds of high risks of default. In the fourth quadrant a positive relationship between real income and repayment is shown because the higher the level of income, the greater will be the repayment of the farmers. Thus we have;

$$r = f(Y, R, \pi)$$

and $\quad f_1 < 0, \quad f_2 < 0, \quad f_3 > 0.$

It is then easy to work out that a low income level like OY_1 will generate low repayment ($= OR_1$), a strong stranglehold of the moneylenders over the farmers ($= O\pi_1$) and a high rural interest rate ($= Or_1$). A high rate of growth in the agricultural sector will raise the real income to OY_2. This, in its turn, will increase repayments to OR_2, reduce the degree of risk and the probability of default, and reduce the monopoly power of the moneylenders to $O\pi_2$, which will then reduce rural interest rate to Or_2.

2.4 Financial policies for agricultural development

At the policy level, it is thus necessary to raise the rate of agricultural growth and real income within the agricultural sector to reduce the usurious rural interest rate. It is also necessary to attack the problem from the institutional standpoint. In many cases, a complex set of socioeconomic and legal frameworks help to perpetuate an age-old system of land tenure which generates an unequal access to resources including the availability of credit. It has now been confirmed that the landlords in LDCs who own large plots of land and earn higher income, pay (relatively) lower interest rates in comparison with the small landowners. It is also known that a very high proportion of credits from the rural banks and co-operative usually gravitates towards the large landowners as they have more sound collateral (generally land and equipment) to be very credit-worthy clients for the organised financial agencies. Clearly, small farmers, tenants and landless labourers are at disadvantage to be considered credit-worthy customers. Here, a more egalitarian distribution of land through a series of land reform programmes should improve the economic standing of the poorer section of the peasantry in the eyes of the organised financial institutions who will then be more willing to lend to this particular group of the rural sector. A more equal distribution of credit by the organised credit agencies will also generate more competition for the loan-sharks and moneylenders, who will then be forced to charge lower interest rates since the organised financial institutions which generally operate in the rural sector of the LDCs charge rates of interest which are much lower than those charged by the moneylenders. Where political will exists, it is extremely important to carry out such land reform programmes whenever the situation permits.

The other policy that may be pursued by the monetary authorities to mobilise financial savings from the rural sector is to *raise* the deposit interest rate offered by the organised financial institutions. It is well known that the rural banks and the credit co-operatives in most LDCs usually offer very low interest rates to the savers. As such, incentives to save in organised financial agencies are lacking. Given the shortage of credit supply in many areas, one would expect the equilibrium interest rate for a rational allocation of resources will be much higher. The system of keeping interest rate very low by the organised financial agencies has been called a regime of 'finanacial repression'. The following diagram illustrates (see Figure 8.3). Interest rate in rural areas is measured in the vertical axis and savings and investment are measured on the horizontal axis. When the interest rate is too low, say Or_1, savings are OS_1 and investment demand will be equal to OI_1 and a gap of excess demand of S_1I_1 will emerge. If the interest rate is raised to Or_2 savings will rise to OS_2 and the excess demand will be narrowed to S_2I_2. A more rational allocation of resources will require the interest rate to be raised to Or_e. A rise in interest rate from r_1 to r_2 and a fall in investment simply irons out inefficient allocation of resources as shown by the shaded area.

On the basis of the available evidence, it has been observed that a policy of very low interest rate in many LDCs has led to a steady attrition of organised finance (Chandavarkar, 1971). A policy of financial reform in such LDCs should then include the programme of

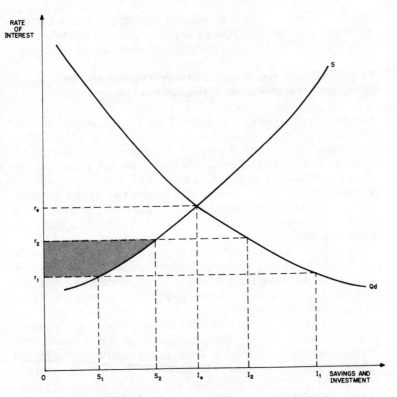

Figure 8.3: *Efficacy of a positive rural interest rate policy*

raising rural interest rates to an equilibrium level (or around it) to attract more savings for investment in agriculture. It is, however, assumed that savings will rise with a rise in interest rates. There is some evidence to confirm that such an assumption is plausible in some LDCs (Gupta, 1970). In any case, the policy of maintaining a very low rate of interest in underdeveloped rural areas has courted such a dismal failure that perhaps it is now necessary to follow a policy of financial liberalisation.

3 Market Imperfections and Marketing Policy

In this section we examine the role of marketing in agricultural development and the objectives of government marketing and price policy, the imperfections of agricultural marketing in LDCs, marketing reform policies, price policy and price stabilisation mechanisms, and the role of the state in agricultural marketing.

3.1 Role of marketing in agricultural development and the objectives of government marketing and price policy

Agrarian societies are characterised by a high degree of self-sufficiency at the village level and even at the level of individual households. Hence, at this stage of development, the demand for commercial marketing services, such as the transport, processing and strorage of food, is extremely limited. But as a consequence of economic growth and development, as signified by urbanisation, rising living standards and an increasing demand for services of all kinds, the volume of economic resources devoted to agricultural marketing inevitably grows.

Marketing and production are interdependent. Producers must be convinced that a remunerative market exists for their products, particularly 'new' products, before they can be induced to produce commercially. An attractive market prospect combines a 'good' price with an assured sales outlet. This simple, self-evident truth has sometimes been overlooked in agricultural development planning. Marketing projects have failed both because an adequate price incentive was not backed up by a neccessary addition to the market infrastructure (e.g. processing facilities) and because infrastructure was provided without a sufficiently attractive price.

Government policy for agricultural marketing may be seen to have four principal objectives: first, to encourage (and occasionally limit) production to meet growing consumer demand; second, to distribute non-farm inputs to agricultural producers; third, to protect consumers from unscrupulous trading practices and unduly high prices; and fourth, to promote 'efficient' marketing.

The types of policy adopted to realise these objectives vary widely in different countries according to their stage of economic development and their government's ideological stance between pure capitalism and pure socialism. A broad choice exists between *regulation* and government *participation*. Where government confines itself to the regulation of markets, trading remains in the hands of private enterprise or producer organisations – government merely prescribes 'fair trading practices', such as the use of standard weights and measures, the prescription of grade standards and the prevention of fraudulent practices (and possibly also the regulation of prices).

Where government also participates in marketing it performs trading activities either in competition with private enterprise, or as the sole (i.e. monopoly) buyer/seller of a particular commodity or commodities. A primary but insufficient reason for government participation in trading is the desire to *control prices*. If the government wishes to regulate prices it can do so without exercising trading powers.

Price policy is an important sub-set of government marketing policy. One objective of price policy is to regulate the agricultural sector's internal *terms of trade* so as to maximise the overall rate of economic growth. But the terms of trade objective may be complemented by a *price stabilisation* objective. The economic rationale of price stabilisation is that, where producers are *risk-averse*, it may promote more efficient resource allocation. Allocation or pricing efficiency requires that the marketing system be responsive both to changes in consumer demand and changes in the conditions of supply, such as factor cost and technological choice. But the full, *economic* efficiency of a marketing system demands *technical* as well as pricing efficiency.

The realisation of technical efficiency in marketing is contingent upon goods moving from producers to consumers at minimum cost, subject to the proviso that all marketing services for which an effective demand exists are actually catered for. Thus an improvement in technical efficiency does not necessarily mean a reduction in the marketing margin; rather, it may involved an improvement in marketing services such that the marginal utility of the extra service (to market users) exceeds the marginal cost of providing it. But if, due to market imperfections, the marginal private and marginal social costs or benefits of a market innovation *diverge*, the *social* benefit: cost ratio is the appropriate efficiency criterion. So, for example, in a country with serious unemployment or underemployment in the agricultural sector, although a policy of reducing the number of marketing intermediaries might be effective in reducing the direct costs of marketing, if the policy also increases unemployment the consequent added social cost may outweigh the incremental benefits to producers and/or consumers of a narrower marketing margin.

3.2 Imperfections of agricultural marketing in LDCs

We have selected three major weaknesses of agricultural marketing in LDCs for brief discussion: (i) infrastructural deficiencies; (ii) the weak bargaining position of producers; and (iii) producers' lack of information.

3.2.1 Deficient Infrastructure

Although a country's general communications infrastructure – its roads, railways, ports, postal services, telecommunications, etc. – serves the economy as a whole, and not just agriculture, the adequacy of these services vitally affects the structure and costs of agricultural marketing. Adequate transport is especially crucial. Without railways or roads suitable for the use of motor vehicles, the cost of long-distance transport becomes prohibitive, thus greatly limiting the size of the market. The development of agricultural marketing may also entail the provision of a more specialised infrastructure. Building are needed for the assembly, buying and selling, and storage of produce and even more specialised facilities, such as slaughterhouses and cold stores, may also be required.

Progress in agricultural marketing is typically constrained by low standards of infrastructural provision in both these areas, the general and the specialised. The problem persists due to underinvestment. The provision of social overhead capital is generally unattractive to private investors who are also deterred from investing in the provision of more specialised infrastructural services for agriculture by what are considered to be unacceptably high risks.

3.2.2 Weak Bargaining Position of Producers

The competitive structure of agricultural markets is invariably such that the number of market intermediaries is greatly exceeded by the number of individual producers. Thus the bargaining position of individual producers tends to be weak. But degrees of 'effective competition' amongst traders buying from farmers varies greatly between countries and even within them. Development economists hold widely divergent views on the position of the competitive norm in LDC agricultural markets between the extremes of pure competitive and pure monopsony, possibly according to where they have gained their field experience. Some take the view that, if not pure competition, then 'effective' or 'workable' competition is the norm, and oligopsony (and even more so, monopsony) the exception in characterising the behaviour of first-hand buyers of agricultural products in LDCs. In other words, according to this view, farmers generally have a real choice of market outlets so that buyers are obliged to compete, by offering a better price or an improved service in order to procure supplies (see e.g. Bauer and Yamey, 1957, pp. 186–7). The opposing view is that, at least in some countries, buying power is so concentrated and collusion amongst buyers so great that sellers do not have an effective choice of market outlets, and are therefore obliged to accept whatever price the buyer offers. It is also held that potential competitors are deterred by the high costs of entry into trading with farmers. The obstacles to entry include the cost of advancing credit to suppliers, the threat of price warfare by established traders, and even intimidation with threats of physical violence (see e.g. Simpson, 1970, on market behaviour in Sudan). In many areas it is customary for the farmer to rely on the final buyer of his crop for an advance of credit to fund crop expenses or even the purchase of food and other consumption necessities to bridge the gap between seed-time and harvest. Where this practice is followed, 'buyer attachment' effectively precludes competition for the purchase of the crop at harvest time.

The available fragmentary evidence on the structure, conduct and performance of the markets serving peasant agriculture in LDCs is conflicting. It suggests the co-existence of differing degrees of competitiveness depending on numerous economic, social, cultural and geographical factors. However, it seems fair to conclude that, as in developed countries, where small farmers are completely unorganised their bargaining position in the market place is weak.

3.2.3 Lack of Market Information
The farmer's bargaining position is further undermined by his lack of information on current market prices, crop prospects and prospective changes in demand, such as export orders. Unlike developed countries, LDCs typically lack a formal market news service and, in any case, many farmers are unable to receive up-to-date news reports by mass media. Although it is claimed that news travels fast by word of mouth, even to isolated areas, traders naturally tend to be better informed than farmers.

3.3 Marketing reform: objectives and implementing policies

The principal objectives of marketing reform follow from the three major market imperfections we have identified:

(1) to improve the quantity and quality of infrastructural provision, especially in transport and other forms of communication;
(2) to strengthen the bargaining power of farmers *vis à vis* the market intermediaries through whom they buy and sell; and
(3) to improve the quantity and quality of agricultural market intelligence.

The means of attaining (1) and (3) can be dealt with very briefly.

Infrastructural development, and the collection, processing and dissemination of market intelligence, are both pre-eminently the responsibility of *government*. Indeed, as we have already remarked, because general infrastructural developments are, in effect, additions to social overhead capital, government funding is virtually inevitable. As for the provision of market intelligence, it is generally easier for government to inspire confidence in its impartiality than it would be for a private intelligence agency. In the rest of this section we concentrate on the choice of means for achieving objective (2) – that of strengthening the bargaining power of small-scale farmers. First, we consider the potential role of voluntary marketing co-operatives, and then that of statutory marketing boards, or 'compulsory co-operatives'. We shall conclude that public policy can play a very important role in promoting marketing reform by both strategies.

The potential role of marketing co-operatives with voluntary membership is to provide their members either with a 'new' (i.e. hitherto unavailable) service, or to replace an existing but inadequate one. The reason for dissatisfaction with the existing service may be that its cost is considered to be unnecessarily high, or its quality unacceptably low, or both. However, the record of voluntary agricultural marketing co-operatives in LDCs is not encouraging. Commonly-encountered problems include an inability to retain the loyalty of members to obtain competent yet incorruptible managers, and to prevent subversion of the co-operative by trading competitors.

Fulfilment of at least three conditions appears to underlie the successful establishment of voluntary marketing co-operatives. They must be capable of:

(1) providing something which is actively wanted by the potential membership and to achieve which members are prepared to sacrifice part of their individual freedom for the common good;
(2) eliciting an active response from their members and retaining their loyalty and confidence; and
(3) making visible progress in reducing marketing costs or improving the quality of marketing services, often in competition with rival traders.

Experience suggests that two further factors are favourable to successful establishment. First, it is easier to achieve success by providing a specialised yet relatively uncomplicated service than by attempting to give a comprehensive set of marketing services. For example, co-operative cotton ginning and transport services (for cotton and other crops) have quite often achieved success. Second, the probability of success is greater for products with a limited choice of marketing channels than for products for which there are many different outlets. For example, an export crop which can be sold only to overseas markets has the advantage that no opportunity exists for more profitable domestic disposal and thus co-operative members are more likely to remain loyal.

Government sometimes attempts to assist agricultural co-operatives not only by according them privileges, such as tax concessions, but also by granting monopoly trading rights. Generally though, this seems inadvisable since the spur of competition encourages efficient marketing by both co-operatives and private enterprise.

Agricultural marketing boards differ from voluntary co-operatives in that they are more directly controlled by government (they exercise delegated powers) and membership is complusory. Technically, marketing boards are producer organisations in the sense that producers or producer representatives share in the management. But, in LDCs, government nominees (officials or politicians) tend to dominate the management, so that boards are under closer and more direct government control than their counterparts in developed countries. The possible functions of marketing boards, and the powers they may exercise, are wideranging. A useful classification scheme (Abbot and Creupelandt, 1966) separates them into six main types. A major distinction exists between *non-trading* types, which leave the existing marketing structure unchanged, and *trading* boards, which transform the structure, either by trading in competition with private enterprise or by assuming monopoly trading powers. In ascending order of degrees of market intervention, the principal functions forming the basis of classification are:

(a) advice on presentation and product promotion,
(b) market regulation (trading standards),
(c) price stabilisation (without regular trading),
(d) competitive trading, and
(e) monopoly trading (export market only, or export and domestic markets).

A primary function of marketing boards is to enhance producers' market returns. This may be achieved either by reducing the costs of marketing, or by selling additional services, or even by curtailing supplies to raise the selling price. Another major function is to reduce market uncertainty through price stabilisation. In addition, marketing boards in LDCs are sometimes used by government as a convenient means of taxing the agricultural sector. It has been objected that, by exercising this fiscal function, marketing boards effectively 'rob' producers of part of their market returns (Bauer, 1954). But if it is accepted that extraction of part of the agricultural surplus may be a legitimate objective of government policy in LDCs (see chapter 3, section 4.1) then this may be an effective means of collection (Blandford, 1979).

In principle, because of their compulsory membership, marketing boards possess much greater bargaining power than voluntary co-operatives. But, in practice, board operations are very difficult to appraise objectively for several reasons (Abbott and Creupelandt, 1967). Relevant information is difficult to obtain due either to the use of inadequate or inappropriate accounting and recording procedures, or to the deliberate secretiveness of governments wishing to avoid political embarrassment. Even if more adequate data were available, valid comparative analysis would still be very difficult because of the widely differing conditions in which boards operate. Due to the lack of objective evidence we can only compare the potential achievements of agricultural marketing boards with the possible costs. Potential achievements include: enhanced producer returns through improved market presentation; economies of scale in selling and other marketing activities; easier access to capital for market development; price stabilisation; and raising returns to producers by controlling supplies and practising market discrimination.

But are these goals achievable by some alternative means at a lower cost? In principle, some of them probably are, *either* by direct government regulation (but *not* trading), *or* by government collaborating with voluntary marketing co-operatives, or even with the private sector. Marketing board achievements depending on the use of monopoly trading powers are, of course, exceptional. There is no substitute for monopoly. However, the hidden social costs of trading monopolies are well known, not only to consumers but also to producers compelled to trade with them. Although the balance between increased efficiency from exposure to competition and from economies of scale need not invariably favour competition, it will often do so even for producers.

3.4 Price policy and price stabilisation mechanisms

The general level of agricultural prices fixes the internal terms of trade between agriculture and other economic sectors. Governments often seek to influence the purchasing power of agricultural products for two reasons: first, to give farmers a bigger incentive to expand output for the market and, second, to accelerate the transfer of capital from agriculture to the remainder of the economy. Since these are potentially conflicting objectives, requiring opposing agricultural price adjustments, finding the 'correct' balance is no easy task. From the A sector's point of view, a *positive* price policy weights the terms of trade in agriculture's favour, whereas a *negative* price policy does the opposite. LDC governments are frequently charged with pursuing a negative price policy; slow agricultural growth is often attributed to this. Whilst such criticisms are doubtless sometimes valid, there is no obvious economic argument to justify giving agriculture a *permanent* terms of trade advantage.

The economic case for government intervention to *stabilise* agricultural prices is inherently stronger than the case for raising them above the mean level or trend dictated by market forces. Buffer-stocks and buffer-funds are widely used as stabilising mechanisms, with agricultural marketing boards frequently acting as the stabilising agency.

Buffer-stocks are built up to raise market prices in periods of over-abundant supplies, and run down to lower prices when current supplies are less than normal. In Figure 8.4, S_L is the market period supply curve when the crop yield is 'low', and S_N and S_H, respectively, represent years of 'normal' and 'high' yields. Thus, without price stabilisation, supplies and prices fluctuate from Oq_1 and P_h in low-yield years, through Oq_2 and P_n in normal-yield years, to Oq_3 and P_1 when yields are high. Assume $Oq_3 - Oq_2 = Oq_2 - Oq_1$. In times of glut the buffer-stock agency removes $Oq_3 - Oq_2$ from the market in order to force the price *up* from P_1 to P_n: in years of relative shortage stocks amounting to $Oq_2 - Oq_1$ are released to force the price *down* from P_h to P_n. Thus P_n is the *stabilised* producer price. Ignoring the costs of storage and handling, and the interest due on deferred income, producers gain in glut years what they lose in years of shortage. These offsetting gains and losses of marginal revenue are respectively represented by the triangular areas A and R. In this hypothetical and idealised example the stocks are *self-liquidating* over a period of time in which there are equal numbers of high- and low-yield years. In practice, however, the *ex ante* determination of, and strict adherence to, the stabilised price, P_n, presents problems. In order to estimate reliably the equilibrium market price in a 'normal' year, the buffer-stock agency needs a complete model of the market which both specifies all the variables

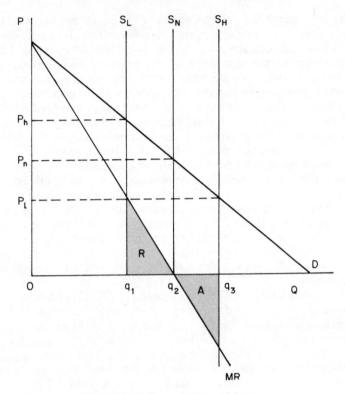

Figure 8.4: *Buffer stock price stabilisation*

affecting supply and demand, and also permits estimation of their parameters together with the probability distribution of stochastic errors.

Even if the agency is able to estimate the market-clearing price in a normal year, producers may well exert pressure for the price to be stabilised above that level. The consequences of yielding to such pressure can be followed through in Figure 8.4. Designate the level around which producers desire the price to be stabilised the 'target price'. Suppose the target price is not set at P_n but at P_h. Then the buffer-stock agency can no longer choose between accumulating and *maintaining* stocks at the current level. Given the supplies of Oq_1, the target price P_h can be obtained from the market without changing stocks. But if supplies are Oq_2 or Oq_3 the agency is compelled to add to its stocks in order to obtain the target price. Thus, over time, stocks can only accumulate, they can never decline; in other words, they are no longer self-liquidating. Moreover, unless stocks are self-liquidating a buffer-stock scheme cannot finance itself. Ever-accumulating stocks require periodic capital in injections to defray the ever-rising costs of holding them.

Buffer-*fund* schemes possess the advantage that the costs of holding physical stocks are avoided. The fund accumulates revenue by 'taxing' producers in high price years, and disburses revenue by subsidising them when market prices are relatively low. Again, the financial viability of the scheme depends on pitching the target price near to the level which equates supply with demand in a 'normal' season.

3.5 The state versus private enterprise in marketing

The case for some government regulation of agricultural marketing is virtually self-evident. Compliance with fair trading practices and adherence to grade standards, for example, must be policed; overseas trade must be monitored and, if necessary, controlled. But how desirable is it that government should actually *participate* in marketing? In seeking an answer to this question we first consider why government may choose to participate (including some non-economic reasons). We then consider a number of *a priori* reasons against participation (FAO, 1969, pp. 94–106).

From observing how governments behave, the possible reasons why they decide to participate in marketing appear to include:

(1) Attempting to obtain more control over national resources and the national economy in the furtherance of national development, by facilitating the control of prices and foreign exchange, and by simplifying the collection of taxes, for example.
(2) Attempting to obtain increased bargaining power for domestic producers in export markets, by setting up monopsonistic export marketing boards.
(3) Attempting to develop new markets and products which the private sector is unable or unwilling to enter or handle.
(4) Attempting to dislodge foreign trading enterprises from domestic markets for ideological reasons, to gain 'local control' in markets where indigenous private enterprise is absent or unable to withstand foreign competition.
(5) Pandering to popular prejudice against private middlemen as a class, regardless of whether they are nationals or foreigners.

Whilst (1)–(3) possess an underlying economic rationale for government participation, (4) and (5), which are non economic, clearly do not.

An important part of the case *against* participation derives from a number of potential handicaps of public enterprise in marketing, these are:

(1) Exposure of such enterprises to political pressure, possibly affecting the appointment of staff, choices of buying and selling contracts and many other policy or managerial decisions. In extreme cases staffing may occur entirely within a system of political patronage.
(2) Overhead costs are excessive and operating costs unnecessarily high because of over-staffing and injudicious investment for reasons of 'prestige' or due to inadequate planning, for example.
(3) The civil servants appointed to manage public enterprises lack the judgement, experience and self-confidence needed to make sound commercial decisions; they are trained to contain public expenditure, within the constraints of a fixed budget, rather than to make profitable investments.
(4) Governments expect public marketing enterprises to show a commercial profit and, at the same time, undertake financial loss-making 'social service' functions, without the benefit of financial subsidy. So, for example, a marketing board or public marketing corporation may be required to accept produce from *all individual producers* on the same terms, regardless of the size, frequency and timing of their deliveries.

Certain potential strengths of *private* marketing enterprise are implied by these potential weaknesses of government or public enterprise. Thus, private entrepreneurs tend to be superior in business acumen, commercial judgement and risk evaluation. They are also less open to political pressures and government interference in discriminating between commercial and non-commercial marketing services. But, despite the shortcomings of public marketing enterprise, a valid case for a measure of government participation remains, based on any or all of the above-listed *economic* reasons for participation, (1)–(3).

Rather than relying exclusively on either public or private marketing enterprise to achieve the goals of agricultural marketing policy, better results may be achieved by adopting a policy of *partnership* between government and the private sector (Fenn, 1972). The economic rationale underlying this approach is a form of comparative advantage. For their part, *private* concerns are frequently capable of lower cost operations and tend to be better able to provide the drive and initiative needed to sustain managerial efficiency and innovation. But *public* intervention is needed to control and direct the activities of private enterprise in accordance with development plans and policies. Also, public enterprise may be needed to function and assume risks for social gains where private enterprise is absent. Moreover, definite policies are needed to reduce market imperfections and stimulate more effective competition in the private sector. The state may also be able to finance capital projects, such as processing plants and storage space, where private investment in such facilities is inadequate.

A precondition of fruitful partnership between the public and private sectors is a willingness to co-operate on both sides. But government officials and politicians in LDCs sometimes display an attitude of indifference (or even hostility) towards the private sector. A remedy may lie in improved education and training to give administrators, and even politicians, a better understanding of the benefits of the economic services provided by private sector market intermediaries.

3.6 Summary and conclusions

The basic objectives of agricultural marketing policy in LDCs include infrastructural development, improving the quantity, quality and dissemination of market information, strengthening the bargaining position of individual producers, and reducing uncertainty by stabilising prices. In pursuit of these objectives general infrastructural development and the improvement of market information services are primarily the responsibility of government. Government is also responsible for enforcing necessary marketing regulations as well as devising and implementing policies to enhance marketing efficiency by reducing market imperfections and making competition more effective. Price stabilisation may be undertaken either by government or by a producers' organisation. The extent of government *trading* activities is likely to depend in part upon the extent and quality of the services offered by private enterprise and partly on government's ideological stance. A good case exists for active collaboration between government and private marketing enterprise in order to provide agricultural marketing services at minimum cost, subject to the proviso that the quality and range of services offered is consistent with the effective service demands of both producers and consumers of agricultural products.

References

Abbot, J.C. and Creupelandt, H.C. (1966), *Agricultural Marketing Boards: their Establishment and Operation*, FAO, Rome.

Abbott, J.C. and Creupelandt, H.C. (1967), Agricultural marketing boards in developing countries', *Monthly Bull. Agric. Econ. & Stats.*, **16**(9).

Adams, Dale W. and Nehman, Gerald I. (1976), 'Measuring farm level credit use: a Brazilian example', *Agricultural Finance Review*, **39**: 77–9, April.

Adams, Dale W. and Tommy, Joseph L. (1974), 'Financing small farms: the Brazilian experience, 1965–69', *Agricultural Finance Review*, **35**: 36–41, October.

Agency for International Development (1973), *Spring Review of Small Farmer Credit*, 20 vols., February–June.

Bardhan, P. and Rudra, A. (1978), 'Interlinkage of hand labour and capital relations: an analysis of village survey data in East Asia', *Economic and Political Weekly*, pp. 367–84.

Bauer, P.T. (1954), *West African Trade*, Cambridge University Press, Cambridge.

Bauer, P.T. and Yamey, B.S. (1957), *Economics of Underdeveloped Countries*, Cambridge University Press, Cambridge.

Berry, R.A. and Cline, W.R. (1979), *Agrarian Structure & Productivity in Developing Countries*, Johns Hopkins University Press, Baltimore.

Bhaduri, A. (1973), 'A study in agricultural backwardness under semi-feudalism', *Economic Journal*, **83**, pp. 120–37.

Bhaduri, A. (1977), 'On the formation of usurious interest rates in backward agriculture', *Cambridge J. Econ.*, **1**, pp. 341–52.

Blandford, D. (1979), 'West African export marketing boards', in Hoos, S. (ed.), *Agricultural Marketing Boards: An International Perspective*, Ballinger, Cambridge, Mass.

Bliss, C. and Stern, N. (1982), *Palanpur: A Study of an Indian Village*, Macmillan, London.

Bottomley, A. (1971), *Factor Pricing and Economic Growth in Underdeveloped Rural Areas*, Crosby Lockwood, London.

Chandavarkar, A.G. (1971), 'Some aspects of interest rate policies in less developed countries: the experience in selected Asian countries', *IMF Staff Papers*, **18**, pp. 48–112.

Dorner, P. (1972), *Land Reform and Economic Development*, Penguin, Harmondsworth.

Fenn, M.G. (1972), 'The entrepreneur in agricultural marketing development', *Monthly Bull. Agric. Econ. & Stats.*, **21** (7/8).

Food and Agriculture Organisation of the United Nations (1969), *State of Food and Agriculture, 1969*, FAO, Rome.

Ghatak, S. (1975), 'Rural interest rates in the Indian economy', *J. Dev. Studies*, **11**, 3, pp. 190–201.

Ghatak, S. (1976), *Rural Money Markets in India*, Macmillan, India.

Ghatak, S. (1983), 'On inter-regional variations in rural interest rates in India', *J. Developing Areas* (forthcoming).

Griffin, K.R. (1976), *Land Concentration and Rural Poverty*, Macmillan, London, (2nd edn.).

Gupta, S.G. (1975), 'Interest sensitiveness of deposits in India', *Economic and Political Weekly*, **20**, pp. 2357–63.

Lewis, S.R. (1967), 'Agricultural taxation in a developing economy', in Southworth, H.M. and Johnston, B.F. (eds), *Agricultural Development and Economic Growth*, Cornell University Press, Ithaca.

McKinnon, R.I. (1973), *Money and Capital in Economic Development*, Brookings Institution, Washington, DC.

Mellor, J.W. (1966), *Economics of Agricultural Development*, Cornell University Press, Ithaca.

Mirrlees, J. (1975), 'A pure theory of underdeveloped economies', in Reynolds Lloyd (ed.), *Agriculture in Development Theory*, Yale University Press, New Haven.

Nguyen, D.T. and Martinez-Saldivar, M.L. (1979), 'The effects of land reform on agricultural production, employment and income distribution: a statistical study of Mexican states, 1959–69', *Economic J.*, no. 355.

Putterman, L. (1983), 'A modified collective agriculture in rural growth-with-equity: reconsidering the private unimodal solution, *World Development,* **11**(2), pp. 77–100.

Rahman, A. (1979), 'Agrarian Structure and Capital Formation: A Study of Bangladesh Agriculture with Farm Level Data', unpublished PhD thesis, Cambridge.

Shaw, E.S. (1973), *Financial Deepening in Economic Development*, Oxford University Press, New York.

World Bank (1975), *The Assault on World Poverty*, Johns Hopkins University Press, Baltimore.

9 Population and Food Supplies

The relationship between population and food supplies has been the subject of major economic debate for at least 200 years. T.R. Malthus published his famous 'Essay on the Principle of Population' in 1789. But Malthus was responding to Adam Smith's more optimistic view of the economic implications of population growth, as expressed in 'The Wealth of Nations' published in 1776. Other classical economists in the British tradition, like David Ricardo and John Stuart Mill, were much influenced by Malthus and his stress on the restrictive influence of diminishing returns.

In what are now termed the developed countries the discipline of economics has now lost some of its reputation as the 'dismal science'. Disbelief in the 'iron law of wages' has been encouraged partly by the declining relative importance of agriculture and other primary industries in national products and partly by the modernisation of agriculture itself, which has diminished the relative importance of land as a factor of production. Moreover, developed countries generally now enjoy the advantage of a relatively stable and low rate of population growth.

A very different view is apt to prevail in many of today's LDCs, where agriculture and other primary industries are still dominant. Moreover, most of agriculture is still 'traditional' (see chapter 2) and a high rate of population growth prevails (higher than historic rates at a comparable stage of development in what are now the developed countries). In such a situation it may well seem that there is much less reason to discard the pessimistic Malthusian view of the prospects for improving the economic lot of the broad masses of a country's population.

This chapter falls into two main sections. First we outline the classical, Malthusian-type model of economic development. We then contrast this with two alternative models of population and food supplies – a contra-Malthusian model in direct antithesis to the classical model, and an 'ecological model', postulating the existence of 'natural' checks on uncontrolled population growth in culturally stable human societies. Section 1 concludes with an attempt to synthesise these various approaches to explaining the relationship between population growth and food supply. In Section 2 malnutrition is defined and its incidence and extent in LDCs is examined. In so doing we distinguish between deficient aggregate food supplies and the maldistribution of food. The causes of famine are also briefly examined. The section concludes with a brief examination of some alternative policy approaches to ameliorating the problem of malnutrition in LDCs.

1 The Classical Model

The basic argument of the classical theory of development, as exemplified by Malthus, can be summarised in five steps:

(1) Increased population causes a parallel increase in the demand for food.
(2) This increased demand for food can be met *either* by bringing 'new' land into cultivation, *or*, by cultivating existing cultivated land more intensively than before, i.e. by applying more labour.
(3) Land is not only scarce (i.e. limited in supply) but also variable in *fertility*. The most fertile land is cultivated first. Then, in response to continuing population pressure, less and less fertile land is successively brought into cultivation. Thus, the least fertile land being cultivated at any particular time is to be found at the 'extensive margin of cultivation'. ('Fertility' is synonymous with 'productivity' in this context.) Thus, as a consequence of bringing additional land into cultivation at the extensive margin, the marginal productivity of agricultural labour is bound to decline.
(4) If additional labour is applied to existing land (at the *intensive* margin) its marginal productivity will again decline due to diminishing returns.
(5) Since diminishing marginal returns to agricultural labour are inevitable and unavoidable, food production will always tend to grow less rapidly than population. In the long run population will always expand to the limit of available food supplies at the

subsistence level. That is, although the rate of growth in food supplies may *temporarily* get ahead of population growth – due, for example, to an unexpected and temporary decline in population or the fortuitous discovery of a more productive method of farming – continual population growth must inevitably force per capita food supplies back to a bare subsistence level in the long term.

More formally, the classical model implicitly assumes that the production function expressing the relationship between population (as a proxy measure of agricultural labour input) and aggregate agricultural output is *continuous through time*. Thus, in Figure 9.1 where the vertical and horizontal axes respectively measure aggregate agricultural product and population, TP and AP are the total and average product functions. The shape of the TP function is consistent with the classical model assumption that aggregate agricultural product is doubly bound by the constraint of diminishing returns at both the intensive and extensive margins of cultivation. The model assumes that agricultural wages are determined by the average productivity of labour. Thus the limit to population increase is at OP_m where AP is at the minimum subsistence level OM. If population exceeds this limit people will starve. The slope of the total consumption function TC_s, is consistent with OM. An agricultural surplus exists provided TP > TC. Such a surplus may be mobilised either for investment to provide employment outside the farming sector or to raise productivity and living standards within agriculture itself (Nicholls, 1963). But, with population at OP_m following unconstrained population growth, the surplus is zero.

The obvious policy implication of the classical model is that in predominantly agrarian societies curtailment of population growth is the sole feasible means of materially improving living standards for the majority of the population. It is easy to criticise this model with the benefit of hindsight. The classical economists did not foresee the extent to which European countries would be able to expand their agricultural production at both the intensive and extensive margins of cultivation, by means of land improvement, improved communications and the adoption of scientific methods of farming. At the time of Malthus, most of the major scientific advances which have revolutionised developed country agriculture during the last 150 to 200 years, including the invention of chemical fertilisers and the selection and breeding of higher yielding crop varieties and strains of

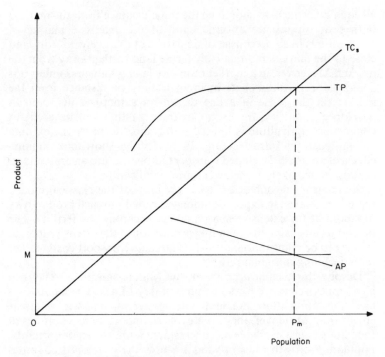

Figure 9.1: *Classical model*

livestock, still lay in the future. Equally important, the extent to which population pressure in the Old World would be relieved by emigration to the New World was not foreseen: nor was it anticipated that an ocean transport revolution would effectively enable much of Western Europe's extensive margin of cultivation to be transferred to the New World.

Although the classical economists can be absolved from failing to foresee the inforeseeable, they are vulnerable to the criticism that it is unrealistic to assume that land at the extensive margin of cultivation must necessarily be of lower fertility than any supra-marginal land.

Economic location theory postulates that the rent of land is jointly determined by its fertility and its distance from the market. If either factor is constant the other – and the rent – are free to vary. Thus, there is no more reason why all plots which are equidistant from the market should be equally fertile (or attract the same rent) than that all equally fertile land should be the same distance from the market: both propositions are absurd. Land at the extensive margin of cultivation is likely to consist of a mixture of relatively infertile land close to the market and relatively fertile land further away from the market. Moreover, in a market economy land is valueless unless it is *accessible*, regardless of its inherent fertility or distance from the market (in the sense of a line drawn on a map). Thus, even in developed countries, the extent of the area that can be used for commercial agriculture is limited in the short term by the existing communications infrastructure. In LDCs the short-term limiting effects of an underdeveloped transport infrastructure on agricultural expansion are likely to be even more significant.

So, contrary to one of the basic premises of the classical model, there is no reason to expect declining marginal physical productivity of labour at the extensive margin due to declining soil fertility. The principal constraint on expanding production at the extensive margin is likely to be rising marginal costs (especially transport costs) rather than declining marginal returns.

Despite this technical criticism, the Malthusian model may still have policy relevance for some present-day LDCs. These countries lack the population pressure safety-valve of massive overseas emigration. Moreover, they are unable to import large quantities of foodstuffs unless available on concessional terms. Most importantly, population growth rates in what are now the developed countries never reached the very high levels of up to 3 per cent annum which characterise today's LDCs. Whereas the developed countries had many years in which to adapt their birth rate to a gradually declining death rate and rising living standards, many LDCs have experienced a later and much more rapid decline in the death rate, whilst the majority of the population (much of it in agriculture) remains very poor. Thus, in the LDCs, birth rates have remained at, or very close to, their traditional high levels.

2 Contra-Malthusian Model

Whereas the classical model postulates that, in the long run, population growth is limited by food supplies, the contra-Malthusian model argues that food supplies increase in response to population growth. These sharply opposed views of the relationship between population and food supplies result from equally contrasting views of the process of technical change in agriculture. The classical (and neo-classical) tradition treats the discovery of improved technology as an exogenous shift variable, i.e. discovery is fortuitous. Since, by definition, *improved* methods of production offer the means of reducing costs of production, discovery is rapidly followed by adoption on the part of profit-maximising producers. In contrast, the contra-Malthusian hypothesis that food supplies increase in response to population growth rests on the assumption that a stock of improved technology is accumulated (again fortuitously) but adoption depends on population pressure. (Boserup, 1965).

Boserup's concept of the dynamics of agricultural technology concerns change in cultivation methods and the choice of tools in relation to changing systems of land use. Population density is the primary determinant of the system of land use, the method of cultivation, and the choice of tools at a particular time. As population pressure increases, progressively more intensive systems of land use are adopted, combined with consequential changes in methods of cultivation and the choice of tools, in order to offset any tendency for food output per capita to decline, due to diminishing returns. In more formal terms, the model postulates that although in the short run there may be diminishing returns to agricultural labour, in the long run the aggregate agricultural production function will always shift upwards in response to population pressure, at whatever rate is required to maintain output per capita. The shift variable is agricultural technology, as expressed by the system of land use, the method of cultivation and choice of tools. Thus Boserup explicitly rejects the classical model notion of a *continuous* agricultural labour productivity function: in her view the true function is discontinuous and the reason for discontinuity is technological change.

The structure of the Boserup model, assuming a Cobb-Douglas-type production function, is formally presented in the appendix.

2.1 Differential intensity of land use systems

Boserup identifies a succession of increasingly intensive systems of land use in primitive agriculture and correlates these with increasing population density. The lowest level of population density is characterised by the system of 'forest fallow', with a fallow period of 20–25 years between successive crops. At a somewhat higher level of population density, forest fallow gives way to 'bush fallow' in which the fallow period is reduced to 6–10 years. Shortening the fallow period raises the labour: land ratio as well as output. In response to further population growth and increase in the agricultural labour supply, bush fallow is successively followed by 'short fallow', with a fallow period of 1–2 years only, and 'annual cropping', with the fallow period reduced to less than a year. The most intensive system of land use – at the opposite extreme to forest fallow – is 'multi-cropping', under which the land is cropped more than once annually with virtually no break between successive crops. Other things being equal, the multi-cropping system is most labour intensive as well as yielding the highest output per unit of land area per unit time period.

2.2 Method of cultivation and choice of tools

Forest fallow farming, the most extensive system of land use, is virtually a 'no-cultivation' system. The land is not broken, the farmer merely fells and burns the forest and plants his crop in the ashes. The essential tool is the axe and little else is needed.

The system of bush fallow calls for the use of the hoe (or digging stick) to help clear the land for planting and also to keep down weeds during the growing season. As the fallow period is successively shortened, so the 'weed problem' intensifies and, by the annual cropping stage, the use of the plough – to bury weeds before planting – becomes essential. The introduction of the plough also entails setting land aside for the grazing of draught animals (motor tractors being strictly outside the 'primitive agriculture' terms of reference). Thus the incremental production resulting from the use of the plough must exceed the production lost by diverting land to grazing. Multi-cropping frequently involves the adoption of irrigation, requiring specialised equipment, in order to break the constraint on production imposed by the natural 'dry season'.

2.3 Land use intensity and labour productivity

A further important feature of Boserup's theory is that, because the necessary number of cultivations and frequency of attention to crops is inversely related to the length of the fallow period, the length of the working day in agriculture *increases* with intensification of the system of land use, i.e. forest fallow farmers work the shortest hours and those practising multi-cropping the longest. This is the exact opposite of the widely-held view that the short working day is most prevalent in densely populated countries, and vice versa. But long fallow farmers are not disguisedly unemployed in the sense of Sen (see chapter 3, pp. 52–4). According to Boserup, their leisure preference is high and they are prepared to sacrifice leisure at the margin of their customary working hours only when forced to do so by population pressure which increases the relative scarcity of food. Only then are they willing to switch to a more intensive system of farming entailing not only changed methods of cultivation but also longer working hours. Underlying this argument is the assumption that in primitive agriculture the motivation for work is the attainment of a fixed *income* target (with the income consisting primarily of food) rather than income (or utility) *maximisation*. Thus, in Figure 9.2, an individual producer's utility surface, as signified by indifference curves I_1, I_2 and I_3, is superimposed on his production possibility surface, as indicated by the curves P_1, P_2 and P_3. Work time is assumed to

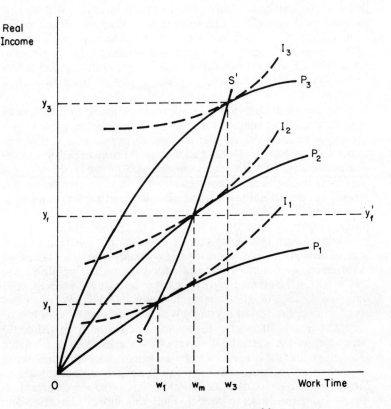

Figure 9.2: *Fixed income target model*

be the sole variable input. The labour time response curve to opportunities for gaining real income through improved methods of cultivation is indicated by SS'. Whereas the neo-classical producer with unlimited wants could move all the way from w_1 to w_3 work time in response to the opportunity of increasing his real income from y_1 to y_3, the producer with a fixed income target of $y_f y'_f$ would not be prepared to exceed w_m work hours unless forced to do so in order to maintain the income target.

A major policy implication of the contra-Malthusian model is that, at least in the long run food output will always tend to keep pace with population growth through changes in the intensity of cultivation and choice of tools. However, food output response to growing population pressure may be inhibited by policy constraints such as a deficient marketing and distribution infrastructure and unfavourable movements in agriculture's terms of trade, (Boserup, 1975). Despite some empirical evidence lending support to the hypothesis that in LDCs technical change in agriculture responds to population pressure, such as the results of a study of rice cultivation in Sierra Leone (Levi, 1976), Boserup's model is clearly open to several criticisms.

First, the model provides a much more plausible explanation of how food supplies might keep pace with population growth in a sparsely populated country than in a densely populated one. Suppose that a country has no idle land so that agricultural expansion at the extensive margin of cultivation is precluded. Suppose further that the transition to the most intensive known system of cultivation has already occurred, but the population is still rising due, for example, to the favourable effect of continuing improvement in standards of public health on the death rate. What options then exist for maintaining per capita food supplies? Within the constraints of the present model the two main options would appear to be either (a) attempting to increase aggregate food output by applying more labour within the existing system, or (b) progressing to an even more intensive system of land use and choice of tools. Whereas the successful application of (a) might well be frustrated by very low or even zero marginal returns to labour, the feasibility of (b) could depend upon the scientific discovery of some hitherto unavailable technology. But the concept of technical progress in agriculture being dependent on scientific discovery and the effective transmission to farmers of the fruits of agricultural research and development, is absent from the Boserup model. Thus this model scarcely offers an effective reply to the argument that the only possible escape route from the Malthusian trap for the most densely populated poor countries is larger food imports provided on concessional terms.

A second serious criticism also concerns the nature and costs of technical change in agriculture. Boserup emphasises the economic incentives needed to induce producers to change their production techniques. However, the model fails to account fully for the costs of adopting new methods. Although the investment costs of intensifying the system of cultivation are not entirely neglected, there is too much emphasis on the special case of farmers utilising their own low opportunity cost labour to create additional farm capital to the neglect of types of investment necessitating the purchase of modern capital inputs from outside the agricultural sector. For example, although hand digging may be the 'best' means of excavating a well if labour is sufficiently plentiful and cheap, it is less likely that manual labour can feasibly be substituted for a mechanical pump with equal effect as the means of extracting the water. But, for a poor peasant farmer, finding the money capital to buy a pump, and possibly other irrigation equipment too, may present a virtually insurmountable obstacle. Such problems are ignored by the Boserup model. Moreover whereas most LDCs now plan to accelerate their rate of agricultural development and *raise* per capita food output, the Boserup model is too long term and insufficiently specified for use as a *planning* model. Furthermore, although rapid population growth in LDCs may have contributed to the urgency of the international search for improved agricultural inputs such as high-yielding varieties, this model does not account for the process whereby the fruits of research conducted mainly in developed countries have been successfully transmitted to and adopted by farmers in LDCs.

A third criticism is that Boserup presents little convincing evidence to support her contention that primitive agriculturists typically exhibit a strong leisure preference. Although it is not irrational for farmers to value their leisure above the disutility of work at the margin of their working hours, the hypothesis lacks plausibility in a context of relative poverty, short working days and ample leisure. Much stronger empirical support is needed to carry with conviction the argument that the hypothesis has general validity.

However, despite these criticisms, the contra-Malthusian model, deserves serious attention for its novel analytical insights into the process of agricultural development as well as for the counter-arguments it provides against the over-gloomy prognostications of neo-Malthusians.

3 Ecological Disequilibrium

The central hypothesis of Wilkinson's ecological model of economic development is that 'changing ecological circumstances, centring on the relationship between population and resources, force societies to exploit their environment in new and often more difficult ways' (Wilkinson, 1973, p. 4). The approach is multi-disciplinary, drawing extensively on biological, anthropological and sociological evidence, as well as upon the literature of economics, in order to support the central thesis and related arguments.

In the natural world *ecological equilibrium* is the norm. 'Rather than over-exploit their resources [natural populations] build up a pattern and a rate of resource use which the environment can sustain indefinitely' (ibid., p. 21). Wilkinson presents anthropological evidence that primitive societies apply cultural restraints upon the rate of resource use. Specifically with respect to the rate of *population growth*, these include abortion, infanticide and taboos upon sexual intercourse during lactation. But, in mankind, ecological equilibrium is not necessarily a permanent state. Ecological *disequilibrium* can occur as and when the constraints which maintain human societies in ecological equilibrium break down. For example, traditional beliefs and customs may be undermined due to contact with alien cultures. Thus, traditional methods of population control, like infanticide, have been abandoned due to the teaching of missionaries that such practices are wicked.

When ecological equilibrium breaks down society will try to find ways of developing its technology in order to increase the yield from the environment – in *agriculture* more intensive systems of land use, frequently involving the sacrifice of leisure, are adopted.[1] Like Boserup, Wilkinson contends that pre-industrial societies are characterised by a pronounced leisure preference; only in industrial societies does the leisure/consumption (or work) trade-off shift markedly in favour of consumption. But again, the empirical evidence supporting this contention is weak.

The apparent policy implication of the ecological disequilibrium model is that in LDCs, where the traditional cultural constraints on excessive population growth have broken down and new agricultural technology is incapable of closing the gap between food supplies and food requirements, a new social consensus is needed to curb population growth. But is such a policy feasible? And how might it be implemented? The logic of the model suggests that re-establishment (and maintenance) of ecological equilibrium might depend on large numbers of individuals either conforming to new forms of social taboo, or behaving altruistically. But in many LDCs, where the pursuit of individual self-interest is encouraged to promote the common goal of more rapid development, and where pluralism may be gaining at the expense of consensus, such an option may be unrealistic. However, although most countries with a population policy rely primarily on the voluntary acceptance and use of birth control techniques to limit family size, some have resorted to direct social and economic pressures to discourage parents from having too many children.

In China, for example, restricting family size to one or two children has long been one of the main planks of social planning. Raising a large family is anti-social since it diverts resources from the more important tasks of self-education and productive employment. Consequently, the Chinese government, as well as advocating family planning, have exerted very strong social pressures to discourage early marriage (which was traditional in pre-communist China). More recently, social pressures to limit family size have been supplemented by economic pressures, at least in some urban areas, where one-child families receive preferential treatment with respect to housing and taxation, whereas families with more than two children are discriminated against. Similar discriminatory policies against large families have been adopted and applied in Singapore, a non-communist country (Birdsall, 1980, p. 60). Barriers to the wider adoption of such policies in LDCs include the implied infringement of parental liberty to determine their own optimum family size and the view that, in any case, children should not be made to suffer for the 'social deviation' of their parents.

4 Synthesis of Population and Food Supply Theories

A principal weakness of the classical model is its underestimation of the potential role of technical progress in offsetting the effects of diminishing returns to agricultural labour accompanying population growth. The classical economists also mis-specified the extensive margin of cultivation by not allowing adequately for the development of communications giving better access to agricultural land both at home and overseas. This mis-specification exaggerates diminishing returns in agriculture. Nevertheless, the Malthusian model may still have policy relevance for some present-day developing countries with a high rate of population growth and severe technical or capacity constraints on either expanding domestic agricultural production or increasing the volume of food imports.

The contra-Malthusian model, as interpreted by Boserup, draws attention to the possibility of a relationship between population growth and the adoption of new agricultural technology (which Malthus missed). But Boserup has nothing to say *either* about the process whereby new agricultural technology is generated – either within agriculture itself or in a separate agricultural service sector – *or* about obstacles to its adoption – such as capital rationing and risk-aversion (as discussed in chapters 2 and 5). Boserup's model is also difficult to square with the 'internationalisation of agricultural R & D' whereby the fruits of research conducted in developed countries are transferred to the LDCs. This process is exemplified by the yield augmenting affects of the 'seeds-fertiliser revolution' which has occurred in certain LDCs, particularly in South-East Asia during the past 25 years (see chapter 6). The thesis that higher population pressure will *always* induce increased productivity of land via the adoption of more intensive cultivation methods is almost certainly invalid for some of today's very poor and most densely populated countries. But, more generally, the long-term *secular* relationship between population and agricultural output in most countries lends more support to Boserup than to Malthus. However, a historical relationship is not necessarily a fool-proof guide to the future.

The theory that population growth is limited by food supplies is also rejected by Wilkinson, except under conditions of ecological disequilibrium where cultural restraints on population growth have broken down. Ecological disequilibrium forces the adoption of new technology as the means of increasing agricultural output. Thus, the theory of ecological disequilibrium is similar to the contra-Malthusian model, except that Wilkinson treats technological innovation as 'second-best' to adherence to cultural restraints on population growth in the race between population and food supplies. Wilkinson's theory poses the problem, for today's over-populated LDCs, of finding new population restraints or institutions to replace outmoded traditional methods which would be unlikely to regain social acceptance in the modern world. Recent developments in China and one or two other LDCs suggest that, at least in some societies, it may be possible to solve this problem.

Both the Malthusian and contra-Malthusian explanations of the relationship between the rates of growth in population and food supplies assume that causation is uni-directional. The difference between them is that the direction of causation between the two variables is reversed. Whereas Malthusians assume that an autonomous food supply growth rate (cause) limits population (effect), contra-Malthusians assume that an autonomous increase in population (cause) induces a corresponding increase in food supplies (effect). Uni-directional causation is open to the criticism that it rules out *feedback*. An alternative hypothesis is that as growing food supplies induce population expansion, so population growth induces further growth in food supplies. It is immaterial whether the causal chain is initiated by an autonomous increase in food supplies or population. The crucial point is that feedback is built into the system so that neither variable is uniquely causal. A resolution of the opposing viewpoints of Malthusians and contra-Malthusians might be found in a model of this kind.

Each of the three models of the relationship between the growth of population and food supplies which we have reviewed is open to criticism. But, despite inherent weaknesses, each model also deserves attention for any positive aspects which add to our knowledge and understanding of a complex problem and which may also inform the deliberations of policy-makers in LDCs.

One further aspect of this relationship deserves mention. It is naive to suppose that if the growth of aggregate food supplies keeps pace with population all individuals comprising the population will necessarily be adequately fed. This is because, except in a society where all households have direct access to land, there is a two-way race between: (a) population and economic participation (or employment) and (b) economic participation and food supply. The unemployed are bound to have difficulty in obtaining an adequate food supply regardless of the aggregate supply situation (Poleman, 1981). (We discuss food redistribution policies, to meet the needs of the unemployed and other socially deprived groups, in section 6 below.)

5 Malnutrition in Developing Countries

Malnutrition refers to a state in which diet is inadequate, either in quantity, quality, or both, for normal health and physical well-being; thus adequate nutrition is a normative concept. Whereas, in a market economy, an individual's food consumption primarily depends upon his income and tastes (assuming constant prices), his food requirements depend upon the application of nutritional norms, including allowances for interpersonal differences in age, sex and occupation, as well as upon factors like climate variations.

The conventional method of assessing the nutritional adequacy of per capita food supplies for a country or region is to construct a 'food balance sheet'. This compares per capita food supplies – including imports, and after making some allowance for wastage – with objectively determined per capita nutrient requirements. The nutritional status of the country (or region) is judged to be inadequate only if per capita supplies are exceeded by per capita requirements.

This approach to the measurement of malnutrition is open to several criticisms (Poleman, 1981). First, the assumption that per capita food requirements are *homogeneous*, even amongst people of similar age and the same sex, is clearly very crude. There must inevitably be a tendency to overestimate 'reference man's' food requirements in order to err on the safe side. Second, estimates of available food supplies are highly likely to *underestimate* the actual amounts available for reasons such as deliberate underreporting by farmers (to reduce taxation, for example) and statistical errors such as incomplete enumeration and underallowance for the volume of subsistence agricultural production (intrinsically difficult to measure). Third, and most importantly, by concentrating attention on what per capita food consumption would be if available supplies were *equally* distributed within the population, the food balance sheet approach ignores the whole problem of the *maldistribution* of food supplies. This requires some elaboration.

5.1 'Overt' and 'silent' food problems

The conventional explanation of hunger focuses on the *supply* of food and *physical* food scarcity as signified by an average or *per capita* shortfall of supplies below consumption requirements. This has been termed the 'overt' food problem (Reutlinger, 1977). But, by assuming that members of the relevant population group are either *all* hungry or *all* adequately fed, this explanation overlooks the possibility of a proportion being adequately fed whilst the remainder are hungry due to unequal distribution of the available food supply. Reutlinger calls this the 'silent' food problem.

The key to 'who goes hungry', regardless of whether aggregate food supplies are 'scarce' or 'plentiful', is the distribution of effective demand, or what Sen (1981) calls 'exchange entitlements'. Any economic change causing a reduction in per capita real income, such as reduced working hours, unemployment, or a sharp rise in the price of food or other necessities, will tend to 'squeeze' the consumption of low income households. From simple comparative statics analysis it can be deduced that, *ceteris paribus*, some reduction in per capita food consumption is the expected consumer response to any real food price increase. But, provided that the price-elasticity of demand is lower for 'rich' than for 'poor' families – a seemingly reasonable *a priori* assumption – the scale of the cut in per capita consumption will be *inversely* related to family income. Whereas rich families may scarcely need to curtail their food consumption, some poor families will be virtually priced out of the market.

The reason for a rise in the food price may be a rightwards shift in the aggregate demand schedule rather than a leftwards shift in aggregate supply. For example, speculative hoarding by high income families could produce this effect. Whether the expected shortage actually materialises or not, aggregate demand must increase in the short term as the hoarders build up their stocks, as well as buying for current consumption, and low income families again suffer from reduced food consumption as the price rises. But poor families can also be forced to curtail food consumption despite food price *stability*. Thus any economic misfortune which reduces the amount of family income, such as falling wages or rising unemployment, will induce a leftwards shift of the demand schedule in the poor families sub-sector of the market. *Ceteris paribus*, this could be expected to induce a parallel shift in the corresponding segment of the *aggregate* demand schedule, and some consequent price *reduction*. However, the effect of this price reduction on family real income may only be marginal in relation to the initial income loss. Of these three situations in which low income families suffer a reduction in food consumption, only the first, in which aggregate food supplies are curtailed, conforms to the conventional explanation for hunger.

5.2 Famine

Carried to its extreme, hunger becomes famine and people die from starvation. The causes of famine have been analysed by Sen (1980a) who uses the term 'food availability decline (FAD)' to signify the most usual and obvious explanation. That is, people starve because per capita food supplies are suddenly reduced to the starvation level by crop failure or some other event which greatly reduces aggregate food *supplies* below their normal level. But Sen finds that when the facts surrounding certain major famines are examined, FAD emerges either as an incomplete explanation, or even as a wholly unconvincing one.

An alternative explanation of famine is that those who starve lack the means to obtain sufficient food to remain alive. Sen terms this the 'entitlement failure' explanation. Entitlement failures are of two kinds: *direct* entitlement failures, relating only to food producers, in which a peasant farmers' crop fails and he is unable to feed himself either by self-supply or by trade; and *trade* entitlement failures, which relate to those who can normally afford to purchase a sufficient quantity of food, but are priced out of the market by a sudden increase in demand. In a 'slump famine' direct entitlement failure predominates, whereas trade entitlement failure is the overriding factor in explaining a 'boom famine'. But, despite its theoretical elegance, the exchange entitlement failure explanation of famine suffers from being much less amenable to measurement than FAD.

It is important to appreciate that the FAD and entitlement failure explanations of famine are not mutually exclusive. Sen cites the notorious Irish potato famine of the 1840s as an example of slump famine. Supplies of other foodstuffs, such as grain, were available in Ireland during the famine despite the devastating failure of the potato crop. But, because of the dominance of the potato in the accustomed diet of Irish peasants, the obliteration of the crop by potato blight *did* represent a substantial FAD, as Sen himself admits. Indeed, FAD may be the dominant cause of some famines. At the same time, the dire consequences of crop failure in Ireland were greatly exacerbated by government's failure to organise more effective measures of famine relief. In Sen's phraseology, such measures could have done much to offset the Irish potato producers' direct entitlement failures.

Thus, explaining the Irish potato famine appears to combine FAD with slump famine conditions brought about by the devastating loss of income suffered by potato producers. Sen cites the Ethiopian famine of 1974, which was concentrated in one province of the country, as an example of a 'pure' slump famine, i.e. famine in a specific geographical area, due to massive entitlement failure for the agricultural population, without substantial FAD in the country as a whole. In contrast to the Ethiopian famine, the Bengal famine of 1943 is cited as an example of boom famine. The statistical record indicates that per capita food availability in Bengal during the famine period was *not* exceptionally low. But the real incomes of agricultural labourers, the prime victims of the famine, were severely curtailed by food price inflation and increased unemployment induced by a combination of adverse weather conditions and disruption caused by India's involvement in the war with Japan.

5.3 Amelioration of malnutrition: policy options

Analysis of the causes of malnutrition in LDCs indicates at least three possible policy approaches to ameliorating the problem:

(1) lowering the population growth rate;
(2) augmenting food supplies (domestic and/or imported); and
(3) redistributing available supplies from the adequately or 'over-fed' to the 'under-fed'

5.3.1 Population Control

A comprehensive analysis of population control is beyond the scope of this book and we confine ourselves to a brief examination of possible reasons for the persistence of high fertility rates in LDCs, particularly in rural areas or where agriculture predominates. Whereas crude birth rates in low income countries still average about 30 per thousand population, compared with only about 15 per thousand in industrial market economies, the crude death rates appear to be virtually the same, on average, at 10 or 11 per thousand population (World Bank, 1981, Table 18).

Fertility appears to decline with increasing urbanisation. But since this, and other socioeconomic trends favourable to falling fertility, such as rising literacy and declining mortality and poverty, tend to be interrelated, it is difficult to isolate the effect of any *one* of them (Birdsall, 1980, p. 42). The urban–rural fertility differential hypothesis states that, other things being equal, fertility in LDCs is lower in urban than in rural areas. The reasons postulated for the differential include not only higher urban incomes and a wider choice of consumption goods in urban areas, but also more advanced standards of education and superior family planning services. The empirical evidence is mixed. The most convincing evidence of a significant differential is provided by Latin America where, during the 1950s and 1960s, fertility was lower in urban than in rural areas in practically all countries. The evidence is less convincing for other continents, and particularly in Africa where 'comparable and reliable measures of fertility differentials are generally lacking' (United Nations, 1977, p. 284). However, this conclusion is as much based on an absence of reliable data as on positive evidence to disprove the hypothesis.

The causation and duration of urban–rural fertility differentials remains obscure. The *general* relationship between fertility and economic growth seems clear enough, i.e. fertility declines as per capita income rises. Thus urban–rural fertility differentials may reflect no more than the *expected transition* from high to low fertility, which is part of the development process, with the urban population merely leading. Long period demographic evidence from some now developed countries, like Japan, suggests that although rural fertility remains higher initially, the gap eventually closes (United Nations, 1977, p. 290).

A high rural fertility rate probably reflects the relative poverty of farmers – and, *a fortiori*, landless labourers – as well as lower literacy, but it may also reflect a stronger economic motivation to invest in children as potential contributors to the family income. The economic motivation for family formation combines consumption with investment aspects, as discussed by Cassen (1976). On the consumption side, the costs of educating and rearing children, plus the opportunity costs of parents' time and earnings possibly forgone, have to be set against the psychic income derived from parenthood. On the investment side, the potential 'yield' includes not only a direct contribution to family income when the child is old enough to work, but also relief labour to cover parents' temporary inability to work (due to illness or accident) and the support of parents in their old age. On balance, Cassen appears to be dubious whether, on a strictly economic calculus, children do typically represent a good investment prospect for poor households in LDCs. Relevant empirical evidence is extremely meagre, but a study in Bangladesh (Eberstadt, 1980) pointed to the relatively optimistic conclusion that the average village boy can be credited with the triple achievements of earning enough to pay for his food by the age of 12; paying for all his past consumption by the age of 15; and paying for the consumption of his sister as well by the age of 21.

We conclude that lowering the rate of population growth, by means such as improved family planning services, is probably a viable approach to improved nutrition in LDCs, especially in the long term. In the shorter term, the decline in fertility may be retarded by the perceived economic advantages of a larger-sized family, particularly in agriculture.

5.4 Augmenting food supplies

This policy option is discussed at length in chapters 5, 6 and 7. Suffice it to repeat here that the constraints on expanding domestic food production in LDCs are complex and include institutional rigidities which are very resistant to rapid change. Thus, like population control, this approach is too long-term for dealing effectively with the most pressing aspects of the problem of malnutrition.

5.5 Redistribution

For the reasons already analysed in this chapter hunger is primarily a function of poverty. If aggregate food supplies are inadequate for all to have sufficient for survival, the relatively well-off are much less likely to starve than the poor. And if, as is much more likely, aggregate food supplies are adequate to meet consumption needs *given equality of distribution*, the poor remain exposed to the threat of starvation.

In an economy consisting exclusively of poor people all might be hungry and the problem of relieving hunger could be resolved only by raising incomes all round. But, coming closer to reality, where the distribution of income is such that only a proportion of households are 'poor' whilst the remainder are 'rich' – i.e. relatively better-off it may be feasible to relieve hunger by redistributional measures alone, i.e. without raising the aggregate or average level of income. Such an approach is open to the critical argument that the best means of curing poverty (and hunger) is by maximising the rate of economic growth in order to make everybody better off. There are two decisive arguments against this criticism and in favour of deliberate redistributional measures to ameliorate hunger. First, due to the uncertain but most likely unequal distribution of its benefits, economic growth does not necessarily cure poverty. Secondly, in any case, the 'trickle down' approach to the relief of hunger is too slow.

Having opted, in principle, for redistribution as the most effective short-term method of relieving hunger, government remains confronted by two major problems: a) identifying the hungry, and b) devising the most effective means of raising their per capita food consumption, particularly in the most needy or 'target' groups. An obvious and commonly used means of identifying the hungry is the 'income approach', assuming that there is a direct and reasonably close correlation between per capita food consumption and household income. Provided appropriate data are available, this relationship can be estimated statistically, most usually in the form of data showing average levels of per capita consumption by household income group. The results of empirical observation in a number of countries confirm the expected systematic increase in average calorie availability per capita with rising household income (or household expenditure) (FAO, 1977, Tables II 1.3–1.7).

The same approach might be used in order to test for a systematic difference in per capita food consumption between urban and rural households. A food consumption differential in favour of urban households might be postulated *a priori* because of the *income* differential which commonly exists in favour of those in urban occupations. We are unaware of empirical evidence with a bearing on this hypothesis. But available evidence does suggest that at the same or similar levels of household income per capita food consumption tends to be higher in *rural* households. This is possibly because whereas farm households usually possess self-produced food supplies, poor urban households typically have much less access to the resources required for subsistence food production. Urban households are consequently more dependent on cash income to purchase food. Lacking access to land, rural landless labour households are also deprived compared with farm households.

A possible alternative to trying to relate actual calorie intake to perceived nutritional needs is to attempt to observe the point where households endeavour to improve the quality of their diet rather than increasing its quantity (Poleman, 1981). This approach is based on the principle that, as a nation becomes richer, the composition of its diet shifts from almost complete reliance on bulky foods rich in carbohydrates, such as root crops and cereals, to the adoption of a more varied diet emphasising less bulky foods with more protein. Bennett's potato/wheat substitution model, based on historical changes in the diet of American immigrants and their descendants, is a well-known application of this principle (Bennett, 1954). Poleman cites more recent developing country evidence, from North-East Brazil for example, suggesting that very poor people do in fact substitute more preferred for less preferred starchy staples as opportunity arises rather than merely consuming more calories.

A major weakness of the household income and consumption approach to identifying the hungry is that it conceals the facts concerning food distribution *within* the family. Yet it is known that young children and pregnant and lactating women are especially vulnerable to protein–energy malnutrition (PEM). Accurate information on the extent of worldwide PEM is lacking though, on the basis of various approximations and assumptions, Poleman estimates that, in 1975, between 62 and 309 million women and children might have been 'at risk'. PEM is related to bad sanitation and health factors, as well as to inadequate food intake as such, so that the 'synergism of malnutrition and infection is difficult to separate into its component elements' (FAO, 1977). But bad health and low food consumption are both linked to poverty.

Reasonably accurate measurement of the extent of malnutrition in a country, and proper identification of the most needy malnourished 'target groups' are important objectives of nutrition policy. If the extent of malnutrition is overestimated or exaggerated not only does the problem appear more intractable than it really is, but scarce resources are wasted in giving relief to people who do not need it (Srinivasan, 1981).

Because of the close link which exists between poverty and hunger, a *general* redistribution of income in favour of the poor represents one approach to the amelioration of hunger. In principle, a choice of redistribution methods exists. First, there is the 'socialist solution' of guaranteed employment for all who are able to work, even though this entails the loss of the individual's freedom to choose the type and place of employment, as in China.[2] Second, there is the possibility of a guaranteed income for all citizens through social security benefits, as in rich 'capitalist' countries, despite the cost to the national budget and the political resistance of vested interests. Third, there is the pursuit of full employment with adequate minimum wages for all, within the context of a free enterprise or 'mixed' economy, despite the well-recognised problems of achieving and maintaining steady growth in the face of inherently unstable market forces.

In spite of the formidable difficulties of achieving significant redistribution of income by deliberate government policy, particularly in market-type economies, evidence is accumulating to show what can be achieved, even in LDCs. So, whereas in Taiwan and Sri Lanka, for example, economic growth has been successfully combined with greater equity in the distribution of income, in other LDCs, such as the Philippines and Brazil, growth has been accompanied by increasing income concentration (Fields, 1980; Sen, 1980b). Thus, although the growth and distribution record of some LDCs appears to confirm the well-known Kuznets curve hypothesis (that, during the process of economic growth, declining income concentration is preceded by a phase of *increasing* income inequality) the evidence is by no means uniform. Some developing countries have actually achieved 'redistribution with growth' (Chenery *et al.*, 1974). However, as a means of alleviating hunger, the general income redistribution approach suffers from the twin problems of its relative slowness and its failure to deal directly with the nutritional problems of especially vulnerable groups such as young children and pregnant and nursing mothers.

More direct approaches to stimulating food consumption of low income families and especially vulnerable groups include at least three possibilities. First, there is nutritional education to encourage the purchase of high-calorie foods at minimum cost. This is difficult to put across effectively, particularly to the poorly educated and, in any case, provides no more than a partial solution to the problem. In many families income may still be too low to buy nutritionally adequate amounts of even the cheapest types of foodstuffs. A second possibility is for the government to grant food price subsidies, as exemplified by India's 'fair price' shops. This approach is limited by the problem of effectively restricting the supply of low price food to low income families: leakages to non-preferential consumers tend to make such programmes very expensive. The third option is special target group food distribution measures, such as the US Food Stamp programme. Although the US is not a developing country, there seems to be no reason, in principle, why LDCs should not adopt this approach. The essential feature is that low income households must be identified and issued with coupons which can be exchanged only for a fixed amount of food. The coupons may either be given free to qualified consumers or sold to them at less than face value. Such a system is clearly open to abuse in that some stamps may get into unauthorised hands: but no scheme is completely foolproof. A more serious objection is that the distribution of coupons benefits only those consumers who purchase food in the market. In particular, low income subsistence farming households are *not* catered for. A different approach, emphasising increased farming productivity, is needed for these. From a strictly nutritional point of view, a main advantage of food stamps compared with open-ended food subsidies or straight income transfers to target groups, is their cost-effectiveness, i.e. the programme cost per unit of extra food transferred to target group consumers is likely to be lower. A simple theoretical model supports this conclusion (Reutlinger and Selowsky, 1976). But a food stamp programme imposes a substantial additional administrative burden on government, and this may be one reason why this approach to the relief of malnutrition has not been more widely adopted in LDCs.

In practice, some combination of policy measures to combat malnutrition in Third World countries, including both short-term and long-term instruments, may best serve national welfare. So, for example, an employment-oriented programme of economic growth (long-term) might be combined with a food stamp programme to give some immediate relief to the most needy families (short-term).

Appendix: Formalisation of Boserup's Model

Consider the Cobb–Douglas production function:

$$q = aL^\alpha, \quad 0 \leqslant \alpha \leqslant 1$$

where q = agricultural output, L = agricultural population (with dependency ratio assumed constant), a = a constant reflecting a given state of agricultural technology, and α = elasticity of output w.r.t. labour (L). The amounts of all non-labour inputs are assumed to be fixed.

Discrete time can be introduced by attaching numerical subscripts to variables. Thus:

$$q_1 = a_1 L_1^\alpha \ldots \tag{1}$$

and

$$q_2 = a_2 L_2^\alpha \ldots \tag{2}$$

It is a condition of Boserup's model that:

$$\frac{q_2}{q_1} = \frac{L_2}{L_1} \ldots \tag{3}$$

Proceeding now to re-expressing condition (3) in terms of (1) and (2), we deduce (from (3)) that:

$$q_2 = \frac{L_2}{L_1} \cdot q_1 \ldots \tag{4}$$

and substituting (1) in (4) yields:

$$q_2 = \frac{L_2}{L_1} \cdot a_1 L_1^\alpha \ldots \tag{5}$$

Substituting (2) in (5), we have:

$$a_2 L_2^\alpha = \frac{L_2}{L_1} \cdot a_1 L_1^\alpha \ldots \tag{6}$$

from which:

$$a_2 = \frac{L_2^{(1-\alpha)}}{L_1^{(1-\alpha)}} \cdot a_1 \ldots \quad (7)$$

and finally, by substituting (7) in (2):

$$q_2 = \left[\frac{L_2^{(1-\alpha)}}{L_1^{(1-\alpha)}} \cdot a_1\right] L_2 \ldots \quad (8)$$

Equation (8) merely expresses the mathematical relationship between the 'required' level of output, q_2, and given values of other model variables and parameters. It is *not* an explanatory relationship in the sense of explaining *how* target output q_2 is achieved. The technological shift factor is readily isolated from (7), i.e.

$$\frac{a_2}{a_1} = \frac{L_2^{(1-\alpha)}}{L_1^{(1-\alpha)}} \ldots \quad (9)$$

But this states no more than that technological state a_2 is sufficiently 'better' than a_1 to enable the proportionate increase in labour of L_2/L_1 to raise output by an equivalent ratio, despite the constraint of α. The means whereby better technology raises output in the required proportion is not specified.

References

Bennett, M.K. (1954), *The World's Food*, New York.
Birdsall, N. (1980), *Population and Poverty in the Developing World*, World Bank Staff Working Paper no. 404, World Bank, Washington DC.
Boserup, E. (1965), *The Conditions of Agricultural Growth*, George Allen & Unwin, London.
Boserup, E. (1975), 'The impact of population growth on agricultural output', *Quarterly J. Econ.*, **LXXXIX** (2).
Cassen, R.H. (1976), 'Population and development: a survey', *World Development*, **4** (11/12).
Crosby, Jr., A.W. (1973), 'New world foods and old world de-

mography', in Crosby, Jr., A.W., *The Columbian Exchange*, Greenwood Press, Westport, Conn.

Eberstadt, N. (1980), 'Recent declines in fertility in LDCs and what population planners may learn from them', *World Development* **8** (1).

FAO (1977), *The Fourth World Food Survey*, FAO, Rome.

Fields, G.S. (1980), *Poverty, Inequality and Development*, Cambridge University Press, Cambridge.

Griffin, Keith (1977), *Inequality, Effective Demand and the Causes of World Hunger*, Queen Elizabeth House, Oxford.

Hayami, Y. and Ruttan, V.W. (1971), *Agricultural Development: An International Perspective*, Johns Hopkins University Press, Baltimore.

Langer, W.L. (1963), 'Europe's initial population explosion', *American Historical Review*, **69** (1).

Levi, John (1976), 'Population pressure and agricultural change in the land intensive economy', *J. Dev. Studies*, **13** (1).

Nicholls, W.H. (1963), 'An agricultural surplus as a factor in economic development', *J. Political Economy*, **LXXI** (1).

Poleman, Thomas (1981), 'A reappraisal of the extent of world hunger', *Food Policy*, **6** (4).

Reutlinger, S. and Selowsky, M. (1976) *Malnutrition and Poverty*, World Bank Occasional Paper no. 23, World Bank, Washington DC.

Reutlinger, S. (1977), 'Malnutrition: a poverty or a food problem?', *World Development*, **5** (8).

Sen A.K. (1980a), 'Famines', *World Development*, **8** (9).

Sen, A.K. (1980b), *Levels of Poverty: Policy and Change*, World Bank Staff Working paper no. 401, World Bank, Washington DC.

Sen, A.K. (1981), 'Ingredients of famine analysis: availability and entitlements', *Quarterly J. Economics* no. 384.

Schultz, T.P. (1969), 'An economic model of family planning and fertility', *J. Political Economy*, **77**: 153–80.

Srinivasan, T.N. (1981), 'Malnutrition: some measurements and policy issues', *J. Dev. Econ.*, **8** (1).

United Nations (1977), *Levels and Trends of Fertility throughout the World, 1950–1970*, United Nations, New York.

Wilkinson, R.G. (1973), *Poverty and Progress*, Methuen, London.

Woodham-Smith, C. (1962), *The Great Hunger*, Hamish Hamilton, London.

World Bank (1981), *World Development Report 1981*, World Bank, Washington DC.

Notes

1. A further aspect of ecological disequilibrium is damage to the natural environment itself, such as soil erosion caused by the cultivation of hillsides, de-afforestation and over-grazing with livestock.
2. Another example is West Bengal where the (Marxist) government has made 'food for work' available to the unemployed as a practical means of redistributing income in favour of the poor by mobilising surplus labour.

10 Agriculture and International Trade

1 Introduction

Trade in agricultural goods can play a significant role in promoting economic development of LDCs. First, exports of agricultural goods can pay for imports of capital goods, technology, manufactured products and other essential commodities for a sustained economic growth of developing countries. It is important to stress that exports of agricultural goods from LDCs like Thailand and Malaysia have helped them considerably to earn valuable foreign exchange for their industrialisation and economic growth.

Second, many LDCs have a comparative advantage in the production of agricultural goods. Given a trade regime which is relatively free from control and regulations, LDCs can use their comparative advantage in producing agricultural goods to raise their standard of living. Indeed, in an export-led growth model of trade, it would be to the advantage of many LDCs to especialise in the production of those goods where they have a comparative advantage and to export the surplus production. Such a policy will lead to the use of trade as an engine of growth apart from ensuring a 'rational' allocation of resources.

Third, even when a developing country successfully raises it standard of living, trade in agricultural goods could still remain an important policy for a number of key industries. Witness the case of Canada, New Zealand, Ireland, Denmark and even the United States of America – in these rich countries, agriculture still performs a very important role as a major export industry for stimulating economic growth.

Fourth, a growth of agricultural trade where LDCs can play an increasingly major role could also be of considerable advantage to the DCs. As the income in the LDCs grows with a rise in their exports of agricultural goods, there will be a corresponding rise of the demand for industrial goods and different types of services in these countries which would be imported from the DCs. In effect, the market for DCs in the LDCs will expand and trade in agricultural goods will be of mutual benefit to both DCs and LDCs. However, we are assuming a trade regime which is free from all types of controls and regulations. In practice, DCs try to protect their agricultural sector by the imposition of different types of controls for keeping out imports. Such controls usually take the form of quotas whereby only a fixed amount of goods from LDCs are allowed to be imported. Also, on a number of occasions, DCs (e.g. France and the EEC) have imposed levies on the imports of agricultural goods. It may be pointed out that LDCs also use tariffs, quotas and other exchange control regulations to protect their 'infant' industries from foreign competition. Some argue that the imposition of these controls like tariff and quotas work against the principles of free trade which is supposed to enhance economic welfare by providing goods at the cheapest price. Developing countries point out that while the DCs preach the virtues of the theory of free trade on the basis of comparative cost advantage in the production of industrial goods, they themselves do not obey these rules while they trade in agricultural goods with the LDCs. A similar point has been made with respect to trading in textile and manufactured products. We shall examine these issues more carefully later.

2 Main Features of Trade in Agricultural Goods

Most LDCs trade in a large number of agricultural goods which have the following important characteristics.

2.1 Commodity concentration in trade

This feature implies that a large number of LDCs tend to export a limited number of agricultural goods. For instance, Brazil tends to depend considerably on exporting coffee; in the case of Ghana, the major export is cocoa; while Zambia's chief export is copper. Bangladesh and India depend substantially on the export of jute; India, Sri Lanka and Kenya depend considerably on the export of tea; rubber is one of the few major exportables for Malaysia; Argentine and Thailand depend very much upon exports of food grains. In other words, trade is heavily dependent upon the exports of a few major agricultural products.

2.2 Geographic or market concentration in trade

Geographic or market concentration implies that most of the major exportables are usually sold in a few markets of the industrialised countries. Examples are the sale of tea, coffee, jute, cocoa, rubber in the markets of Europe, North America, Australia and Japan. Such a geographic concentration has the implication that the economic fortunes of the LDCs are strongly related to the rise of fall of the domestic and economic activities of some industrialised countries. Let us assume that India and Sri Lanka depend significantly upon exports of tea to the British market. Now, if there is severe recession in Britain for any reason, then the market for Indian and Sri Lankan tea will also be adversely affected.

2.3 Fluctuations in primary commodity prices

It has been argued that the degree of fluctuations in trading of agricultural and primary goods is significantly high or at least higher than fluctuations in prices of manufactured goods which are traded in the world market. The figures for Brazilian coffee illustrate the point: *monthly* fluctuations in prices in the New York trading market is very considerable. Many other agricultural goods which are exported by the LDCs show similar fluctuations. Given such fluctuations, it is easy to understand why export *earnings* (i.e. price x quantity of exports) of LDCs will also fluctuate under certain assumptions.

The point has been made that such fluctuations in export earnings are detrimental to the economic growth of LDCs. The theory is quite easy to illustrate. Assume that the producers of primary goods are generally risk-averse. If risk can be approximated by an index of instability of export earnings, then a high degree of instability in such earnings will induce producers to withdraw resources from the production of such goods and, under such a circumstance, output and employment will fall. If the producers were planning to commit investible resources in the production of those goods whose earnings oscillate substantially, then again production plans may be cut back and an increase of employment will not take place, and once again, economic growth will be stifled.

Primary commodity prices may also fluctuate in the export market due to *demand* factors, two of which deserve special emphasis:

(i) The *income* elasticity of the demand for agricultural products is usually less than unity. This implies that 10 per cent rise in the level of income will raise the demand for the product by less than 10 per cent. Hence with economic growth and a rise in per capita income, the rise in the demand for agricultural products will tend to decline in proportional terms. This will have an adverse impact on the export earnings of the LDCs.

(ii) The *price*-elasticity of the demand for agricultural product is usually less than unity. Thus, an increase in supply of agricultural products in the international market will reduce the price. But the additional demand generated by a decline in price of agricultural exports will bring less revenue than before due to price inelasticity of products. Once again, the LDCs will be hard hit if their combined effort to expand production (following the principle of comparative advantage) brings down the price by such an extent that total earnings will actually fall.

The income and price inelasticities of agricultural exportables seem to suggest that the LDCs are in a 'Catch 22' situation where they cannot win! One argument that can be advanced to escape the impasse is to *raise* rather than to reduce prices of primary exports because if the demand is price-*inelastic*, then an *increase* in price will raise export revenue. Unfortunately, the LDCs are generally price-takers rather than price-makers in the international market and as such they cannot raise the price unless they can exercise monopolistic control in the supply of a particular primary product, like oil. One of the major reasons why the Organisation of Petroleum Exporting Countries (OPEC) has been so successful in raising the export earnings from oil is that the cartel of oil producers has been able to collude in production and price planning to take full advantage of their monopoly gains as a group. But as far as other primary products are concerned, it is difficult to see how other LDCs can form such cartels to create such monopolies and 'exploit' the market.

3 Trade Policies in Developed Countries and their Impact on Agricultural Trade

It is generally acknowledged that some trade policies followed in the DCs are not really conducive to the growth of exports of agricultural goods from LDCs. We shall demonstrate here the impact of two main instruments of commercial policy which have been successfully used by the DCs to keep out imports of agricultural goods from the LDCs. The first instrument that has been generally used in the DCs is an *import levy*. This has the effect of a tax on imports. The other commercial policy has been *export subsidy*.[1]

Consider first the case of an import levy. Figure 10.1 illustrates the point of taxing imports very clearly. On the vertical axis, price level (= P) is measured and quantity (= Q) demanded and supplied is measured along the horizontal axis. The domestic demand is shown by D_d and the domestic supply is given by S_d. The world price of the product is given by OP_W. At this price, domestic supply is given by OS_W and the domestic demand is OD_W. Under the import levy scheme (as it has been used by the Common Agricultural Policy (CAP) within the EEC) a lower limit price is given by OP_L. Let us say, no country within the EEC will be allowed to import agricultural

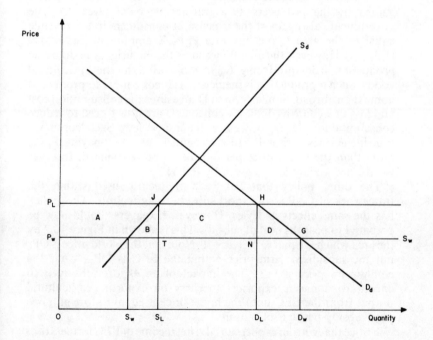

Figure 10.1: *Welfare cost of an import levy*

goods at a price below OP_L from the LDCs. Within the EEC as a result consumers are forced to pay OP_L as the domestic market price. Notice that when the domestic price rises from OP_w (the world price) to OP_L (the market price), $P_L P_w$ is the amount of levy that has been imposed on the amount of import ($= S_w D_w$) within the domestic market. The size of this levy will actually depend upon the farm price policies of the EEC countries.

The impact of such a levy is easy to illustrate. Domestic supply rises from OS_w to OS_L and domestic demand falls from Od_w to OD_L. The decline in the balance of trade deficit is shown by $S_L D_L$ instead of $S_w D_w$. This may be regarded as the balance of payment effect of import levy which is now 'successful' in keeping out the imports of primary goods from LDCs. Production effect within the EEC is clearly observable as output rises by $S_L S_w$.

However, the welfare cost of such an import levy is mainly borne by the consumers. This is shown by the reduction of consumer surplus by the area $P_w GHP_L$. It is not a cost for the community as a whole because what the consumers have lost as a group, is partly gained by the producers as a transfer payment (area A). The governments also gain at the expense of consumers in tax revenue equal to the area C (i.e. tax rate $P_L P_w X$ amount of import i.e. $H_L J_L TN$). However, the area B measures the inefficiency in domestic production from producing $S_w S_L$ more of domestic agricultural goods. Such a production is inefficient as it costs more to produce at home than abroad. Similarly, area D measures the consumption cost and loss of welfare as domestic consumers are now forced to reduce consumption by $D_w D_L$ because of the import levy. Such 'consumption inefficiency' arises because consumers are now paying $P_L P_w$ more than the world price per unit of output consumed. In other words, net welfare loss is given by $B + D$.

The other policy that has been frequently used within the framework of CAP is an export subsidy to agriculture. This policy has the same effects as levies. However, taxpayers should now be prepared to bear the cost of subsidy. This is shown in Figure 10.2 by the area which is equal to $N + R + T$. Note that the world price is OP_w but the subsidised farm price within the EEC is OP_{su}, while the equilibrium price is OP_e. The impact of an export subsidy is to encourage domestic exporters to carry on exporting agricultural output from the EEC until the home price is equal to foreign price plus subsidy per unit of output.

Notice that with an export subsidy, net income of EEC farmers rises by $M + N + R$. Although there is a rise in exports because of the

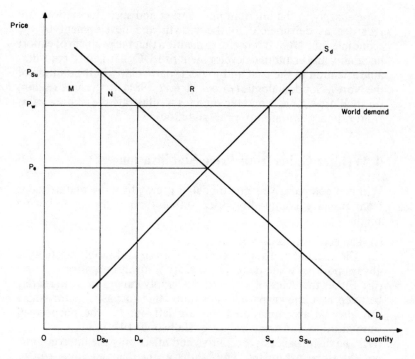

Figure 10.2: *Common agricultural policy and the impact of an export subsidy*

subsidy, an increase in home price reduces home demand by $DS_{su}D_W$ because of the rise in prices to OP_{su}. The loss of welfare due to 'consumption inefficiency' is clearly shown by the area M + N. Also notice that farm producers from LDCs are now compelled to face extra competition due to subsidies in the world market.

It is clear from the above analysis that the use of protectionist policies for sheltering agriculture in DCs has almost certainly reduced the trading welfare of the LDCs. As most LDCs are generally exporters of agricultural products, the point has been made that the imposition of trading embargos, even if they are indirect, should be removed not only for the sake of LDCs, but also to remove the types of 'consumption inefficiencies' in DCs that we have just now observed.

There is an additional problem about the trade in primary products. It is often claimed that prices of primary products fluctuate quite sharply in the international market and such fluctuations are regarded as detrimental to the growth and development of the economies of LDCs. It is useful to mention that the problem of export instability and economic development of LDCs has gained considerable attention in the past and present, particularly in the context of the North–South dialogue (*Brandt Report*, 1981). In the next section, we shall use simple economic theories to illustrate the welfare gains from primary commodity price stabilisation.

4 Welfare Gains from Price Stabilisation

Within a comparative static model, it is easy to show welfare gains from primary commodity price stabilisation with the following simple

(i) The economy is competitive.
(ii) The demand and supply curves are linear and supply elasticity is always positive while demand elasticity is always negative.
(iii) Either the demand curve or the supply curve shifts alternately between two known parallel positions (the stochastic disturbances can also be accommodated but are left out for the purpose of simplification; for details, see e.g. Hallwood, 1979).
(iv) The disturbances are additive and affect only the intercept and not the slope parameter. Thus shifts in the demand curve can be regarded as the product of a 'business cycle' in the exporting country and the shifts in the supply curve by a 'production cycle' in the exporting country.
(v) Supply and demand respond instantaneously to price changes. In other words, adjustment lags are assumed away.
(vi) The buffer-stock scheme can operate without any cost and it can stabilise market price on a unique intervention price (once again, the positive costs of operating buffer-stocks can be included, but are left out to facilitate exposition).

In Figure 10.3, we show the movements of prices due to shifts in demand (from D_1 to D_2). The stabilising price P^* stabilises the total revenue which is equal to total earnings from exports regardless of price elasticities of demand and supply. In Figure 10.3 OHGP* is the stabilised total revenue. This is more stable than the oscillating free market revenues at prices P_1 and P_2, i.e. OJIP$_1$ and OLKP$_2$.

In Figure 10.4, we illustrate the case of price instability due to shifts in supply from S_1 to S_2. Here, stabilisation may either stabilise or

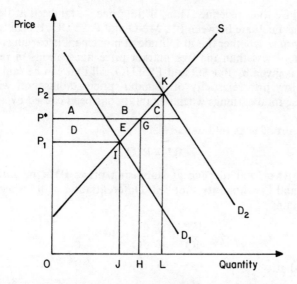

Figure 10.3: *Welfare gains from price stabilisation: demand shifts*

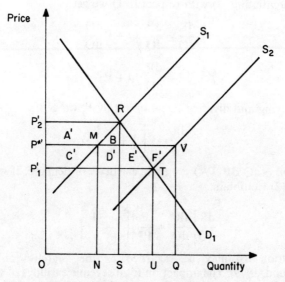

Figure 10.4: *Welfare gains from price stabilisation: supply shifts*

destabilise total revenue. Thus, if the price is stabilised at P^*, total earnings fluctuate between $P^{*'}$, MNO and $P^{*'}$ VQO. It is difficult to say, *a priori*, whether this amplitude of movement in earnings will be greater or less than the free market price fluctuations in revenue, which is given by P'_2RSO and $P^{*'}_1$TUO. All that can be said is that given low price elasticity of demand, price stabilisation will also stabilise total earnings when instability has been caused by shifts in supply.

The proof is as follows:

Let
$$R^* = P^*Q \ldots \quad (1)$$

where $R^* \equiv$ total revenue at stabilising price, $P^* \equiv$ the stabilising price, and $Q \equiv$ quantity supplied. Differentiating with respect to Q, we obtain:

$$\frac{dR^*}{dQ} = P^* \frac{dQ}{dQ} = P^* \ldots \quad (2)$$

Note that:
$$R = PQ \ldots \quad (3)$$

where R = total revenue without price stabilisation.

Differentiating (3) with respect to Q we get:

$$\frac{dR}{dQ} = \frac{dP}{dQ} \cdot Q + P \frac{dQ}{dQ}$$

$$= \frac{dP}{dQ} \cdot Q + P \ldots \quad (4)$$

Multiplying and dividing equation (4) by P, we get:

$$\frac{dR}{dQ} = P\left[1 + \frac{1}{\eta_d}\right] \ldots \quad (5)$$

where $\eta_d = dQ/dP \cdot P/Q < 0$ price elasticity of demand. If we divide (5) by (2) we obtain:

$$\frac{dR}{dQ} \bigg| \frac{dR^*}{dQ} = \frac{P}{P^*} \bigg| \left[1 + \frac{1}{\eta_d}\right] \ldots \quad (6)$$

Equation (6) may be regarded as the 'stabilisation factor' (Hallwood, 1979). The impact on total revenue earnings of stabilisation price at P^* will depend upon η_d. It is clear from the left-hand side of equation (6) that if the standard conversion factor > 1, price stabilisation will also stabilise total revenue because R is more stable with than without price stabilisation. When SCF < 1, price stabilisation will lead to a revenue destabilisation. When the SCF = 1, the effect on revenue is neutral. (Readers may wish to work out the different kinds of relationships between price stabilisation and revenue stabilisation by assuming different values of price elasticity.)

The analysis has so far been carried out in a micro context. It is possible to highlight some macro issues regarding export instability and economic growth.

5 Export Instability and Economic Growth: some Macro Issues

It has been sometimes argued that a high degree of instability in export revenue would usually mean a larger fluctuation in the income of the export earners and as long as such income comprise a significant proportion of the gross national product (GNP) of the economy, the GNP would be subject to large oscillations. Many argue that such an instability in expot income would have harmful effects on the economic development of a country. If there is a fall in export income, then there could be a 'foreign exchange crisis', and the capacity to import (particularly food and capital goods) will be adversely affected. Thus, there could be a serious balance of payment crisis. A decline in the imports of capital goods will have an adverse effect on the rate of growth of investment, which, in its turn, will reduce the overall rate of economic growth.

So far we have assumed away the presence of uncertainty. It is possible to take into account the role of uncertainty explicitly. Let us assume that uncertainty in export income exists; let us also assume risk-aversion by the producers, then, if the producers expect a large fluctuation in their export income, they may divert funds from the export to some other sectors of the economy where the return from capital investment is expected to be fairly stable. This type of speculation could clearly jeopardise the long-run growth of export industries in the economy.

There is another problem with respect to wage rigidity. Assume that wages are rigid downwards, then, it can be shown that a rise in instability in export earnings could imply an increase in the rate of inflation. The argument may be understood in a very simple way if it is acknowledged that prices and wages tend to increase during a boom, but would not fall much during the 'trough'. Note that many countries, particularly LDCs, suffer from low supply elasticities. If the Philips curve-type trade-off is assumed to have some validity in the short run, then, a rise in the rate of inflation will affect employment. Further, a high rate of inflation can adversely affect the level of savings as people suffer the erosion in the real value of financial savings. There could be a movement away from financial assets with adverse consequences on the process of monetisation of the economy and its beneficial effects.

Some economists have argued that sometimes export instability can have a favourable impact on the economic growth of the country. The argument runs as follows. Given a curvilinear consumption function, the higher the income instability, the lower will be the average propensity to consume and the greater will be the average propensity to save. This proposition rests on Friedman's theory of consumption function where he regards consumption as a function of *permanent* rather than transitory income. A transitory rise in income will thus leave consumption unchanged and savings will rise. This is supposed to raise the level of investment and the rate of economic growth. In other words, instability in export income will favourably affect the rate of economic growth.

It is important to remember that the above analysis does not take into account the risk-aversion by the investors in the face of instability in export earnings. Clearly, given uncertainty and risk-aversion by the producers, investment may not always be positively correlated with an increase in instability in export income.

6 A Survey of the Literature

The array of past studies can be broken down into three categories: those that tried to explain export instability, testing to see whether it was indeed related to such variables as commodity concentration; those that tested whether any significant difference could be detected in the degree of export instability experienced by developing and developed countries; and those who explored the repercussions of export instability in domestic economies. A review of past studies must first include the work of Coppock (1962) who developed a log-variance index of instability using the total exports of eighty-three countries for the period 1946 to 1958. Coppock tried to explain the different indices for countries by means of single and multiple correlations. He found export earnings instability most closely related to instability in the volume and price of exports and imports, and the terms of trade. He also found that instability was negatively correlated with regional concentration and had a low positive correlation with commodity concentration.

Michaely (1962) studied the relationship between export *prices* and commodity concentration for thirty-six countries for the period 1948 to 1958. His results showed a significant positive relationship between export price instability and commodity concentration. Massell used a sample of thirty-six countries and data from 1948 to 1959. He regressed instability of earnings on commodity concentration. Two measures of each of the variables were used but no coefficients were significant at the 5 per cent level. After a geographic concentration index was added commodity concentration became significant. And when the percentage of primary products of total exports (% XP) was added as a variable to an equation including a variable for geographic concentration, % XP was significant. The coefficient of the geographic concentration index in this equation was negative and insignificant. Massell's results were inconclusive. His empirical evidence did not support the hypothesis that concentration of exports were significantly related to export earnings instability (Massell, 1964, 1970).

The data compiled by Coppock and Michaely was used by MacBean (1966). His findings suggested that export instability experienced by LDCs and DCs did not differ greatly, and tried to explain the variance in instability between countries with three variables: the ratio of primary products in total exports, and commodity and geographic concentration indices. All these variables succeeded in explaining only 25 per cent of the total variation of export instability between countries. It can be shown that Coppock's log-variance (LVC) instability index was determined only by the first and last observations in the sample.

$$LVC_1 = \text{antilog}\left[\frac{1}{n-1}\sum_{t=1}^{n-1}\left(\log\frac{V_{t+1}}{V_t} - r\right)^2\right]^{1/2}$$

or

$$r_t = \frac{1}{n-1}\sum_{t}^{n-1}\log\frac{V_{t+1}}{V_t}$$

or

$$r = \frac{1}{n-1}(\log V_n - \log V_1)$$

The log variance index formulated by Coppock was only one of four instability indices in MacBean's study. Twelve countries emerged as having unstable exports earnings by all four criteria. But by further examining each of these countries, MacBean attributed the instability to very specific local and political factors. Only five countries exhibited instability due to the type of produce exported. However these products proved to be primary products which had a history of instability. His general conclusion was that:

Such theoretical proposed general factors as specialisation in primary products or commodity concentration *per se* may have some slight systematic tendency to produce export instability but their explanatory value in particular cases is very small.

MacBean also explored the influence of export instability on investment in LDCs. He defined the rate of growth of investment as the dependent variable and the rate of growth of import capacity (the total value of exports plus net invisibles and net capital transfers, divided by an index of import prices), the instability of importing power as merchandise exports, and the rate of change (increase or decrease) in reserves of gold and foreign exchange over the period as independent variables. The instability of the importing power of merchandise exports was insignificant at the 5 per cent, denying a terms of trade effect. However the rate of growth of import capacity was positive and significant. Evidence did not support the case that export instability had a detrimental effect on investment.

Several other authors, including Erb and Schiavo-Campo, point out the danger of generalising for all periods from evidence of export earnings instability drawn from one time period. After computing Coppock's instability index for countries from 1954 to 1966, Erb and Schiavo-Campo (1969) found the mean instability index for LDCs was 13.4 with a standard deviation 6.2. Developed countries had a means instability index of 6.2 with a standard deviation of 2.2.

Knudsen and Parnes (1975) constructed a transitory income index of export instability, derived by dividing export income into permanent income and transitory income components. They proposed that instabilities in export earnings could not be explained by simple comparisons with the nature of exports, concentration indices, or degrees of political instability. Knudsen's and Parnes' sample consisted of twenty-eight countries and data from 1958 to 1968. Their evidence suggested that there was no general relationship between instability in export income and the type or concentration of exports. Knudsen and Parnes also drew in interesting conclusion with regards to instability and investment. Empirical work revealed that lower propensities to consume measured under higher levels of instability resulted in increased aggregate investment, that is, instability stimulated investment.

Finally we can turn to Hallwood's study (1979). For the purposes of his study, Hallwood combined later work of Coppock with a study done by UNCTAD. A new method had been used by Coppock to compute an instability index for 109 countries and observations from 1959 to 1971. UNCTAD dealt with data from 147 countries and the years 1950 to 1970 in five years sub-sets. Hallwood observed that a marked difference existed in the export instability experienced by developed and developing countries:

dispersion of export earnings instability of the LDCs was greater than that of the DCs. Export earnings instability was on average 1.5 times higher for LDCs with a per capita income below $250 than those with a per capita income equal to or greater $250.

This conclusion seems to bear out the observation regarding criterion for disaggregating developed and developing countries. Our review of some of the literature concerning export instability can now be concluded. The studies have not led to any definitive consensus on differing magnitudes of export instability in DCs and LDCs or causes of effects of instability.

7 Some Measurement Problems

Some other points can also be mentioned in connection with the measurement of the instability indices. For example, Coppock uses the 'percentage increase in GNP per year' adjusted for price changes. MacBean, on the other hand, uses the compound annual growth rate of GDP, 1950–51 to 1957–58 in current prices. But problems arise because LDCs generally experience higher rates of growth of population than the DCs and hence the use of the total GDP rates of growth rather than the respective per capita growth rates introduces an upward bias into the rates of LDCs. It should also be borne in mind that since the LDCs usually experience a higher rate of inflation than the DCs, the use of income data in current prices for measuring the rates of growth of income implies an over-estimation of the growth rates of LDCs. Coppock's work seems to have suffered from other methodological defects. Coppock uses export instability indices for one period (1946–58) while his data for income growth rates covers a different period (i.e. 1951–7), (see Gleazakos, 1973).

8 Prebisch's Hypothesis

Prebisch (1959) was one of the first economists who has drawn attention to the problem of a secular downward trend in the terms of trade of the LDCs (where the terms of trade are defined as the ratio of export to import price paid by the LDCs in their trade with DCs). Prebisch's findings implied that the LDCs are losing out in their trade with DCs and the only way to develop their economies is to resort to industrialisation, protection and import substitution (for details, see Ghatak, 1978). Free trade has failed to be a vehicle of economic growth for the LDCs.

Prebisch's conclusion has been criticised by many. It has been shown that the final judgement about a net fall (or rise) in the terms of trade against the LDCs is critically dependent on the choice of base periods. Lipsey (1963), for instance, has shown that the terms of trade has actually moved in favour of the LDCs *vis à vis* Europe and the USA. On the other hand, Kindleberger's study (1956) does not confirm Lipsey's findings.

In the 1960s, doubts have been cast by some empirical studies on the assumed harmful effects of export instability and economic development of the LDCs. The major conclusions (eg. MacBean, 1967) of these studies can be summarised as follows:

(a) The correlation between instability in export earnings and the proportion of primary products in the export mix has been found to be very weak.
(b) The correlation between the degree of primary commodity concentration in exports and the instability in export income has been very poor.
(c) The association between instabilities in export earnings and the geographic concentration of the export market has been very tenuous.

Studies during the 1970s have not, however, always confirmed the 1960s' finding. Many recent studies have shown that instabilities in export incomes for LDCs are significantly higher than for DCs. Indeed, some of these studies report higher instabilities in export earnings of LDCs even when the commodity boom of the early 1970s has been excluded. In some recent empirical work, a high degree of positive correlation has been observed between export instability and export concentration. Indeed, Glezakos (1973) concludes:

> we feel that the empirical evidence presented here (in contrast to that presented earlier by Coppock and MacBean) supports the *a priori* arguments that export instability is generally larger in the LDCs than in the DCs and this instability is detrimental to economic growth in the former but not in the latter.

The adverse effects on trade on economic welfare has been well articulated by Bhagwati. This phenomenon has been labelled as 'immiserising growth' and can be best explained with a diagram. In Figure 10.5 let us measure industrial goods on the vertical axis and agricultural goods on the horizontal axis. The domestic production possibility curve is initially given by TT_0, and the terms of trade by the line AB. Initially the country produces at the point P and consumes at the point C where the indifference curve i_1 is tangent to the price line AB. Exports and imports can be measured by PP_0 and P_0C respectively.

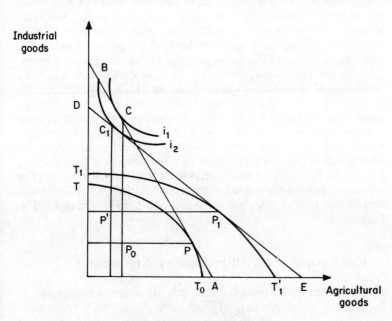

Figure 10.5: *Immiserising growth*

Suppose a LDC has a comparative advantage in the production of agricultural goods and with trade its production possibility curve shifts to $T_1 T_1'$. But the terms of trade of the country is now given by ED. The country now produces at P_1 and consumes at C_1 as the $T_1 T_1'$ curve is tangent to the price line at P_1 and the indifference curve is tangent to the price line at C_1, respectively. Clearly, the country's welfare has been reduced as C_1 is on a lower indifference curve than C.

The following conditions must be satisfied for 'immiserising' growth to occur:

(a) It is imperative that the economic growth of a country should be biased in favour of the export industry.
(b) Global demand for the home country exports must be inelastic.
(c) It is important that the economy of the home country should be closely tied to international trade in the export commodity in question.
(d) There should be substantial mobility of resources among different sectors.

It is possible to argue that all these conditions may not be satisfied for the immiserisation effects of trade on growth to occur. Nevertheless, it is difficult to reject the spirit of the model or its internal logical consistency, particularly when one looks at the primary commodity exporting LDCs and the decline in their terms of trade *vis à vis* the DCs in the last twenty-five years. If it is difficult to reject the possibility of immiserising growth, then the LDCs may be better-off, as some argue, by *withdrawing* productive resources from the primary agricultural commodity sector and channelling them into the industrial sector. Hence many, in line with the Prebisch–Singer hypothesis, would like to pursue 'inward looking' import substitution industrialisation policies for the long-run development of the LDCs via protection. However, protection has its benefits as well as costs. (For a good analysis of the impact of protection on LDCs, see, Bhagwati and Desai, 1970; Little, Scitovsky and Scott, 1970; Balassa, 1971; Ghatak, 1978)

9 The Agricultural Self-Sufficiency Argument

It has sometimes been argued that it is important for countries to achieve self-sufficiency in food production. Such an argument has a special appeal for the LDCs as many of them are heavily dependent upon imported food to meet excess home demand. An increase in domestic food production will obviously mean a fall in food imports and an easing of the balance of payment constraint. It will also imply that a developing country can use foreign exchange to import capital and other essential goods for domestic development. Further, a rise in domestic food production will raise the level of nutrition where malnutrition is a severe problem, assuming of course that a greater volume of food production will be equally available to all. Such a rise in nutrition level is supposed to raise labour efficiency and productivity. Some argue that self-sufficiency in food production is necessary to make a country truly independent of foreign governments internationalist policy. Hence we hear the slogan: 'No country should rely on the world market for its supply of food.'

In order to discuss these issues, it is important to define the term self-sufficiency. Self-sufficiency ($=$ SS) may be defined as the ratio of domestic production ($=$ PR) to domestic consumption ($=$ C). Hence, we have:

$$SS = PR : C$$

The SS ratio can be transformed into percentage terms easily. However, this ratio can be misleading ideas about independence from imports because of the use of imported inputs in domestic agriculture. Thus the SS argument 'involves not only the proportion of the consumption of a particular product, but also the proportion of the inputs used in the production of the product which are themselves domestically produced' (Ritson, 1977). Further problems arise in the estimates of 'overall SS' and here it is important to find a common measure to aggregate across commodities. In most studies, market prices have been used as weights. However, the use of a composite 'nutritional index' or 'calorie intake' index could be quite fruitful.

The welfare effects of a rise in agricultural terms of trade can be explained easily. In Figure 10.6, we measure industrial goods (I) on the horizontal axis and agricultural goods (A) on the vertical axis. PP is the production possibility curve and with TT_1 as the terms of trade (assuming only two products, A and I, the exchange ratio between A and I is also the terms of trade for that country) the country produces at K and consumes at C. The country's welfare is maximised when it produces ON of industrial goods and OE of agricultural goods. The country imports HE of food for QN of industrial products.

If the terms of trade rise in favour of A goods (for example, because of sudden supply shock due to crop failure, war or flood) then the TT_1 will move to T_0T_2. Welfare clearly falls as the country consumes at J

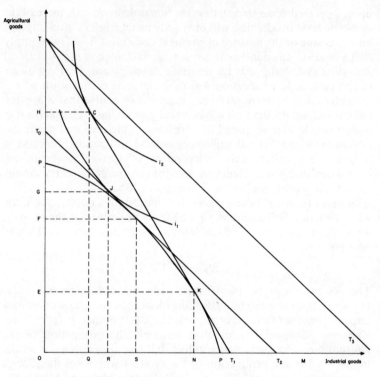

Figure 10.6: *The welfare effect of a rise in agricultural terms of trade*

(at a lower indifference curve i_2). Another way of describing the loss of welfare is to say that the country now needs to exchange OT_3 of I against OT of A goods (note that the slope of the terms of trade line TT_3 is the same as that of T_0T_2). Clearing the welfare of the food importing country diminishes. In the next period, to optimise welfare the food importing country should raise the production of A goods (food) from OE to OF with a reduction of food imports from EH to FG.

It seems that food importing countries should be better off by introducing policies to stimulate domestic food production if it expects a rise in global agricultural terms of trade. Imports may be controlled to allow domestic price to rise, and supply should rise given a normal supply response curve. However, free-traders argue

AGRICULTURE AND INTERNATIONAL TRADE 299

that once the effects of the rising global food prices are felt by the producers and consumers, appropriate adjustments will be made without resorting to government interventionist policies. On the other hand, there is no guarantee that the desired adjustments will be made very quickly and, as such, there remains a case for an intervention by government to promote desired adjustment.

Where a rising term of trade for food articles has been anticipated and future benefits have to be set against present costs, government may also intervene to affect land transfer. The problem here is that land cannot be transferred easily from one market to the other (rural and urban) with every turn and twist in the agricultural terms of trade. It is thus necessary to take a decision regarding optimum land use in agriculture with respect to a specific terminal period of planning. Here planning techniques and sometimes cost-benefit analysis could be used fruitfully (for a discussion of the cost benefit analysis and planning, see chapter 11).

10 Cartels in Commodity Trade and Welfare Gains and Losses

Following Viner, we can describe three major conditions for a successful cartel agreement among the primary commodity exporting countries:

(i) It is necessary for the cartel to control a very large proportion of the total commodity supply in the global market. If this condition is not met, demand curve which an individual cartel member faces could be elastic even when the final demand from the consumer is inelastic.

(ii) It is also important to have inelastic demand in the consuming countries in both the short and long term. It is useful to remember that higher commodity prices today may lead to the growth of the substitutes in the long run.

(iii) In both the short and long term, the supply of relevant primary commodities by the *non-member* countries should be rather inelastic in both the short and long term, otherwise, the profits of the members of the cartel will fall as non-member countries raise their supplies due to a rise in prices.

Apart from these conditions, it is important to note that, for *any* cartel to hold, the political will to unite must be there. If any country wants to follow an independent policy it may be tempted to sell more than its quota and this may lead to a feud within the cartel and eventually it may break up.

A policy of trade embargo may also raise welfare if a country's trade associates do not take up their choice to alter the terms of trade back to their own advantage. For LDCs exporting tropical products, this could be the case. Another way of looking at the situation would be to accept the exploitation of monopoly power as a just weapon to encourage development resource flows. Indeed, in the meeting of the Commonwealth prime ministers and presidents in 1975, the following resolution was adopted: 'We accept that the emergence of producer's associations is a reality born out of historical experience. In an unequal world it is understandable that such a development should take place' (Commonwealth Secretariat, 1975).

It is necessary to point out, at this stage, that cartels and trade restrictions could lead to welfare gains through market power exploitations. But such welfare gains can diminish or even evaporate if non-cartel countries retaliate. For example, Ritson (1977) observes that during the 1960s, 'the European Economic Community established minimum import prices for most agricultural products which were very high relative to global market prices.' The policy reduced the level of EEC agricultural imports beneath the level which would have applied with free trade and this undoubtedly had a depressing effect on world agricultural prices compared with their free trade equilibrium level. The result was that some food exporting countries raised their subsidies on exports of agricultural goods and this aggravated rather than reversed the initial movement of the terms of trade.

The 'inevitability' of the collapse of a cartel has been questioned recently (see Osborne, 1976). It has been shown that if members of a cartel can devise a quota rule, then cheating by some 'naughty' members of the cartel may be prevented. The rule is very simple. If any member of the cartel is found to be cheating, then members which are not cheating can always retaliate by increasing their production and profit. This will lead to a fall in the profit of the 'cheating' member. In other words, as long as the threat of retaliation persists against the 'cheating' members, self-interest will maintain stability within the cartel. However, the LDCs have now shown a marked preference for an International Commodity Agreement (ICA) scheme rather than for cartels in the interests of both producers and consumers.

11 Integrated Commodity Agreement (ICA) Schemes

In the United Nations Conference on Trade and Development (UNCTAD), a Nairobi (1976) (known as UNCTAD IV), a resolution was adopted for an 'Integrated Programme for Commodities' (IPC). This resolution had three major components: aims, coverage of commodities, and global measures of the programme.

Two important aims are: the stabilisation of global commodity markets by 'avoidance of excessive price fluctuations', and the rise in the real income that LDCs obtain from exports of their commodities by ensuring a remunerative and a just price for the producers in LDCs that take full account of world inflation and revenue stabilisation 'around a growing trend' value of earnings from exports.

As regards the *coverage* of commodities UNCTAD specified ten 'core' commodities: cocoa, coffee, copper, sugar, cotton, jute, rubber, sisal, tea and tin. This list was supplemented by seven other goods: bananas, bauxite, beef and veal, iron ore, rice, wheat and wool. The first ten products are branded as 'core' commodities because they account for about 75–80 per cent of the export earnings of LDCs of all seventeen goods. Since these products can also be stored for stockpiling, they are regarded as suitable for promoting a buffer-stock scheme.

The heart of the IPC programme was the establishment of a common fund to finance the entire project. Initial calculations show that it would cost about US $6 billion and such an amount should be raised by subscriptions from the exporters and importers of these commodities. It was agreed that the least developed countries should be exempted from paying the subscriptions. The argument for setting up a common fund is to share and minimise risks and the principle is the same as the one followed by any insurance company which seeks to minimise its risks. Further, it is believed that the establishment of a common fund will help LDCs to have more bargaining power *vis à vis* the global capital markets than could a set of individual funds for the same goods. 'It would also require smaller financing than the aggregate of a set of individual funds because of differences in the phasing of cycles across commodity markets' (Behrman, 1979).

It is important to look at the empirical evidence of the rates of growth and fluctuations in prices and values of the UNCTAD core and other commodities. Behrman has worked out rates of growth and fluctuations in prices of such goods for a period that varies between 1953 and 1972 and these figures are cited in Table 10.1. Considerable evidence of price fluctuations has been confirmed in this table, and

Table 10.1: *Secular trends in deflated prices for UNCTAD core commodities UNCATD other commodities, and additional commodities of possible interest 1950–1975*[1]

Core commodities			
Coffee	−0.035	Bauxite	0.019
Cocoa	−0.024	Iron ore	−0.017
Tea	−0.030	Maize	0.029
Sugar	−0.004[3]	Tobacco	−0.015
Cotton	−0.038	Lumber	−0.005
Rubber	−0.058	Hides and skins	−0.030
Jute	−0.018	Groundnut oil	−0.033
Sisal[2]	−0.004[3]	Olive oil	−0.010
Copper	0.004[3]	Coconut oil	−0.023
Tin	0.004[3]	Palm oil	−0.028
		Linseed oil oils[4]	−0.035
Other commodities			
Wheat	−0.021	Soybean oil[4]	−0.015
Rice	−0.008[3]	Cottonseed oil[4]	−0.016
Bananas	−0.037	Palm kernel oil[4]	−0.006
Beef and veal	0.026	Lead	−0.028
Wool	−0.041		

[1] Calculated from UNCATD price indices in United Nations and OECD GDP price deflator.
[2] 1954–75.
[3] Not significantly different from zero at standard 5 per cent level.
[4] 1954–74.
Source: Behrman, (1979).

clearly it suggests that a number of ICA concluded in the 1960s and early 1970s have not succeeded in lowering instability. As a crosscheck, if fluctuations in the World Bank index of primary commodity terms of trade (based on unit values of LDCs' exports of thirty-four non-petroleum primary commodities) are observed between 1954 and 1975, then again Behrman's finding of large fluctuations can be confirmed. The figures in Table 10.1 are more instructive as they reveal a disaggregated picture of price fluctuations of individual commodities which could be concealed in an aggregated World Bank index.

In Table 10.1. Behrman's calculations of the secular trends in deflated prices for UNCTAD core and other commodities have been reproduced. The important conclusions which Behrman draws are as follows:

(a) Negative secular trends have been generally observed for most products except for copper and tin (non-agricultural products) beef and veal (dairy products) and maize.

(b) Behrman fails to observe any correlation between the amplitude of price instability and the size of the secular trends in prices for seventeen commodities as listed by the UNCTAD.
(c) Somewhat surprisingly, a positive correlation between income elasticities and secular trends in prices has not been found. Differential supply shifts (along with demand shifts) and different supply price elasticities could have been responsible for such a finding.

Under the auspices of the UNCTAD during the 1960s and 1970s, several commodity cartels have been set up to promote the interests of the LDCs, for instance, An International Sugar Agreement was concluded in 1968. An International Wheat Agreement led to the negotiation of the International Grains Agreement in 1967, which then led to the establishment of an International Wheat Agreement alone in 1971. Likewise, an International Tea Agreement was reached in 1969 and was followed by similar agreements in sisal, cocoa and henequen in 1970. It is useful to point out that the International Coffee Agreement collapsed in 1972, but a new agreement on coffee was reached in 1975. Eckbo (1975) observes that, for a sample of fifty-one international commodity cartels, the mean duration of the formal agreements has been 5.4 years and the median has been only 2.5 years – a finding which seems to confirm the view that a large number of the International Commodity Agreement schemes cannot be regarded as great successes.

One of the major reasons for such a lack of success has been the failure to achieve price stabilisation. Law (1975) has shown that in the case of coffee, the mean annual price changes has been about 50 per cent greater during the years of agreement (i.e. 1964–72) than for the pre-agreement years (i.e. 1950–63). In the case of sugar, according to Law's calculations, the mean price oscillations has been about 75 per cent per annum larger for twelve recent years of price agreements than during the eleven other non-agreement years. The cocoa agreement failed when prices were pitched too high in 1970. It is interesting to observe that in the case of rubber, the International Rubber Agreements actually destabilised the market. It is only in the case of wheat and tea that commodity agreements have led to an increase in price stability – but here again, thanks mainly to stockpiling decisions made outside the agreements by Canada and the USA.

The other objective of UNCTAD IV has been to raise the export income of LDCs by raising the prices of some selected commodities. In some cases, these attempts have been successful. For example, in the case of coffee, recent agreements might have transferred funds from consumers to producers at the rate of $500 to $600 million per year. In the case of some other commodities, agreements have failed to raise revenue, except in the very short run. The interesting point about the commodity agreements that has emerged recently is that, even for the 'successful' commodity agreements, the median duration is only four years and the mean life is about 6.6 years. Behrman (1979, pp. 64–5) has neatly summed up the reasons for a 'successful' rise in prices in commodity agreement schemes as follows:

(a) high income elasticities of demand;
(b) high price elasticities of demand;
(c) high concentration of production;
(d) high concentration of foreign trade;
(e) a high share of foreign trade controlled by members of the commodity agreements;
(f) low cost differences among producers in the agreement; and
(g) less intervention by the government.

The most common causes for the failure of the commodity agreement schemes are supposed to be intense competition among the members of the agreement schemes; and strong competition from those who are not members of the agreement schemes.

The important question that has almost continuously been asked in connection with the commodity agreement schemes is: Will the gains in the short run outweigh the costs to be incurred in the long run because of the growth of the substitutes and a rise in production from 'fringe' producers? The answer is probably yes, even when the goods concerned do not always possess the same attributes which have accounted for the short-run success of commodity cartels in the past. As Behrman concludes, after a rigorous and thorough analysis,

> even short-run market regulations may have a reasonable probability of success only if the consumer nations can be persuaded to co-operate in order to enforce discipline. The growing strength and cohesion of the developing nations and the more accommodating posture of the developed nations probably lead to a higher probability of at least short-run success for an UNCTAD-type proposal than past experience alone might suggest. (Ibid., p. 67)

In Table 10.2 Behrman's simulations of buffer-stock price stabilisation for UNCTAD core commodities and basic foodgrains have been shown. These simulations suggest quite clearly that the gross

Table 10.2: *Summary of simulation of buffer-stock price stabilisation schemes for eight of UNCTAD core commodities and two basic foodgrains*[1]

Commodity	Mean percentage changes as compared to base simulation			Maximum buffer-stock (1000 metric tons)[2]	Longest continuous period (in years) of buffer-stock activity		Present discounted value of real revenue: stabilisation-base simulation, millions of 1975 dollars at 5 per cent discount rate	Ratio of real revenue standard deviation in stabilisation/basis simulation	Present discounted value of buffer-stock activity, millions of 1975 dollars at 5 per cent real discount rate	
	Price	Quantity supplied	Value of producers' revenues		Without buying	Without selling			Excluding final stock	Including final stock
UNCTAD core commodities:										
Coffee	4.7	0.6	5.2	21738[2]	5	7	2662	1.0	−4	324
Cocoa	10.0	1.2	11.1	546[2]	4	10	1115	0.5	−126	351
Tea	0.8	0.1	0.9	161	8	9	114	0.9	−5	243
Rubber	9.3	0.0	9.2	3432	7	4	2243	0.5	−879	777
Jute	0.0	0.0	0.0	0.0[2]	10	10	0.0	1.0	0.0	0.0
Sisal	8.9	0.5	10.2	232	5	5	−1	0.4	−53	−16
Copper	−0.9	0.0	−0.9	507	7	2	−1339	0.8	−730	−454
Tin	−5.2	−1.0	−6.2	19	10	1	−656	0.9	−95	−95
Core commodities (sum or average)	3.4	0.2	3.7		7.0	6.0	4140		−1892	1130
Basic foodgrains:										
Wheat	0.1	0.1	0.2	1200	4	7	0.0	0.9	−2997	−1865
Rice	11.1	0.5	11.8	1600	3	3	33440	1.0	−5054	−641
Basic foodgrains (sum or average)	5.6	0.3	6.0		3.5	5.0	33440		−8051	−2509

[1] Details of the underlying econometric models are given in Behrman (1979). The buffer-stock simulations are identical to the base simulation except that the buffer stock purchases or sells whatever is necessary to keep the deflated price within the indicated brand-width of the secular trend for 1950–74. The simulations are all for the decade. 1963–72. The initial and final stocks are valued at the 1962 and 1973 prices, respectively, which overstates the present discounted values of buffer-stock operations.
[2] Units are: thousands of 60 kg bags for coffee; thousands of long tons for cocoa; millions of pounds for jute.

revenue gains to the LDCs from the operation of these specific programmes may be large, though these revenue gains seem to vary significantly across commodities. However, this is not a surprising result given the difference in the basic characteristics of these commodities.

12 The Compensating Financing Schemes

Another way of stabilising the foreign exchange earnings of LDCs is to adopt a compensatory financing scheme. This idea gained considerable attention during the 1960s and it has still retained some of its appeal. According to this method, a country's earnings from foreign exports can be stabilised by setting up a stabilisation fund from which countries should be entitled to compensatory drawings for any shortfall in their export income whenever it falls below a predetermined, mutually agreed level. The definition of export income could follow the principle which has been adopted by the International Monetary Fund, i.e. current account receipts in the balance of payments. Alternatively, such income can be defined according to the principle followed by the European Economic Commission under the Stabilisation of Exports scheme (STABEX), i.e. only *commodity* export income, or predetermined commodity export income. However, problems cropped up in the late 1970s when the benchmarks against which shortfalls in export income are measured were fixed at a level which was considered too low. Clearly, under such a rule the developing countries failed to earn substantial accounts of foreign exchange. In the case of the IMF, the shortfalls in export income needed for obtaining compensations were too large. Also, conditions for drawing from the stabilisation fund were very restrictive. In 1975, some of these severe restrictions were partially eased. For instance, 'forecasting restrictions' (the clause that stipulated that exports in the post-shortfall year, which had to be forecast, should not exceed 110 per cent of the nominal level of the two pre-shortfall years), have been rejected and the quotas for drawings have been raised. However, given the limitations of the CFS, UNCTAD is now advocating an Integrated Programme for Commodities under which a buffer-stock will be established as an important part for setting up a New International Economic Order. The problem, however, is the financing of the buffer-stock scheme which will cost about US $6 billion.

The other criticisms of the CFSs can be stated briefly. It has been argued that the commodity agreement schemes will intervene into the free operation of the market and hence the efficiency in resource allocation will be adversely affected. Also, an export quota system will tend to protect high cost producers and result in a sub-optimal level of production. Further, the costs of stockpiling some agricultural goods may be a lot larger than the benefits of such stockpiling.

In defence of the CFSs, it needs to be mentioned that the international markets in agricultural products are not perfect and there is also some imperfection in the flow of information between the principal agents of the markets – producers, consumers and speculators. It is precisely because of this sort of imperfection that price instability occurs and institutions like the CFS and STABEX are needed to prevent market failures. Where it is very difficult to manipulate the two principal instruments for stabilising commodity export earnings, i.e. buffer-stock and export quota, it is necessary to use the compensatory finance scheme to aid the LDCs.

References

Balassa, B. (1971), *The Structure of Protection in Developing Countries*, Johns Hopkins University Press, Baltimore.
Behrman, J. (1979), *Development, The International Economic Order and Commodity Agreements*, Addison-Wesly.
Bhagwati, J. (1959), 'Immiserizing growth: a geometrical note', *Rev. of Econ. Studies*, 25, 201–5.
Bhagwati, J. and Desai P. (1970), *India: Planning for Industrialization*, Oxford University Press, OECD and Oxford.
Brandt, W. (1980), *North-South: A Programme for Survival*, Pan.
Coppock, J.D. (1962), *International Economic Instability*, McGraw-Hill, New York.
Eckbo, P.L. (1975), 'OPEC and the experience of previous international commodity cartels', Cambridge, Mass., Energy Laboratory Working Paper.
Erb, G. and Schiavo-Campo, S. (1969), 'export instability, level of economic development and economic size of less developed countries', *Bull. Oxford Inst. Econ. & Stats.*, XXI, November, 263–83.
Ghatak, S. (1978), *Development Economics*, Longman, London and New York.
Gleazakos, C. (1973), 'Export instability and economic growth: a statistical verification', *Economic Development and Cultural Change*, 21, 670–9.

Hallwood, P. (1979), *Stabilization of International Commodity Markets*, JAI, CT.
Kindersen, O. and Parnes, A. (1975), *Trade, Instability and Economic Development*, Lexington, Mass.
Kindleberger, C.P. (1956), *The Terms of Trade: A European Case Study*, New York.
Law, A.D. (1975), *International Commodity Agreements. Setting, Performance and Prospect*. Lexinton, Mass.
Lipsey, R.E. (1963), *Price and Quality Trends in the Foreign Trade of the United States*, National Bureau of Economic Research, Princeton, New Jersey.
Little, I. Sktovsky, T. and Scott, M. (1970), *Industry and Trade in some Developing Countries, A Comparative Study*, Oxford University Press, OECD, Paris and Oxford.
Macbean, A.C. (1966), 'Export Instability and Economic Development, George Allen & Unwin, London.
Massell, B.F. (1964), 'Export concentration and export earnings', *Amer. Econ. Rev.*, **54**, 47–63.
Massell, B.F. (1970), 'Export instability and economic structure', *Amer. Econ. Rev.*, **60**, 618–30.
Michaely, M. (1962), *Concentration in International Trade*, North-Holland, Amsterdam.
Osborne, D.K. (1976), 'Cartel problems', *Amer. Econ. Rev.*, **66**, 835–44.
Ritson, C. (1977), *Agricultural Economics, Theory Policy*, Crosby Lockwood, London,

Notes

1. Barriers to trade such as the import levy and export subsidy do not apply to tropical products such as tea, coffee, etc. but they do create barriers when LDC exports compete with temperate zone crops.

11 Planning Agricultural Development

1 Introduction

In this chapter we shall be concerned, first, with the concept of planning and its potential for accelerating development in LDCs. Next we discuss alternative macro-planning models with particular reference to experience in the Soviet Union and China. We then consider the choice of planning strategy for the agricultural sector. This is followed by a brief consideration of the commonly used planning techniques of input–output analysis, linear programming and activity analysis. Finally, we undertake a detailed examination of agricultural project planning covering the underlying principles of project analysis and problems of application in LDCs, including some peculiarities of agricultural projects.

2 Meaning and Objectives of Economic Planning in LDCs

Planning, in general, means a deliberate attempt on the part of a country or its government to follow a specific pattern of economic growth and structural change in the economy. Frequently, planning implies intervention by government in the normal working of the market. However, the degree of intervention very much depends on government ideology. France, for example, follows a system of indicative planning whereby the degree of intervention by government is minimal. Indeed, government simply sets some targets for the private sector, outlines the major directions of economic policy, and provides information to the market economy so that the desired targets can be achieved. Here the state does not control all the *means* of production. On the other hand, in Soviet Russia, the state controls most of the means of production, dictates the pattern of demand and supply, and owns the 'commanding heights' of the economy. As a consequence, the degree of state intervention in the market mechanism is the greatest. The market mechanism, in fact, hardly operates to allocate real resources within the Soviet economy.

In between these two extremes, we have a large number of intermediate degrees of government intervention in the operation of the market. However, prices are not suppressed completely. Indeed, in many countries, prices play important roles in the allocation of resources. Such countries operate within the framework of a 'mixed' economy, where both planning and the market coexist side-by-side. Many LDCs have mixed economies and, given market imperfections, government interventions are taken as attempts to rectify these. The main reasons for the introduction of planning in LDCs are derived from the Soviet experience which can be stated as follows:

(a) Planning is supposed to speed up the rate of economic growth of a country. If the market is allowed to operate without any intervention then it may take LDCs a long time to attain a reasonable standard of living.
(b) The history of the operation of markets in LDCs does not seem to suggest that allocation of resources will always be very efficient or optimal in an unregulated economy. In many cases, markets operated with considerable distortions and economies of most LDCs stagnated.
(c) In a market economy, private investors usually try to maximise the short-run *private* profits rather than the long-run *social* benefits. As a result, marginal private net benefits and marginal social (net of costs) benefits will not be equal and resource allocation will be less than socially optional.
(d) Given the lack of entrepreneurial skill, initiative and leadership in private business and commerce in many LDCs, it is sometimes argued that government via planning should provide dynamic leadership and enterprise to stimulate investment in order to attain a higher and better standard of living.

Against these arguments in favour of planning, it has been pointed out that government intervention in the market has considerable drawbacks. Here we shall briefly state three major criticisms. (More detailed discussions of planning can be found in Heal (1971); Ellman (1979) and Healy (1971).)

(1) If the marginal private and social net benefits are not equal to one another, then government should provide information and other services (e.g. the creation of social capital like roads, schools, hospitals) so that they can be equalised. This would, however, imply a greater role for government in the economy than the supporters of a pure market economy may favour. But it does not imply the introduction of planning on a grand scale.

(2) Planning – either for the entire economy or for a particular sector, such as agriculture – needs considerable information to match supply with demand at a price which is mutually acceptable to producers and consumers. In many countries, such data are very difficult to obtain. Even if information is available, its nature and quality (particularly for the rural sector) are not always reliable. Building large input–output or econometric models for planning the agricultural sector on the basis of rather dubious information may yield very misleading conclusions.

(3) Many LDCs do not have enough trained manpower to design and implement planning very effectively. As a result, in a number of LDCs where planning has been introduced, the implementation of planning has hardly been very efficient. Moreover, the use of planning in many LDCs has led to the growth of enormous and sometimes corrupt bureaucracies. The objective of achieving a satisfactory balance between demand and supply by planning (when the market mechanism has been suppressed) has consequently not been achieved in many countries. In particular, the task of disseminating information to obtain a proper coordination for an optimal use of resources has remained unfulfilled in a large number of cases.

3 Macro-Planning: comparison of Soviet and Chinese Models

Despite these and other criticisms, many LDCs have tried to use planning partly to achieve a faster rate of economic growth to raise standards of living and partly to achieve industrialisation via diversification of a largely agrarian economy. A number of LDCs have tried to emulate the Soviet planning example in this respect where agriculture had to pay for the growth of industries. This type of planning of the agricultural sector has been described as the 'tribute' model.

3.1 The Soviet 'Tribute' model

Just after the Bolshevik Revolution of 1917 Soviet politicians and planners argued that to achieve a rapid rate of industrial growth, 'surplus' resources must be transferred from the agricultural to the industrial sector. The term 'surplus' to this context means the excess of sectoral production over sectoral consumption. (For a full discussion of the concept of 'marketable surplus' and its mobilisation for economic development, see chapter 4.)

The Soviet leaders decided that agriculture should pay for industrialisation and the rapid growth of the whole economy, by a 'tribute' levied on agriculture to finance the cost. The mechanism for imposing such a tax is described in Figure 11.1. Let us assume that there are only two sectors in the economy: agriculture and industry. Let us also assume that the industrial sector produces only two commodities: machinery and cloth by using labour and capital (machinery). Wages of all labourers are assumed to be paid in grains.

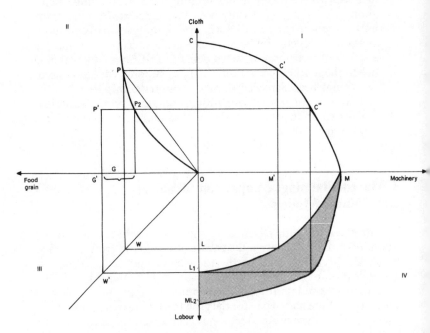

Figure 11.1: *The tribute model*

These grains are produced by labourers in the agricultural sector who receive cloth in exchange for grains.

The supply of grains is a function of the 'offer' curve, i.e. farmers supply grain for cloth. Such supply of grains also determines the employment of labour and the wage rate in the industrial sector.

Consider quadrant I in Figure 11.1 which describes the industrial production possibility curve CM. Assume that line is discrete in the model. The planner at period 1 can choose a certain combination of cloth and machinery. He may choose to produce only cloth ($=OC$) but no capital i.e. machinery; or he may decide to produce maximum possible machinery only ($=OM$) for the industrial sector, but no cloth at all in period 2. Should the planner decide to produce OM' of machinery, then the production of cloth will be equal to M'C' which can be exchanged for grains produced by farmers in the agricultural sector. The amount of grains which will be available in exchange for cloth is given by OG in quadrant II as indicated by the offer curve of farmers OP. Notice that the slope of OP shows the terms of trade (i.e. price ratio) between grains and cloth. In quadrant III the slope of the line OW determines wages (measured in grains) for the employment of labour in the industrial sector. Thus the supply of OG amount of grains determines the employment of OL amount of labour in quadrant III.

Once we have all the necessary information about total machinery, cloth, grains and labour i.e. C', P, G, W and L), we can determine, in quadrant IV, the production possibility curve like ML_1.

It is clear from the above analysis that factors which will constrain the growth rate are the initial endowments of capital, the terms of trade and the wage rate. If it is decided by the Central Planning Board that the growth rate should be raised then, in a command economy, farmers can be forced to give up more grains for the same amount of cloth, i.e. a move from a point on the farmers offer curve such as P_2 to a point like P'. Such a 'tribute' or tax ($=GG'$) imposed on the agricultural sector could then be utilised to employ more units of labour in the industrial sector. As such, the production possibility curve shifts to the right to ML_2 and a higher rate of economic growth for the whole economy can be attained. In other words, the policy of collectivisation is very desirable since it enables the planners to obtain the necessary 'surplus' from the agricultural sector to pay for the industrialisation of the economy.

A number of criticisms can be levelled at the 'tribute' model. First, while it may be applied to a totalitarian economy where farmers are forced to give up a part of their product, most LDCs – with the exception of China and Vietnam – do not have a planned economy. Indeed, most LDCs have 'mixed' economies where both the private and the public sector co-exist and market prices are not entirely suppressed. Such prices in LDCs still play important roles in allocating resources. Hence, it is not clear to what extent the transfer of the 'tribute' model in an underdeveloped economy would be successful.

Second, there is the problem of providing farmers with an adequate incentive to produce grains when the terms of trade are shifted against the agricultural sector. It could be pointed out that when the terms of trade move against agriculture (i.e. when the farmers have to give up more grain for obtaining the same amount of cloth), then their incentive to produce more will obviously diminish. A fall in the production of food grain will also lead to a fall in the amount of 'tribute' or surplus that can be raised in the agricultural sector as it is well known that marketed surplus and production are directly related. In a free market, a reduction in agricultural price ($=Pa$) *vis à vis* the industrial price ($=Pi$) (i.e. an adverse movement of (Pa/Pi) could easily lead to a fall in the production due to the operation of the substitution and income effects. Figure 11.2 illustrates this.

The budget lines for the farmers are given by AB and CD, in terms of agricultural output, respectively, prior to and after the transfer of surplus labour from the agricultural sector. Hence, AC is the additional per capita of agricultural population after the surplus labour transfer since in the figure we measure agricultural output per capita. Where the terms of trade line, AB, is tangent to the indifference curve, the initial equilibrium is obtained at point E. A relative change in terms of trade against agricultural sector is reflected by the change in the slope of CD which is steeper than AB. (e.g. *difference* between slopes of AB and CD represents the change in the terms of trade against agriculture). The new terms of trade line CD passes through E since it necessary to absorb the whole of the potential savings. Since the indifference curves cannot intersect, it follows that the new equilibrium on line CD will occur at a point like E′, and this indicates a fall in marketed surplus. A more drastic change in the terms of trade (advocated by Stalin in Soviet Russia in the 1920s and 1930s) may, however, raise the marketed surplus (Readers can verify this easily by manipulating the figure.) However, such a policy may involve significant adverse economic, social and political consequences. (For a further elaboration of this point see Ghatak, 1978, ch. 6.)

It is also important to remember that the Soviet coercive model for agricultural planning may largely explain the high costs and low productivity in Soviet agriculture. A forcible rise in the flow of wage-goods from the agrarian sector must be viewed against the fall in

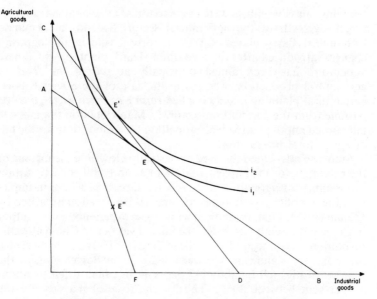

Figure 11.2: *Terms of trade effect on agricultural surplus*

livestock production and a decline in the flow of industrial goods – including investment goods – to the agricultural sector. In the long run, a fall in the income and purchasing power of the peasantry must lead to a fall in productivity in the agricultural sector due to a fall in investment in the agrarian economy, which in its turn is very likely to affect the production of agricultural output (the main determinant of agricultural 'surplus') adversely. Indeed, after Stalin's death the coercive nature of Soviet agricultural planning was gradually reduced, chiefly due to its adverse impact on agricultural production. Yet, despite such relaxation, Russia still needs to import large quantities of grains – chiefly from the US – even at the beginning of 1980s. In the light of Soviet experience, it is not surprising that following the Chinese Revolution in 1948, Mao decided to follow a different agricultural planning model.

3.2 The Chinese model

There are important differences between the Chinese and the Soviet models for agricultural planning. First, in Chinese agricultural planning, massive efforts were made to utilise all available resources for the growth of the agricultural sector. Second, the Chinese Communist Party played a very important role in the agrarian changes introduced after the revolution. Third, planning of Chinese agriculture has been aimed principally at *raising the level* of agricultural production, whereas in the case of the Soviet Russia, agricultural planning mainly implied *raising as much wage-goods as possible* from the agricultural sector. In Mao's opinion the massive tax burden imposed on Soviet agriculture was responsible for the fall of agricultural production.

Mao also advocated the need for a steady growth in the income of the peasantry for the implementation of a successful collectivisation programme for agriculture. The relative success of China in implementing its collectivisation programme has been well summarised by Ellman (1979). First, in contrast with Soviet experience, grain output in China did not fall. Secondly, the fall in livestock in China after the revolution was quite modest, while in the case of Soviet Russia such a fall was very significant (between 1929–30 in Soviet Russia, the number of pigs fell by about 47 per cent; in China it fell by about 18 per cent between 1956). Thirdly, the scale of massacre of the peasantry in Soviet Russia during the 1920s and 1930s was significantly higher than in China during the 1950s. Finally, China did not deport some of its best farmers to achieve its goal of changing the rural society through collectivisation.

However, it is useful to note that much of the gain from the collectivisation programme in China was thrown away during the period of the Great Leap Forward (1958–62) as both the crop and livestocks output fell significantly due to the bad management of the economy. Nevertheless, during the 1950s, substantial progress was made with irrigation. Further, the system of income distribution was made much more egalitarian in the rural sector. Capital formation within the agricultural sector registered a steady rise as resources were diverted from property income to the accumulation of capital.

Since 1961 (i.e. after the Great Leap Forward), agricultural output began to rise with the increasing application of chemical fertilisers and improved seed varieties. A general rise in the proportion of irrigated land also helped enormously to expand output. Even so, with population rising steadily, Howe (1978) estimates that the mean grain consumption in China in per capita terms in the 1970s was only 25 per cent higher than that of Bangladesh. According to another estimate, Chinese agriculture still suffers from some basic problems such as a low growth rate, low yield and productivity, and low level of total production (Padoul, 1975). Indeed, one estimate suggests that the growth rate of grain output in China has only been two-thirds that in India during 1952–73 (Bandyopadhyaya, 1976).

It is, however, important to remember that Chinese agricultural growth estimates are highly sensitive to the base period, as well as the output figures chosen for the base period (Ellman, 1979). Ellman claims that the mean annual grain output growth rate is 'respectable' 3.6 per cent per annum when 1949 is taken as a base; but it goes down to 2.4 per cent per annum if 1952 is chosen as base. As regards the official grain production figures, it is useful to point out that many of these estimates suffer from a number of inaccuracies. Hence, a good deal of caution is necessary in drawing final conclusions. As Sinha (1975) points out:

There is substantial *qualitative* evidence that the level of food consumption has improved [in China], which in itself indicate that food production has risen faster than population. However, with the present state of quantitative information one cannot say whether the levels of food consumption in China today are better or worse than in the mid-1930. But one can easily say that because of the egalitarian policies the level of food consumption of the poorer people is much better now than it was in the 1930s. For the same reason, it can be said that poorer people in China are eating better than India or that the Chinese can withstand food scarcity better than Indians. (emphasis added)

The other point that needs to be emphasised is that the argument for collectivisation, based on its alleged efficacy of raising net transfers from agriculture to industry, is wrong (Ellman, 1979). It is difficult to reach a definitive judgement about the direction of financial and real resource transfer from agriculture to industry for all the periods in the Chinese economy since 1949. Some preliminary investigations suggest that in some periods there has been a steady shift of resources in *favour* of the agricultural sector (Paine, 1976).

4 Choice of Planning Strategy for Agriculture

The Soviet and the Chinese experiences are useful in understanding the nature of problems that might arise in the planning of agriculture in socialist economies. Most LDCs, however, belong to 'mixed' economies. It is, therefore, necessary to adopt those strategies for planning agriculture in these countries which will be conducive to the long-run development of the agrarian sector.

The formulation of such strategies is of crucial importance for the growth of output and employment within the LDCs. Some argue that development plans should aim at the application of the latest in science and technology and condemn 'the placing of restrictive import on farm implements of proven productive performance such as tractors and power tillers as they are thought to be "labour saving" and might cause serious unemployment in the rural economy' (Hopper, 1968). On the other hand, although it is important to emphasise the role of modern science and technology in the agriculture of LDCs, for some it may be 'patently impossible to apply the latest science and technology except in a highly selective manner' (Johnston, 1972). In this context, it is useful to discuss the uni-modal and bi-modal strategies for agricultural developments, bearing in mind that the following aims are particularly relevant:

(a) The contribution to the overall rate of economic growth and the process of structural transformation.
(b) The realisation of a reasonable rate of rise in agricultural output at a minimum cost by helping sequences of innovations which 'exploit the possibilities' for technical progress most suitable to a country's factor endowments.
(c) The realisation of a general increase in rural standard of living and welfare.
(d) 'Facilitating the process of social modernisation', (i.e. controlling fertility, advancement of rural education and improving the entrepreneurial abilities, etc.

One of the most important problems that LDCs face is the choice between a bi-modal or uni-modal strategy. In the first resources are concentrated within a sub-section of a large and capital-intensive unit, while under a uni-modal strategy, policy formulators try to induce a gradual and wide rate of adoption of technical progress, the nature of which is generally geared to the nature of factor proportions of the whole sector. Johnston (1972) has put it very clearly:

> The essential distinction between the two approaches is that the uni-modal strategy emphasises sequences of innovations that are highly divisible and largely scale-neutral. These are innovations that can be used efficiently by small-scale farmers and adopted progressively. A uni-modal approach does not mean that all farmers or all agricultural regions would adopt innovations and expand output at uniform rates....
>
> Although a bi-modal strategy entails a much more rapid adoption of a wider range of modern technologies, this is necessarily confined to a small fraction of farm units because of the structure of economies in which

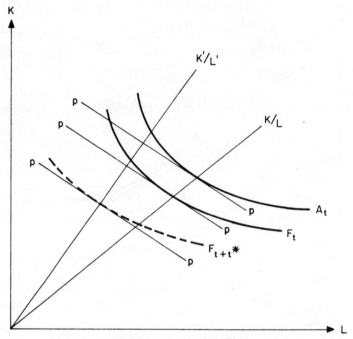

Figure 11.3: *A bi-modal strategy*

commercial demand is small in relation to a farm labour force that still represents some 60 to 80 per cent of the working population. (pp. 36–7)

Owen (1971) argues that 'it may be posited as a basic condition of economic growth in all countries that most of the available land resources should be incorporated in the commercial sub-sector.' However, it should be recognised that, from the point of view of income distribution and equity, a uni-modal approach may be preferable to a bi-modal one. Following the analysis of Farell, it can be shown that 'total efficiency' (as the sum of price, technical and economic efficiency) of a farm under uni-modal strategy will be greater.

Consider a firm using capital ($=K$) and labour ($=L$) to produce a given amount of output shown by the isoquants in Figures 11.3 to 11.5. A technological frontier is a point of output which can be produced by the minimum amount of capital and labour. A firm is

320   AGRICULTURE AND ECONOMIC DEVELOPMENT

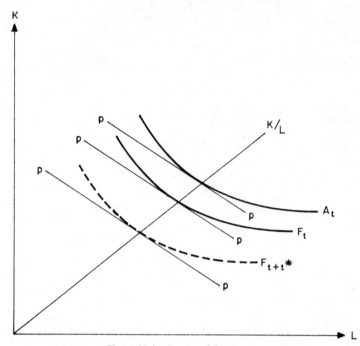

Figure 11.4: *A uni-modal strategy*

technically efficient if it has an input combination that lies on the frontier isoquant (= Ft). Firms which use 'average' technology (i.e. not the most efficient) have input combinations that lie on an 'average frontier' (= At).

By using the simple principles of micro-economics we show by these figures that firms are price-efficient where price lines (PP) are tangent to the isoquants. Notice that a bi-modal strategy should involve 'capital-augmenting' technical progress within a modernised sub-sector. A uni-modal strategy should involve a gradual technical progress and a smooth rise in capital intensity which involves the whole agricultural sector. These discrepancies are shown in the figures.

In Figure 11.3, F_{t+t^*} shows a new frontier isoquant after a capital-augmenting bi-modal strategy has been introduced; whereas in Figure 11.4, F_{t+t^*} shows the frontier given by the wide use of improved inputs like seeds and fertiliser.

PLANNING AGRICULTURAL DEVELOPMENT

Given the same relative factor prices (PP), under a bi-modal strategy a higher K/L will emerge for those (usually large) firms that have access to land and capital. As such, a bi-modal strategy implies a considerable difference in the factor proportions used by 'best' and 'average' firms as shown by slopes of the K/L and K'/L' rays. Now if resources are concentrated within a small sub-sector to promote bi-modal strategy, then there will be a general decline in the ability of the firms outside the sub-sector to use new inputs. It will also imply less foreign exchange for the firms outside the sub-sector to the extent that capital has been imported. Further, given the underpricing of capital (as rates of interest are low and currency is overvalued in the foreign exchange market) and overvaluation of labour in LDCs (since wages are higher than marginal productivity of labour), the divergences in K/L caused by biased technical progress will be accentuated. As such, if a country adopts a bi-modal strategy the difference between the

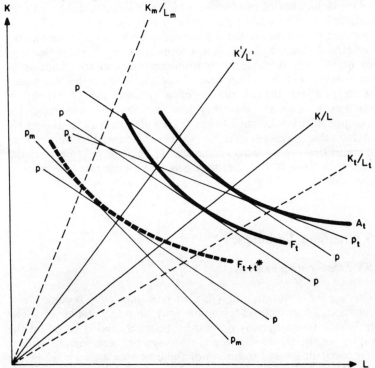

Figure 11.5: *Biased technical progress under a bi-modal strategy in the presence of distortions*

factor intensities and the technical efficiency of the 'best' and 'average' firms is likely to widen as the agricultural sector undergoes a process of structural change. Indeed, given the types of distortions in the factor and the foreign exchange markets we have described, these differences are likely to be accentuated even further. The aggravated divergences in the presence of distortions (underpricing of capital and overpricing of labour) is shown in Figure 11.5 by the rays, K_m/L_m and K_t/L_t for the modern and traditional sectors, respectively.

A bi-modal approach will also be inadequate to absorb 'surplus' labour, and hence the strategy will fail to promote a more egalitarian system of income distribution. On the other hand, a uni-modal strategy emphasises divisible and scale-neutral innovations which generally imply that the factor intensities will be fairly similar for both the most efficient and the average firms. Also, under a uni-modal strategy, most firms within the agricultural sector will adopt techniques of production which will be compatible with the economic structure and factor endowments. Such a choice of factor intensity will minimise the foreign exchange needs of a developing country. As more firms adopt new technology, a process of 'learning by doing' will generate its own momentum and the diffusion of innovation will be widespread. Needless to say, the uni-modal strategy offers the possibility of a greater use of labour in relation to capital – via irrigation and multiple cropping, for example. If the objective is to increase employment as far as possible and to achieve a more egalitarian system of income distribution, then it seems that a uni-modal system should be preferred to a bi-modal one, even though the latter may generate a high rate of economic growth.

5 Planning Techniques

5.1 Input–output analysis

The use of input–output (IO) analysis has now become quite common as a sectoral planning tool in many countries). The technique was originally devised by Leontief, and is based on the input–output tables showing the linkage between various sectors (e.g. agriculture and industry) of the economy and providing the foundation of the economy-wide general equilibrium model that focuses on production. In IO analysis the main assumptions are:

(a) inputs are needed in *constant* proportion to produce output in each sector;
(b) each activity or sector can produce only one product; and
(c) joint products cannot be produced by any sector.

Note that (a) clearly implies that the returns to scale are constant.

It is necessary to separate prices and quantities if we wish to develop a model from the IO tables. Assuming a closed economy, i.e. no exports and imports, the rows of IO tables can be written as follows:

$$P_i X_i = \sum P_i X_{ij} + P_i F_i \ldots \quad (1)$$

where X_i = production in sector i, P_i = price of output in sector i, X_{ij} = flow of intermediate goods from sector i to sector j, F_i = final demand for sector i. Let a_{ij} = intermediate needs from sector i per unit of output of sector j.

As the IO coefficient is fixed, we have:

$$a_{ij} = X_{ij}/X_j$$

Hence, in any period, we have:

$$\frac{P_i a_{ij}}{P_j} = P_i X_{ij}/P_j X_j$$

With an IO table for a given year, it is useful to state the units of real flows in such a way that prices will add up to one.

If we now divide equation (1) by the price, we have:

$$X_i = \Sigma a_{ij} X_j + F_i \ldots \quad (2)$$

In matrix form, this material balance equation is:

$$X = AX + F \ldots \quad (3)$$

and hence we can solve for X as follows:

$$X = (I - A)^{-1} F \ldots \quad (4)$$

Note that the solution indicates that given the final demand, production for different sectors can be determined so that supply is equal to demand in different sectors.

This model can be extended to include foreign trade. Assume that E = exports, and M = imports. We then have:

$$X + M = AX + F + E \ldots \quad (5)$$

which gives

$$X = (1-A)^{-1}(F + E - M)\ldots \quad (6)$$

Notice that the IO technique is very useful if planners wish to promote a balanced growth between agriculture and industry. Notice also, that many LDCs are net importers of food and as such the use of equation (6) raises problems as the net final demand turns out to be negative.

A solution has been suggested by Chenery and Clark who recast the problem in terms of total demand and supply. (See Chenery and Clark (1959); and also Dervis, de Melo and Robinson (1982), who give a very good account of the use of IO technique and a computable general equilibrium model.) It should be pointed out that IO analysis is not very helpful in finding the 'optimal' technique of production from a variety of technologies that may be available for producing a given commodity. In order to identify the 'optimal' (most efficient) technology, it is necessary to use linear programming models.

It is important also to note that the IO analysis is a macro-planning tool, and its relevance to sectoral planning *for agriculture* is largely confined to testing for the *feasibility and consistency* of sectoral targets.

5.2 Linear programming

Some planning models have been developed in both the socialist and the non-socialist countries for improved functioning of agriculture at farm, local or national levels. These models are generally used to illustrate the objective function; identify the constraints; evaluate problems of efficiency in the allocation of resources; and derive policy implications in the light of such evaluations. It is useful to remember that some form of planning the agricultural sector is unavoidable in both socialist and market economies. In the case of market economies, examples of agricultural planning can be easily cited by noting the extent to which public policies play a role in determinig production quotas, support prices, subsidies for output and the money to be spent on research and development for agriculture, irrigation, fertiliser use, etc. Indeed, some problems of farm planning are common to both market and non-market economies as all farms have to face several constraints – land, capital and labour, for example – to maximise a given objective function (e.g. profit or income). In addition, many face institutional constraints – transport, credit and marketing facilities, say. Farms also face problems associated with risk and uncertainty – the impact of weather and the use of new types of seed, pesticide or fertiliser, the effects of new technology, and so on.

5.3 The 'cattle-feed' problem of cost minimisation

Corresponding to the solution of the maximisation problem in the linear programming analysis, there exists a method to solve a minimisation problem. For students of agricultural economics, such a solution can best be illustrated by analysing the minimum-cost of cattlefeed (the 'diet') problem.

Let us assume that the daily feed of an animal should be given by minimum-cost subject to certain nutritional constraints. To simplify the problem, we assume that there are only four kinds of raw materials – oats, barley, sesame flakes and groundnut meal. Let us further assume that there are two nutritional constraints: (i) the feed should have at least 20 units of protein; and (ii) the feed should have at least 5 units of fat.

In Table 11.1, the protein and fat content per unit of each raw materials and units costs are shown (this example is adapted from van de Panne). Let us write the system of equations as follows:

$$12x_1 + 12x_2 + 40x_3 + 60x_4 \geqslant 20$$
$$2x_1 + 6x_2 + 12x_3 + 2x_4 \geqslant 5 \qquad (7)$$

Let us introduce the 'slack' variables (i.e. y_1 and y_2) which are non-negative, indicating an excess of protein and fat needs. We can then transform the inequalities as follows:

$$12x_1 + 12x_2 + 40x_3 + 60x_4 - y_1 = 20 \ldots \qquad (8)$$
$$2x_1 + 6x_2 + 12x_3 + 2x_4 - y_2 = 5 \ldots \qquad (9)$$

If the aim is to minimise the costs of the feed, then we can write the

Table 11.1: *Protein and fat content and costs of raw materials*

	Protein content	Fat content	Cost per unit
x_1 Barley	12	2	24
x_2 Oats	12	6	30
x_3 Sesame flakes	40	12	40
x_4 Ground nut meal	60	2	50

equation for such an objective function as follows:

$$F = 24x_1 + 30x_2 + 40x_3 + 50x_4 \ldots \quad (10)$$

Rearranging equation (8)–(10), we have:

$$0 = -24x_1 - 30x_2 - 40x_3 - 50x_4 + F$$
$$-20 = -12x_1 - 12x_2 - 40x_3 - 60x_4 + y_1$$
$$-5 = -2x_1 - 6x_2 - 12x_3 - 2x_4 + y_2 \ldots \quad (11)$$

In Table 11.2 the simplex tableaux for the minimum cost cattlefeed problem is described. It is obvious that the solution in tableaux is not feasible as y_1 and y_2 (the basic variables) are negative. We know that the simplex method needs a feasible solution to begin with; hence we must find a basic feasible solution. Note that half the sesame flakes should meet the protein need and have 6 units of fat, which is more than what we need. We now find the corresponding basic solution in which the amount of sesame flakes, x_3, is not-zero. Thus a non-basic variable in which the slack variable for protein y_1, is reduced to 0 and this becomes a basic variable. We find this solution is tableaux 0 by pivoting on the element in the row of y_1 and the column of x_3. Since cost-minimisation is our aim, the variable corresponding with the *most positive* element in the first row is chosen as the new basic

Table 11.2: *Simplex tableaux for minimum-cost cattlefeed problem*

Tableau	Basic variables	Values basic variables	x_1	x_2	x_3	x_4
0	f	0	-24	-30	-40	-50
	y_1	-20	-12	-12	-40	-60
	y_2	-5	-2	-6	-12	-2
			x_1	x_2	y_1	x_4
1	f	20	-12	-18	-1	10
	x_3	$\frac{1}{2}$	$\frac{3}{10}$	$\frac{3}{10}$	$-\frac{1}{40}$	$1\frac{1}{2}$
	y_2	1	$1\frac{3}{5}$	$-2\frac{2}{5}$	$-\frac{3}{10}$	$\frac{16}{11}$
			x_1	x_2	y_1	y_2
2	f	$19\frac{3}{8}$	-13	$-16\frac{1}{2}$	$-\frac{13}{16}$	$-\frac{5}{8}$
	x_3	$\frac{13}{32}$	$\frac{3}{20}$	$\frac{21}{40}$	$\frac{1}{320}$	$-\frac{3}{32}$
	x_4	$\frac{1}{16}$	$\frac{1}{10}$	$-\frac{3}{20}$	$\frac{3}{160}$	$\frac{1}{16}$

Source: van de Panne (1976).

variable. This is shown by the following equation:

$$20 + 12x_2 + 18x_2 + y_1 - 10x_4 = F$$

It is shown in the first row of tableaux 1. Note that x_4 now enters the basis as y_2 departs. We now obtain tableaux 2. It is evident that the first row does not contain any positive number, and surely we have obtained the solution for cost minimization! As tableaux 2 shows, the minimum costs is $19\frac{3}{8}$ and the values for x_3 and x_4 are given by 13/32 and 1/16, respectively.

Sometimes, *imputed* values can be assigned to the raw materials used and generally these imputed values are derived from the costs of the raw materials of the cattlefeed. If markets for such raw materials are very imperfect – as they could be in many LDCs – then shadow prices are usually worked out to obtain optimal solutions.

A typical cost-minimisation problem can then be summarised as follows:

Minimise $\qquad F = C_1x_1 + C_2x_2 + \cdots + C_nx_n \ldots \qquad (12)$

subject to:

$$A_{11}x_1 + A_{12}x_2 + \cdots + A_{1n}x_n \geqslant b_1$$
$$A_{m1}x_1 + A_{m2}x_2 + \cdots + A_{mn}x_N \geqslant b_m \ldots \qquad (13)$$

All the x variables are non-negative. (For further discussion of the use of artificial variables, the saddle point and the dual method, see Dorfman, Samuelson, Solow, 1959; and Van de Panne (1976).)

5.4 Activity analysis and the choice of production technique

Linear programming is a form of *activity analysis* in which the optimum combination of activities is selected subject to the objective function and one or more linear resource constraints. However, the technique of activity analysis *without* resource constraints can also be used to select the optimum activity from a limited choice of activity vectors, where the objective function is to minimise total resource costs. An interesting application of this method of analysis relates to rice milling in Indonesia (Timmer, 1974).

Five rice milling techniques were simultaneously available, distinguished by different capital/labour ratios. The most labour-intensive technique, hand pounding, had a zero capital requirement. At the other extreme, the large bulk rice mill technique was characterised by a high K/L ratio and a relatively high absolute capital requirement.

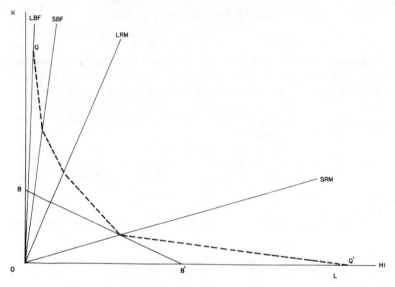

Figure 11.6: *Choice of technique*

The K/L ratios of the remaining three milling techniques – the small rice mill, the large rice mill, and the small bulk rice mill–lay between the extremes. In Figure 11.6 where required capital and labour inputs are measured along the two axes, the fixed input coefficients of the various techniques are signified by the slopes of the vectors HP, SRM, etc. The objective function is to minimise the costs of milling relative to a given 'output' (strictly, value added) of milled rice.

In the figure this output constraint (arbitrarily determined) is represented by the linear-segmented isoquant QQ'. A 'corner' occurs on QQ' wherever the isoquant is intersected by an activity vector. The cost-minimising activity or technique is located at the corner where the relevant 'budget' or 'isocost' line is tangent to QQ'. In the figure the slope of the line BB' reflects the relative costs of labour and capital, and the point of tangency with QQ' identifies the small rice mill (SRM) as the cost minimising technique.

Empirical evidence published by Timmer shows that in Indonesia, where wages, interest rates and rice prices have been determined by market forces, without undue interference from the government, small rice mills have indeed become the dominant rice milling technique, largely at the expense of hand pounding. However, hand pounding is still practised on farms with surplus labour, and especially for self-consumption. Despite their ready availability on the market, large capital-intensive rice mills have been largely ignored by profit conscious millers, even though large mills offer some technical advantages such as a higher extraction rate and a lower proportion of broken grains. But at prevailing prices of both rice and resources used in rice milling, these technical gains are not worth pursuing. Moreover, sensitivity analysis, involving variation of the labour/capital price ratio represented by the slope of the isocost line BB′, indicated substantial stability of the corner solution corresponding with the choice of the small rice mill. However, as may be readily perceived from Figure 11.6, large rice mills may well become the optimal cost-minimising technique at some time in the future as labour becomes more expensive and the relative cost of capital declines. In the meantime, adherence to small and relatively labour-intensive rice mills as the dominant technique must confer very substantial employment benefits on society. Had millers chosen to adopt one of the more capital-intensive methods, very large numbers of jobs in the rice-milling industry would have been lost, with poor prospects of early alternative employment for a high proportion of the workers concerned.

This example also serves to illustrate how market price distortions, or misguided government intervention in the market, can lead to socially inappropriate technological choices. In the Indonesian case any number of market imperfections or government interventions could have combined to alter the K/L price ratio sufficiently to shift the optimum choice to one of the more capital-intensive techniques of rice-milling. The social costs of such a shift, in terms of increased unemployment, could have been very heavy indeed.

6 Agricultural Project Planning

This section falls into three sub-sections. First, the basic principles of project appraisal are outlined and discussed; second, special problems of application in LDCs are considered; and finally, questions of practical application and choice of appraisal methods are reviewed.

6.1 Principles

6.1.1 Objectives of Project Appraisal

Project appraisal is a form of applied welfare economics or cost-benefit analysis. Its central objective is to estimate the social net benefits of an investment project on the *economy as a whole* in contradistinction to merely estimating its financial prospects as viewed by private investors.

The earliest examples of cost-benefit analysis were concerned with the appraisal of major flood control, irrigation and hydro-electric schemes in the USA, where it was realised that there were many 'hidden' costs and benefits associated with such schemes. The US government accordingly sought a standardised appraisal procedure for use in separating viable from non-viable projects, and for placing the viable ones in priority order where these were in competition. The Food Control Act 1936 laid down that for projects to be authorised the benefits must exceed the costs 'to whomsoever they may accrue' (Eckstein, 1965, p. 2). Cost-benefit analysis represents an attempt to apply this criterion.

6.1.2 Choice of Investment Criteria

Two of the most comprehensive decision criteria used in project appraisal are the net present value (NPV), and the internal rate of return (IRR). A third useful criterion, the benefit-cost ratio (B/C), is similar to NPV.

Net present value. This is the difference between the PV of total benefits and the PV of total costs. The following definitions and notation are borrowed from Eckstein (1965, ch. III). Fomally we have:

$$PV_{TB} = \sum_{t=1}^{T} \frac{B}{(1+r)^t} \dots \quad (14)$$

where B = total *annual* benefits. Similarly:

$$PV_{TC} = K_o + \sum_{t=1}^{T} \frac{V}{(1+r)^t} \dots \quad (15)$$

where K_o = fixed (initial) investment, and V = total *annual* costs of operation and maintenance.

For simplicity, we assume that both B and V remain constant for all t, and that there is no further fixed investment after time t = 0. It follows that:

$$\text{NPV} = \text{PV}_{\text{TB}} - \text{PV}_{\text{TC}}$$
$$= \sum_{t=1}^{T} \frac{B}{(1+r)^t} - \left[K + \sum_{t=1}^{T} \frac{V}{(1+r)^t} \right] \ldots \quad (16)$$

Equation (16) expresses the difference between the PVs of two time-streams, the one representing benefits, the other costs.

In use, the condition NPV > 0 must be satisfied to justify the project. *Competitive* projects are ranked by the magnitude of NPV, subject to the previous condition unless there is an overall capital constraint. The discount rate is exogenous, i.e. it is part of the input data.

Internal rate of return. Continuing with the same notation, this is the discount rate satisfying the condition NPV = K_o, where $K_o \geqslant 0$. Thus, we have:

$$K_o = \sum_{t=1}^{T} \frac{B - V}{(1+i)^t} \ldots \quad (17)$$

where i = internal rate of return:

In use, the expected IRR must reach some positive minimum value to justify the project. More formally, justification requires satisfaction of the condition IRR $\geqslant i^t$, where i^t is the opportunity cost of capital, the social time preference rate, or other 'qualifying' discount rate. *Competitive* projects are ranked by the magnitude of IRR, subject to the previous condition. The discount rate i is endogenous, i.e. it emerges from the analysis. However, the test discount rate, i^t, still has to be determined independently.

Benefit: cost ratio. The components of the calculation are identical with those of the NPV criterion. But instead of taking $\text{PV}_{\text{TB}} - \text{PV}_{\text{TC}}$ we take $\text{PV}_{\text{TB}}/\text{PV}_{\text{TC}}$. More precisely:

$$B/C = \sum_{t=1}^{T} \frac{B}{(1+r)^t} \left[K_o + \sum_{t=1}^{T} \frac{V}{(1+r)^t} \right]^{-1}$$

In use, a project is justified provided B/C $\geqslant$ 1. *Competitive* projects are ranked by the magnitude of B/C subject to the first condition. Unlike the NPV criterion, this one does not discriminate between large and small projects.

6.1.3 Critical Problems

Two of the most critical problems of project appraisal are the choice of discount rate, and allowing for risk and uncertainty.

Choice of discount rate. As already noted, a test discount rate is implicit in all NPV and B/C ratio estimates. It also defines the minimum acceptable rate or return for comparison with IRR estimates. Pragmatically, if the discount rate is set too high a bias is created in favour or projects with relatively low initial capital expenditure and with benefits accruing earlier rather than later. So in agriculture, for example, a rapidly maturing extension project might be favoured at the expense of an irrigation project with a larger initial capital requirement and a longer lead time. If the discount rate is set too low the converse argument applies. Additionally, the choice of a low rate will tend to result in too many projects being justified.

Theoretically, the optimum discount rate is the one which equates the marginal social opportunity cost of capital (SOC) with the marginal social time preference rate (STPR). Whereas SOC defines the rate of trade-off between present and future consumption, according to the level of current saving and investment, STPR shows the rate at which society collectively trades off present against future consumption (Pearce and Nash, 1981, ch. 9). Since it is economically rational to rank investment projects in descending order of their expected rate of return, an inverse relationship between the levels of saving/investment and SOC is assumed. The reward for 'waiting' goes down with the declining social profitability of ranked projects. The relationship between the levels of S/I and STPR is assumed to be direct, reflecting the increased marginal opportunity *cost* of waiting at higher levels of S/I. The two functions are shown in Figure 11.7 where, given a socially sub-optimum level of investment I_s, the SOC is q whereas the STPR is lower at i. The condition SOC = STPR defines the optimum level of investment I_o and the discount rate r. Thus, where the coordinates of SOC and STPR meet at I_o and r, the 'gap' between q and i is effectively eliminated.

In the real world a government may be unable to induce as much investment as I_o due to a 'savings gap'. The government is more 'far-sighted' than private investors and taxpayers, particularly regarding investment for the benefit of future generations. Although private citizens as taxpayers may be prepared to sacrifice more, i.e. accept a higher cost and lower reward for 'waiting', than as voluntary savers, the total level of saving and investment is nevertheless subject to political constraints. Thus, if the 'first-best' level of saving and

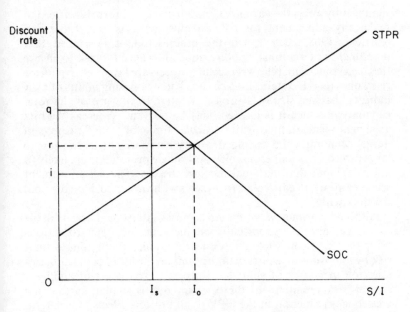

Figure 11.7: *Choice of second-best discount rate*

investment and the corresponding social equilibrium discount rate are I_o and r, it is easy to see from Figure 11.7 that a compromise, 'synthetic' discount rate may be a 'second-best' solution. So, if I_s is the highest feasible level of total saving and investment, q and i, respectively, represent the upper and lower limits of the synthetic rate, d (i.e. q > d > i). Regarding d as some weighted average of q and i, it can be argued that the relative weights depend upon national economic priorities and political judgement. Being fairly closely related to the opportunity cost of capital in the private sector, q is likely to be easier to estimate by observation than i (which is much more subjective). Thus, in practice, choosing d may boil down to little more than answering the question: 'How much less is d than q?' (We return to the choice of discount rate for further discussion in section 6.2.2.)

Risk and uncertainty. Substantial disagreement exists in the literature on project appraisal about the need to allow for the inherent riskiness of individual projects. On the one hand, there is the argument that since the success or failure of a single project does not significantly affect the variance of national income there is no need to make any adjustment for risk on most projects (e.g. Little and Mirrlees, 1968, ch. xv). On the other hand, it is argued that investment criteria must embody some allowance for risk to ensure that, *ceteris paribus*, relatively secure projects are favoured over more risky ones (e.g. Eckstein, 1965 ch. IV). These opposing points of view indicate that the size of a project, relative to the size of the total economy in which it is located, may be relevant. Whereas the first argument seems to hold with particular force for 'small' projects in 'large' economies the second one may carry weight in relation to large projects in small economies. As a single project is more likely to affect regional than national income, the case *for* risk adjustment seems to apply, in particular, to projects with an emphasis on regional development.

Although various *ad hoc* methods of risk adjustment are available to project appraisers, virtually all methods are open to serious criticism (Eckstein, 1965, ch. IV). For example, attempting to limit risk by imposing an arbitrary upper limit on the life of projects biases selection in favour of short-lived projects. The effect of adding a special 'risk premium' to the discount rate is similar, in that the consequent reduction in the NPV of all projects affects longer-term projects disproportionately. The deduction of an arbitrary flat rate 'safety allowance' from the project's NPV is clearly inappropriate as a means of adjusting for risks which *do* grow with time. Because of the numerous defects of these methods of attempting to adjust for risk and uncertainty it is better to use the technique of *sensitivity analysis* in order to evaluate the problem.

Sensitivity analysis involves appraising and re-appraising the project on the basis of differing sets of assumptions regarding the values of critical variables and parameters – such as future prices, crop yields and supply response co-efficients. By examining the effects of relatively 'optimistic' and 'pessimistic' assumptions on the range of NPV estimates about the mean or 'expected value', it may be possible to judge whether acceptance of the project in its present form entails an unacceptable degree of risk. If the risk is judged to be too high, it may be possible to reduce it by redesigning the project, by the substitution of crops in an agricultural project, for example. But notice that sensitivity analysis leaves decision-makers to judge how much risk is acceptable.

A simple hypothetical example will serve to clarify the rationale of sensitivity analysis. Suppose that the expected NPV of an agricultural project, based on the best available forecasts of crop yields and prices, is £100,000, but that the forecasting error is such that the risk of the actual NPV falling below £50,000 is 20 per cent. By altering the crop mix (to substitute less risky for more risky crops) the project can be redesigned to yield an expected NPV of £80,000, with the risk of falling below £50,000 reduced to 10 per cent. Those responsible for project selection must judge whether they prefer (a) the original higher-risk version of the project, which also offers a higher expected NPV, or (b) the redesigned lower-risk but less socially profitable alternative.

Because of space limitations, this review of the basic principles of cost-benefit analysis is incomplete. Other aspects, such as the problems of defining and measuring social benefits and costs, including external economic effects (externalities), are comprehensively covered by sources such as Pearce and Nash (1981), Mishan (1972) and Eckstein (1965), to whom interested readers are referred.

6.2 Application in less developed countries

6.2.1 Reasons for Treating LDCs as a 'Special Case'

The restrictive conditions under which private profit reflects social profit – such as the existence of perfect competition and full employment and the absence of externalities – are never fully satisfied in the real world, even in *developed* economies. Project appraisal has consequently been widely used in economically advanced countries, particularly in areas such as water resource development, transport and defence planning.

In turning from developed to *developing* economies, there are numerous reasons for expecting market prices to diverge even further from the values of social costs and benefits (Little and Mirrlees, 1968, vol. II, ch. II: UNIDO, 1972, ch. 2). Some of the more important reasons for expecting such a divergence in LDCs are:

(1) Price controls and other measures to combat inflation distort relative prices.
(2) Overvaluation of the currency – often resulting from inflation with an inflexible exchange rate – makes imports 'cheap' and domestic products 'dear' at market prices.
(3) Market prices reflect the existing distribution of income within the economy, which may be socially sub-optimal. They also fail to reflect externalities.

(4) Market values represent only *minimum* levels of consumer satisfaction, i.e. they do *not* measure consumer surplus.
(5) Wages fail to reflect the real social costs of employment, especially in agriculture, because of disguised unemployment.
(6) Contemporaneous interest rates vary by more than is necessary to cover differential risks because capital markets are underdeveloped and imperfect.
(7) A divergence exists between the market rate of interest and the social rate of discount due to an aggregate 'savings gap' (as already explained in the previous section). The gap tends to be wider in LDCs due to the greater prevalence of tax evasion or less efficient collection.
(8) The argument that large projects can have important repercussions on profits elsewhere in the economy is especially relevant in LDCs with comparatively undiversified economies and small national products.
(9) Provided that the demand for its exports is less than perfectly elastic, an export tax may yield a net social benefit in the country where it is imposed (an application of optimum tariff theory). This particularly applies to exporters of primary commodities without close substitutes, who also enjoy a substantial share of the total export market.
(10) Protective measures (tariffs, quotas, etc.) are widely employed by LDCs to foster domestic industry; like currency overvaluation, this tends to make domestic manufactured products expensive relative to their world market prices. Inter-industry competition for resources is distorted by differing degrees of protection afforded to different industries, so an industry which is potentially capable of yielding a net social benefit may appear to be unprofitable due to excessive protection of one of its intermediate good suppliers – for example cement used by the building industry, or purchased feeding stuffs used in intensive poultry or pig production.

For all these reasons, and because the development strategy of LDCs tends to be heavily reliant on public investment and public projects, there is an especially strong case for LDCs to use investment appraisal techniques which show the difference between private (or commercial) profitability and social (or national economic) profitability. We now proceed to consider briefly the rationale and methodology of 'shadow pricing' which tends to be an especially important feature of project appraisal in LDCs.

6.2.2 Shadow Pricing

Since market prices cannot be relied on to measure social benefits and costs in LDCs, cost-benefit analysis uses real rather than money prices for project appraisal. 'Shadow' or 'accounting' prices reflect the real costs of project inputs and the real benefits of project outputs to society.

Shadow prices can be broadly defined as the *social* values of economic goods and services, as perceived by government on behalf of the nation. These social values usually diverge from the *market prices* of the same goods and services due to market imperfections (causing resource misallocation), market failures (causing unemployment and the existence of externalities) and the pursuit of government policy objectives such as income redistribution. It is incorrect to think of shadow prices as being the market prices which would prevail after the removal of market imperfections and government intervention. Shadow prices will equate with equilibrium market prices only under the restrictive assumptions of perfect competition, no unemployment, no externalities and a socially optimum income distribution. Each one of these assumptions tends to be unrealistic in the real world and the combination of all four is clearly unattainable.

In summary, then, shadow pricing serves the twofold purpose of correcting for market distortions and market failure, and re-allocating resources in response to government policy objectives (including redistributional objectives). Shadow prices are synthetic in the sense that, unlike market prices, they cannot be observed directly. Therefore, the estimation of shadow prices is an important function of project appraisers. Provided a good or service has an observable market price, this serves as a starting-point in estimating the shadow price, despite the need for adjustment. However, in the case of 'untraded' goods and services (like family labour in subsistence agriculture) it can be misleading to regard the shadow price as an 'adjusted market price', and an alternative method of estimation must be used. Methods of estimation can best be clarified by specific consideration of the shadow prices of labour, investment and foreign exchange. Our treatment of this topic closely follows the methodology of shadow pricing presented in 'Guidelines for Project Evaluation' (UNIDO, 1972, chs. 14, 15 and 16).

Shadow price of investment. The shadow price (or social opportunity cost) of investment, P^I, may be defined as 'the ratio of the social value of investment to the social value of consumption'. In this context social value signifies 'the value of the relevant time-stream of aggregate consumption benefits discounted back to the present at the social rate of discount' (UNIDO, 1972, p. 341). A premium on investment typically exists due to government's inability to induce the socially optimal level of saving and investment where SOC = STPR, as discussed above. Thus, whereas unity is a limiting value of P^I consistent with the condition SOC = STPR, the value of the ratio of the value of investment to the value of consumption is normally expected to exceed 1. As well as being a function of SOC and STPR, P^I is also a function of the marginal rate of reinvestment of business profits. Representing this by s, and SOC and STPR by q and i, respectively, P^I can be determined from the expression:

$$P^I = \frac{(1-s)}{i - sq}$$

Clearly, in the limiting case where $i = q$, $P^i = 1$. Moreover, it is obvious that the magnitude of P^I relates *directly* both to the size of s and the size of $q - i$ (given $q \geqslant 1$).

Thus, instead of assuming that 1 money unit of investment is worth the same as 1 money unit of consumption in project analysis, use of the shadow price of investment enables the differential social values of investment and consumption to be reflected in the appraisal.

Shadow wage. In an LDC the market wage may fail to measure the social cost of labour for three reasons. First, in the traditional sector, labour is allocated and rewarded not in conformity with the 'rules of competition', but according to a customary norm or convention such as a subsistence wage. Thus, even without visible unemployment, there is a gap between the MP of labour in the traditional sector (i.e. its direct opportunity cost there) and the modern sector wage (as paid by private capitalist employers and government). If there is visible unemployment in either sector, the argument is the same but stronger. Since the MP of the unemployed is zero, by definition, the modern sector wage must obviously exceed the direct opportunity cost of labour.

Secondly, where increased public sector employment is financed by taxation there is an income transfer from capitalists to workers which reduces aggregate investment and expands aggregate consumption (assuming that workers consume more and save less of their marginal income than capitalists do). Provided the shadow price of investment, P^I, exceeds 1, this transfer creates additional *indirect* costs to be added to the direct opportunity cost of labour in calculating the shadow wage.

Thirdly, the same income transfer changes the distribution of workers' consumption over time. Present consumption increases, but future consumption declines due to the consequent fall in the current rate of investment. If, in pursuit of government policy, a special premium (or weight) is attached to increasing workers' present consumption, the present value of workers' future consumption forgone for every additional public sector job created in the present must be included in the calculation of the social cost of labour.

More formally, the shadow wage can be defined at three successive stages of approximation. At Stage 1, accounting only for direct opportunity cost:

$$SW = W_s$$

where W_s = direct opportunity costs of labour (usually in the traditional sector). At Stage 2, allowing also for indirect costs, through the effect on saving;

$$SW = W_s + S_c(P^I - 1)W_c$$

where S_c = capitalists' MPS and W_c = capitalist sector wage. At Stage 3, allowing also for the redistribution of workers' consumption in time;

$$SW = W_s + S_c(P^I - 1)W_c + m[W_s + (S_cV_w - 1)W_c]$$

where m = redistributional premium $(0 \leqslant m \leqslant 1)$ and V_w = PV of workers' future consumption forgone.

The significance of the Stage 2 approximation obviously depends upon the value of P^I (at the limit where $P^I = 1$, the indirect cost is zero) whereas the added cost at Stage 3 depends upon m (at the limit where m = 0, there is no case for allowing for the redistribution of workers' consumption in time). Stages 2 and 3 pose difficult problems of parameter estimation for project analysts and decision-makers and, in practice, estimation of the shadow wage may not proceed beyond Stage 1.

Shadow price of foreign exchange. According to one view, the real price of foreign exchange is no more than the ratio of the free market price to the 'official' price set by government. However, the authors of *Guidelines for Project Evaluation* (UNIDO, 1972, ch. 16) argue that this is unsatisfactory since it is not only countries whose currencies trade in the free market at a substantial discount who are concerned about their balance of payments. The basic reason why foreign currency is worth more than its face value at the official exchange rate is that the supply is usually rationed by government. There is consequently a relative scarcity of imported goods, and domestic consumers are prepared to pay more than the landed price of imports expressed at the official exchange rate. For example, suppose that tractors of a particular size and specification can be imported at a landed price of £5000 each. Thus, given an official exchange rate of £1 = RS 20, the landed cost (in rupees) is RS 100,000 per tractor. But if, at the margin, large-scale farmers are prepared to pay RS 150,000 per tractor, each RS 1.0 worth of foreign exchange (at the official rate) provides goods worth RS 1.5 in terms of domestic willingness to pay. In this case, then, the shadow price of foreign exchange is 1.5.

Generalising to any number of commodities, the shadow price of foreign exchange, P^F, can be written:

$$P^F = \sum_{i=1}^{n} f_i \frac{P_i^D}{P_i^L}$$

where f_i = fraction of (rationed) foreign exchange allocated to commodity i at the margin by purchasers of imported goods, P_i^D = domestic market clearing price of i, and P_i^L = cif price of i in domestic currency at official exchange rate.

Choice of numeraire. Consideration of the shadow price of foreign exchange leads to the projects appraiser's choice of unit of account or numeraire. If the practise is followed of choosing the price of freely convertible currency as the numeraire, as recommended by Little and Mirrlees (LM) (1968, ch. VIII) then the shadow pricing of foreign exchange is redundant, since all project costs and benefits are valued at their 'world prices'. In this case, shadow pricing is transferred from foreign currency *per se* to all *traded* goods and services either purchased or sold by the project. One difficulty of adopting this procedure is that of valuing non-traded goods and services – such as indigenous building materials and power supplies – at world prices. This is one reason for preferring the UNIDO alternative of valuing all project goods and services at their *domestic* prices (market or shadow) combined with the application of a shadow exchange rate to imports and exports.

6.2.3 Classification of Projects: the Production and Distribution Trade Off

Development projects are designed to achieve economic and social goals. Even with specifically agricultural projects, the objectives differ considerably. Thus, the emphasis may be on technological innovation, improvement of the rural infrastructure, the reform of agricultural institutions – such as the land tenure or marketing system, the economic and social betterment of either a particular subsection of the rural population (such as farmers with relatively little land) or a particular geographical region of the country (Benjamin, 1981, ch. 3). A broad distinction can be made between two project types:

(1) Projects emphasising increased aggregate production where the distribution of incremental production and income is a secondary consideration.
(2) Projects which differentially increase production or income in one or more 'target groups' of the rural population where increased aggregate production of the entire agricultural sector is of secondary importance.

In a sense, all agricultural projects are designed to relieve rural poverty through a general improvement of agricultural incomes. However, projects emphasising increased aggregate production and income rely on the so-called 'trickle down effect' to transfer part of the benefits to socially disadvantaged sections of the rural population. It may be considered that this indirect method of benefiting the poor is too uncertain and too slow, and there may be a case for making the redistribution of income a primary project objective in order to attack rural poverty more directly and more rapidly.

We have drawn a sharp dichotomy between the objectives of increasing aggregate production and income redistribution to serve an analytical argument. In practice, however, governments sometimes endeavour to design a project to serve multiple objectives. Because transferring income to the poor inevitably raises aggregate consumption, this poses the familiar problem of a trade-off between consumption and investment, or between present and future consumption.

The orthodox approach to project analysis has been to assume that the primary project objective is to benefit the *national* economy by raising net aggregate production and consumption. The redistributive consequences of the projects, or income transfers, are a secondary consideration which may be ignored altogether. But this approach is clearly open to the objection that it neglects the distinction between a gain in national income and a gain in national welfare.

Suppose that two agricultural projects with identical costs are in competition. Project X is a technological-type project aimed at boosting aggregate agricultural output without a major transfer of income to the agricultural sector in the long run. Project Y is a rural development-type project which is less concerned with expanding the marketed surplus of agricultural products and more concerned with improving farm incomes as a means of advancing rural welfare. If the orthodox approach to project appraisal is applied, emphasising national project benefits to the neglect of regional or group benefits, project X must inevitably be preferred to project Y. But if, in pursuit of its agricultural policy and contrary to the orthodox approach, government wishes to pursue production and income redistribution objectives simultaneously, how can such a conflict of 'project benefits' be resolved?

One alternative is to subject the comparison of national economic benefits to an *ad hoc* 'social interpretation' (Benjamin, 1981, p. 154). Thus, in our hypothetical example, the government uses subjective judgement to decide whether the social benefits of transferring income to project Y farmers outweigh the social costs of forgoing the additional national aggregate consumption benefits associated with project X.

Another and more sophisticated method of appraising projects with multiple objectives is to incorporate different types of benefits in an appropriately weighted aggregate benefit function (UNIDO, 1972, ch. 3). Aggregation involves the conversion of different measures of benefit – additional consumption, employment, etc. – into a common set of units by establishing equivalences between different types of benefits. Suppose W_1 units of benefit type B_1 is considered to be equivalent to W_2 units of B_2 and also to W_3 units of B_3. Then the aggregate benefit, $\bar{B}$, can be written as:

$$\bar{B} = W_1 B_1 + W_2 B_2 + W_3 B_3$$

where W_1, W_2, W_3 are weights assigned by the planners.

More generally:

$$\bar{B} = \sum_{i=1}^{n} W_i B_i$$

where W_i are the complete set of national weights.

Extending our earlier simple hypothetical example will serve to illustrate the use of the method. Suppose that, for both projects, B_1 represents national aggregate consumption benefits, whereas B_2 represents regional consumption benefits (income transferred to farmers). Suppose further that with a given test discount rate and identical costs for both projects we have:

$B_1 = 1000$ and $B_2 = 0$ for project X

and
$$B_1 = 800 \text{ and } B_2 = 800 \text{ for project Y.}$$

Suppose also that, for both projects:
$$W_1 = 1.00, \quad W_2 = 0.50$$

This implies that the planners judge 50 pence worth (cents) of money income redistributed to farmers to have the same social welfare value as £1 ($1) worth of money income available to the nation at large in the form of additional national aggregate consumption.

Hence, for project X:
$$\bar{B}_x = (1.00)(1000) + (0.50)(0)$$
$$= 1000$$

whereas for project Y:
$$\bar{B}_y = (1.00)(800) + (0.50)(800)$$
$$= 1200$$

Clearly, if B_1 was the sole criterion of selection, to the exclusion of B_2, project X would be selected. But when B_1 and B_2 criteria are combined, using the weights assigned by government to each type of benefit, project Y is the rational choice. It is easy to see from this example that even if the net benefits of a project at the *national* level were zero (or even negative) the project might still be justified by its redistribution benefits alone if the weight assigned to these was large enough.

The assignment of weights to different types of project benefits is subjective depending on the government's value-judgements. But a major difference between the aggregate benefit function and the *ad hoc* approach is that, in using the former, the same set of weights can be used in appraising different projects, so achieving greater consistency of treatment. It may not always be feasible to express all types of project benefit in terms of a single unit of account such as money. Thus, for example, extra employment (which may be valued for its own sake rather than for its contribution to national income) is measured in physical terms. However, the problem of aggregating benefits measured in different units can be circumvented by deriving a set of equivalences or conversion coefficients with a common base (see UNIDO, 1972, p. 34)

6.2.4 Project Formulation Process
Economic appraisal is but one aspect of a much broader process of project formulation. The starting-point of most projects is

a set of technical proposals based – in the case of agricultural projects – largely on the work of agronomists, livestock specialists and engineers. The technical options must be identified before the accountants can conduct a financial appraisal and the economists an economic appraisal. Thus the preparation of a development project is a complex multi-stage process: it is also a multi-disciplinary activity in which the economist builds upon the feasible technical options, whilst the agronomist or engineer must work within the bounds of economic feasibility.

A 5-stage project cycle is discussed in the literature (Benjamin, 1981, ch. 1). *Stage 1* is 'project concept definition' in which government accepts development through projects as a feasible means of promoting development. *Stage 2* is 'identification' in which the outline proposals for a project are delineated. *Stage 3* is 'preparation' in which proposals identified at the previous stage are expanded and examined in detail. At this stage the full technical and organisational details of the project are presented, together with the expected financial results and the economic justification. Preparation is both time-consuming and costly. *Stage 4* is 'appraisal' to check further the economic viability of the project, including its economic justification in competition with other projects. Since the external financing of projects is often conditional upon meeting a standard of economic justification which is acceptable to the lending or aid agency, appraisal may be undertaken independently of preparation by an external agency. Negotiations for external project finance (loan or grant) are inextricably linked with appraisal. *Stage 5* is 'supervision' which is concerned with all the details of project *implementation*, including the monitoring of progress reports and even modifying details of the project itself during implementation if this is deemed to be necessary in response to unforeseen contingencies.

Like other stage theories, this one is open to criticism. It is admitted that 'in practice many of the phases overlap' (ibid., p. 1), as is all but self-evident with respect to identification, preparation and appraisal. More interestingly, Benjamin maintains that, in practice, appraisal tends to be 'employed *post facto* to justify a project already selected and designed' (ibid., p. 155). This is, of course, contrary to the economic ideal of using appraisal to discriminate between alternative projects.

Despite the possibility of disagreement between economists and other practitioners of project planning about the purpose of economic appraisal, the proposition that project formulation and implementation consist of much more than economic appraisal is beyond dispute. Most projects have a complex technical structure, as well as being based upon a set of interrelated economic assumptions, so much so that if the technical assumptions are unreliable, but the degree of possible error is unknown (which is often the case), little credence can be given to the results of economic appraisal. This is so regardless of the reliability of the *economic* assumptions and the professional skill of the economists making the appraisal.

6.3 Practical application: appraisal methods

In this section we discuss the application of comprehensive models of project appraisal to agricultural projects, using the UNIDO model as an example. We then consider some peculiarities of agricultural projects and the case for using simpler appraisal technique, with examples. Finally, we can contrast *ex ante* appraisal with *ex post* evaluation and the feedback from project experience to project design.

6.3.1 Comprehensive Models: Application of the UNIDO Method of Appraisal to an Agricultural Project

In this section we describe and summarise the elements of a case study borrowed from the UNIDO *Guidelines for Project Evaluation* (1972, ch. 21) to the authors of which we make due acknowledgement.

Description of project. This is a case study of the appraisal of an irrigation project in a particular geographical region of an LDC. The project is designed to achieve three principal objectives:

(1) The intensification of agricultural production in an arid river valley by damming the river and constructing a water distribution network.
(2) Increasing the number of farmers in the valley by the settlement of farmers from outside the region following redistribution of the land.
(3) The social betterment of the region's inhabitants through the construction of a road network, an urban centre and satellite villages.

The project would be administered by a statutory water authority formed by the government specifically to develop the region's water resources. Regarding finance, the water authority would apply to the government and to the World Bank for long-term loans to cover the capital costs of the project. The World Bank would be asked to cover the foreign exchange component of the total investment and the home government the remainder.

This case study exemplifies the appraisal of net benefits at the levels of: (i) the national economy, (ii) the region, and (iii) a particular 'target' social group (small farmers).

Methodology. We briefly consider the following aspects.

(a) framework of government social objectives,
(b) data requirements,
(c) appraisal procedure (in 3 stages),
(d) presentation of results,
(e) interpretation of results, and
(f) summary view of the method.

(a) Statement of the government's social objectives. Consistency demands that detailed project proposals should be prefaced by an explicit statement of the government's social objectives. In the present case those are:

(1) Increased aggregate consumption to raise the average standard of living in the country as a whole (national welfare objective).
(2) Redistribution of income to the region in which the project is located (regional welfare objective).
(3) Redistribution of income to small farmers (group welfare objective).
(4) Creation of new employment opportunities.
(5) Provision or improvement of basic welfare facilities, such as housing.
(6) An improved balance of payments.

Analysis of the project actually revolves around its expected contribution to realising the first three objectives since, in the words of the authors, 'the last three items do not so much represent separable objectives as observations on the limited ability of market prices – the wage rate, the price of social services and the foreign exchange rate – to reflect true social benefits and costs with respect to the aggregate consumption objective'. Thus the realisation of objectives (4)–(6) is sought indirectly by substituting shadow prices for market prices, where the latter are not thought to reflect true social values.

(b) Data requirements. Two basic types of data are required. First, data are needed to construct a *table of resource flows* generated by

the project. This table shows the expected monetary values at market prices of all project benefits and costs, grouped and totalled by major categories on a year-by-year basis for the life of the project. Also shown in the table are the expected monetary transfers between social classes, or between citizens and government. At this stage of the analysis the expected values of all *future* resource flows remain *undiscounted*. The form of presentation is exemplified by Table 11.3 reproduced from UNIDO (1973). The bare contents of the table conceal the amount of detailed work needed to prepare it. First, a vast quantity of data on input availabilities, feasible input mixes, product choices, and product mixes, physical input–output coefficients and prices, must be collected. Second, these basic data must be analysed and processed in order to produce estimates of the expected levels of output (or benefit) by time period, together with the corresponding costs. In order to suit the form of the subsequent analysis, a detailed breakdown of each year's costs is needed, the main distinctions being between capital and current (or operating) costs, and costs borne by the principal 'actors' of government, farmers and taxpayers.

As far as economic appraisal *per se* is concerned, the table of resource flows might be regarded as technical data which the economist responsible for appraisal must accept without question. This view probably exaggerates the division of labour between the economist and specialists from other disciplines who are also involved in project analysis. The important point is that, as one of the most basic documents used in the process of appraisal, the results of appraisal are to an important degree predetermined by the table of resource flows. Most important of all, the reliability of the final result is critically dependent on the quality of the information available to the analysts responsible for preparing this table and their skill in using it to produce trustworthy estimates.

The second basic data requirement is a set of '*national parameters*' for use in evaluating the project within the context of the government's social objectives. Examples include the social rate of discount, the shadow prices of investment, labour and foreign exchange, and the weights assigned by government to dissimilar project benefits such as increased per capita consumption and income redistribution (regional and social). We have already discussed the welfare economics rationale of using subjectively determined weights in project analysis. The set of national parameters may also include certain more objectively based coefficients, such as the marginal

Table 11.3: *Benefit, cost and transfer flows by year (at market prices) (million pesetas)*

		Year										
		1	2	3	4	5	6	7	8	9	10	11–54
Benefits												
(1)	Agricultural output	—	—	—	—	65.80	78.96	92.12	105.28	118.44	131.60	131.60
(1 – D)	Domestic currency	—	—	—	—	49.80	59.76	69.72	79.68	89.64	99.60	99.60
(1 – F)	Foreign exchange	—	—	—	—	16.00	19.20	22.40	25.60	28.80	32.00	32.00
(2)	Housing and social services	—	—	—	—	2.80	2.80	2.80	2.80	2.80	2.80	2.80
Costs												
(3)	Construction costs	64.00	75.00	188.00	152.00	—	—	—	—	—	—	—
(3 – L)	Unskilled labour	20.30	24.90	69.40	64.40	—	—	—	—	—	—	—
(3 – S)	Skilled labour	11.70	13.20	31.19	25.30	—	—	—	—	—	—	—
(3 – D)	Domestic materials	9.70	12.40	33.50	19.90	—	—	—	—	—	—	—
(3 – F)	Foreign exchange	22.30	24.50	54.00	42.40	—	—	—	—	—	—	—
(4)	Operating costs	—	—	—	—	4.00	4.00	4.00	4.00	4.00	4.00	4.00
(4 – L)	Unskilled labour	—	—	—	—	1.60	1.60	1.60	1.60	1.60	1.60	1.60
(4 – S)	Skilled labour	—	—	—	—	0.92	0.92	0.92	0.92	0.92	0.92	0.92
(4 – D)	Domestic materials	—	—	—	—	1.00	1.00	1.00	1.00	1.00	1.00	1.00
(4 – F)	Foreign exchange	—	—	—	—	0.48	0.48	0.48	0.48	0.48	0.48	0.48
(5)	Farmer agricultural costs	—	—	—	—	24.16	28.99	33.73	38.66	43.49	48.31	48.31
(5 – LF)	Family (unskilled) labour	—	—	—	—	8.73	10.48	12.22	13.97	15.71	17.46	17.46
(5 – LH)	Hired (unskilled) labour	—	—	—	—	2.91	3.49	4.08	4.66	5.24	5.82	5.82
(5 – D)	Domestic materials	—	—	—	—	4.37	5.24	6.12	6.99	7.86	8.73	8.73
(5 – F)	Foreign exchange	—	—	—	—	8.15	9.78	11.41	13.04	14.68	16.30	16.30
(6)	Ministry agricultural costs	—	—	—	—	40.50	10.50	10.50	10.50	10.50	10.50	6.50
(6 – S)	Extension workers	—	—	—	—	3.00	3.00	3.00	3.00	3.00	3.00	0.50
(6 – D)	Working capital	—	—	—	—	7.50	1.50	1.50	1.50	1.50	1.50	—
(6 – F)	Foreign exchange	—	—	—	—	30.00	6.00	6.00	6.00	6.00	6.00	6.00
(7)	Agricultural income forgone	—	—	—	—	0.68	0.68	0.68	0.68	0.68	0.68	0.68
Transfers												
(8)	Compensation to landowners	—	5.00	—	—	—	—	—	—	—	—	—
(9)	Irrigation fees	—	—	—	—	20.00	20.00	20.00	20.00	20.00	20.00	20.00
(10)	Rental and interest payments	—	—	—	—	4.94	5.93	6.91	7.90	8.89	9.89	9.89

opportunity cost of capital in the private sector (q), the marginal rate of reinvestment (s), and the marginal propensity to save (MPS) amongst different categories of savers.

An example set of national parameters is shown in Table 11.4, reproduced from the same source as Table 11.3. Adequately reliable data at an acceptable cost is again critical. Highly inaccurate or unrealistic national parameters, applied at a later stage of the analysis, must inevitably undermine the credibility of the final results.

(c) *Appraisal procedure.* In this case study appraisal of the project is based on three criteria: (i) net national aggregate consumption benefits, (ii) regional redistribution benefits, and (iii) social group redistribution benefits. In explaining the method of appraisal it is convenient to deal with each criterion separately before proceeding to discuss overall assessment of the project from values inserted in an aggregate benefit function.

(i) Appraisal of net aggregate consumption benefits. This part of the analysis is best explained in three successive stages. In *Stage 1*, all the component elements of the net aggregate consumption function are valued at *market* prices, on the simplifying assumption that these adequately reflect social opportunity costs. Thus, we have the identity:

$$\text{MC} = (1) + (2) - (3) - (4) - (5) - (6) - (7)\ldots \tag{a}$$

Table 11.4: *Values of general parameters*

(1) Foreign exchange premium	$\varphi = +1.0$
(2) Unskilled labour premium	$\lambda = -1.0$
(3) Extension worker premium	$\lambda^E = +1.0$
(4) Marginal rate of return on investment in private sector	$q = 0.20$
(5) Marginal rate of reinvestment of profits	$s = 0.20$
(6) Social rate of discount	$d = 0.05; 0.075; 0.010$
(7) Associated social price of investment	$P^I = 16.00; 4.57; 2.67$
(8) Marginal propensities to save:	
(8-a) Farmers	$S^F = 0.2$
(8-b) Unskilled workers	$S^L = 0.0$
(8-c) Government	$S^G = 1.0$
(8-d) Taxed public	$S^T = 0.8$
(9) Marginal propensity to (re-) spend in region	$\gamma = 0.2$
(10) Weights on objectives:	
(10-a) Aggregate consumption	$\theta^C = 1.00$
(10-b) Redistribution to region	$\theta^{RM} = 1.00$
(10-c) Redistribution to small farmers	$\theta^{RSF} = 0.50$

Source: Adapted from UNIDO *Guidelines for Project Evaluation*, Table 21.20

where MC = net aggregate consumption at market prices, and (1), (2), etc. are resource flows specified and quantified in Table 11.3.

In *Stage 2* the shadow prices of labour and foreign exchange are substituted for their market prices, except that no allowance is made for the premium on investment (on the simplifying assumption of a zero savings gap). The identity of components of the net aggregate consumption function now becomes:

$$SC = (1) + \varphi(1\text{-}F) + (2) - (3) - \lambda(3\text{-}L) - \varphi(3\text{-}F)$$
$$- (4) - \lambda(4\text{-}L) - \varphi(4\text{-}F) - (5)$$
$$- \lambda[(5\text{-}L^F) + (5\text{-}L^H)]$$
$$- \varphi(5\text{-}F) - (6) - \lambda^E(6\text{-}S) - \varphi(6\text{-}F) - (7)$$

where (1), (1-F), (2), etc. are resource flows from Table 11.3, φ = shadow price of foreign exchange, λ = shadow of wage of unskilled labour, and λ^E = shadow wage of agricultural extension personnel.

Re-arranging terms:

$$SC = MC + \varphi FE + \lambda L + \lambda^E E \cdots \quad (b)$$

where FE = (1-F) − (3-F) − (4-F) − (5-F) − (6-F),
L = − (3-L) − (4-L) − (5-L^F) − (5-L^H) and E = − (6-S).
Note that whereas the shadow wage of unskilled labour, λ, is expected to be at a discount compared with the market wage rate, the shadow wage of extension personnel, λ^E, is expected to be at a premium. All shadow prices are expressed *relative* to a market price of unity.

In *Stage 3* the substitution of shadow prices for market prices is completed by also bringing in the shadow price of investment. Assuming that the marginal propensity to save differs amongst social classes, as well as between citizens and government, assessment at this stage entails the allocation of all SC benefit and cost flows amongst four groups of gainers or losers, as follows:

$$SC = SC^F + SC^L + SC^G + SC^T$$

where the superscripts F, L, G and T identify four classes of net consumption benefit recipients, farmers, unskilled workers, government and taxpayers. Moreover:

$$SC^F = (1) + (2) - (5) - (6\text{-}S) - (7) + (8)$$
$$- (9) - (10) - \lambda(5\text{-}L^F) - \lambda^E(6\text{-}S)$$

$$SC^L = -\lambda[(3\text{-}L) + (4\text{-}L) + (5\text{-}L^H)]$$
$$SC^G = -(3\text{-}F) - (4) - (6\text{-}D) - (6\text{-}F)$$
$$\qquad -(8) + (9) + (10) + \varphi(FE)$$
$$SC^T = -(3\text{-}L) - (3\text{-}S) - (3\text{-}D)$$

Then, if the average farmer saves a proportion S^F of his marginal gains, the (adjusted) social value of the net consumption benefit going to farmers is:

$$C^F = [S^F P^I + (1\text{-}S^F)]SC^F$$

where P^I = shadow price of investment.
Similarly:

$$C^L = [S^F P^I + (1\text{-}S^L)]SC^L$$
$$C^G = [S^G P^I + (1\text{-}S^G)]SC^G$$
$$C^T = [S^T P^I + (1\text{-}S^T)]SC^T$$

Thus, with full adjustment for divergencies between the market prices and social opportunity costs of labour, foreign exchange and investment (= current consumption foregone), we have:

$$C = C^F + C^L + C^G C^T \ldots \qquad (c)$$

which can be re-written as:

$$C = SC + (P^I\text{-}1)[S^F SC^F + S^L SC^L + S^G SC^G + S^T SC^T]$$

(ii) Appraisal of regional distribution benefits. There are two steps in the assessment procedure. First, we assess the net aggregate consumption benefits redistributed to the region of the project in any given year. Thus:

$$DR = (1) + (2) + (3\text{-}L) + (3\text{-}S) + (4\text{-}L) + (4\text{-}S)$$
$$\qquad -(5) + (5\text{-}L^F) + (5\text{-}L^H) + (6\text{-}S)$$
$$\qquad -(7) + (8) - (9) - (10)$$

We note that the project benefits (1) and (2), which were previously counted as benefits to the nation, are now treated as benefits to the region. Of even more interest, some items previously treated as costs to the nation, including the costs of all *labour* employed by the project – (3-L), (3-S), (4-L), (4-S), etc. – are now counted as benefits to the region. Second, we adjust DR for the marginal propensity to respend in the region (= regional income multiplier).

$$R^M = \frac{DR}{1-\gamma} \ldots \qquad (d)$$

where γ = marginal propensity to respend and $1/1 - \gamma$ is the regional income multiplier.

(iii) Appraisal of social group redistribution benefits. Small farmers are the target group to whose improved welfare redistribution is directed. The total value of net consumption benefits provided to small farmers by the project is given by:

$$R^{SF} = A[(1) - (5) + (5-L^F) - (9) - (10)] + B(2) - C(7) \ldots \qquad (e)$$

where A = small farmers' aggregate share of the project's cultivated area, B = number of small farmers as a proportion of the total number affected by the project, and C = small farmers' aggregate share of the land taken over by the project.

Small farmers enjoy a share of the agricultural output (1) and employment benefits $(5-L^F)$, as well as part of the benefits of housing and social services (2). But they also bear a share of the non-labour agricultural costs, and the agricultural disbenefits of the transfer of farm land to non-agricultural uses (houses, roads, etc.) by the project.

Estimates of the project net benefits accruing to the nation, the region and small farmers during each year of the project's life are now complete. The final stage of the appraisal procedure is to discount and sum all annual net benefits in order to find the NPV of the project as follows:

(1) Take the values of the social rate of discount and other national parameters (assumed constant over time for simplicity) from Table 11.4.
(2) Use the SRD to find the NPV of each time *unit* of benefit, cost or transfer shown in Table 11.3, and sum to find the NPVs of the corresponding time *flows*.
(3) Substitute the resulting NPVs together with the appropriate national parameters where required, in equations (a)–(e).

The results of following this procedure are summarised in Table 11.5.

(d) Presentation of results. The results are shown at three different social rates of discount as well as separately for each of the three

separate appraisal criteria of national, regional and social group redistribution benefits. Looking first at *aggregate consumption (national) benefits*, alone, the results show that with full adjustment for divergencies between market prices and the social opportunity costs of labour, foreign exchange and investment (appraisal criterion C) the project is 'justified' only at the lowest of the three test social rates of discount (5 per cent). The effects of full adjustment are brought out by comparing the 'no adjustment' criterion MC, and the partial adjustment parameter SC with the corresponding values of criterion C. Generally, SC > MC > C in this case for two reasons. First, market prices are judged to overvalue unskilled labour and foreign exchange. Second, the project redistributes income in favour of 'low savers' despite a relatively high P^I. Turning now from aggregate consumption to redistribution benefits (regional and social group), the project is justified on both redistributional critieria at all three SRDs. It is apparent, then, that if the test SRD is judged to be at either of the two higher rates the project cannot be justified economically *unless* the benefits of redistribution (regional and group) are judged to outweigh the disbenefits of the expected loss of net aggregate consumption benefits. This brings us back to the aggregate benefit function, $\bar{B}$, and the weighting of its components.

The weighted sums of project benefits, or the aggregate benefit, $\bar{B}$, are shown in the bottom line of Table 11.5. Although there is a strong

Table 11.5: *Net present value of regional irrigation project in year 0 (million pesetas)*

Item	Equation	Social rate of discount		
		5%	7½%	10%
Aggregate consumption[1]				
MC	a	572	219	39
SC	b	1030	506	234
C	c	467	−164	−276
Regional redistribution				
R^M	d	1701	1158	855
Social group redistribution				
R^{SF}	e	887	555	374
Aggregate benefits				
$\bar{B}$	(2)	1336	403	125

[1] MC, SC and C are defined by equations (a), (b) and (c) pp. 349–51.
[2] $\bar{B} = C\theta^C + R^M\theta R^M + R^{SF}\theta^{SF}$, with weights on objectives θ^C, θR^M and θR^{SF} taken from Table 11.4.
Source: Adapted from UNIDO (1972), *Guidelines for Project Evaluation*, Table 21.22.

inverse relationship between D̄ and SRD, the project is justified by a narrow margin even at an SRD of 10 per cent. Thus, in the case of this project, national *disbenefits* at higher rates of SRD are judged to be outweighed by positive redistributional benefits.

(e) Interpretation of results. The method of appraisal we have outlined so far is to find the NPV of the project as a whole, subject to given values of the national parameters, i.e. the SRD and the weights on objectives. An alternative approach is to find combinations of national parameter values at which the project NPV is zero. Recalling that the condition NPV = 0 defines the internal rate of return, the approach is to find combinations of the IRR and other national parameters which 'switch' the project from being viable to being nonviable (Pearce and Nash, 1981, p. 174).

More formally, define:

$$\text{NPV} = \sum_j \sum_i \frac{\theta_{ij} B_{ij}}{(1+d)^j}$$

where B_{ij} is net benefit type i in year j, θ_{ij} is the weight given to B_{ij}, and d is the social rate of discount.

The first approach is to find NPV given d and the sets θ_{ij} and B_{ij}. The alternative approach is to substitute 0 for NPV on the LHS of the above identity. Then, given the B_{ij}, solve for *either* θ_{ij} *or* d (= internal rate of return) on the RHS. A supposed advantage of this approach is that decision-makers can learn to judge, by experience, whether the internal rate of return implied by a given set of weights (or the set of weights implied by a given IRR) is acceptable. For practical reasons, the size of the weight set that can be conveniently handled by this method is only quite small. The 'switching values' of the SRD and other national parameters can also be presented graphically for ease of interpretation (UNIDO, 1972, p. 352).

Even if the project passes the test of economic viability at accepted levels of the national parameters, this does not necessarily imply that it should be undertaken in the precise form of the original proposal. Relevant alternative forms need to be explored. So, for example, the present case study project could be made to yield a higher expected level of net national aggregate consumption benefits by means of two specific changes in project design, as follows:

(1) changing the land distribution pattern in favour of large farmers (and to the deteriment of small farmers), and
(2) changing the cropping pattern to increase the ratio of cash crops to subsistence crops.

The same changes in project design would also yield an improvement in regional redistribution benefits. The cost of project redesign in this case would be borne by small farmers (the target social group in the original project design) who would inevitably suffer a substantial loss of redistribution benefits. However, despite a lower R^{SF} (possibly negative), because C and R^M are both higher, the *overall* net benefits of the redesigned project could still be better than the original, depending upon the relative magnitudes of (weighted) positive and negative effects.

The foregoing discussion is limited to only one of many possible project redesign options. The total number of options which it is feasible to compare is limited by practical considerations. (Our discussion of project risk and uncertainty (section 6.1.3) included some more general comments on forms of 'sensitivity analysis'.)

(g) UNIDO method of project appraisal: summary view. Compared with simpler or more *ad hoc* methods of project appraisal, comprehensive methods, like the UNIDO model, offer at least two potential advantages:

(1) their greater emphasis on the distinction between the financial and economic merits of projects through the use of relatively sophisticated methods of shadow pricing; and
(2) their ability to combine measures of *economic* (or aggregate income) and *social* (or redistribution) benefit in the appraisal procedure systematically.

However, comprehensive methods are more costly than simpler ones in terms of their technical complexity. Also despite their sophistication, the application of comprehensive methods remains substantially reliant on subjective judgements by those responsible for selecting projects.

The UNIDO method is remarkable not only for its technical complexity and its dependence on political value-judgements, but also for its pragmatism. The method's techniques of shadow pricing as described in Section 2.2 typify its technical complexity. The dependence on political value judgements is a consequence of the method of assigning weights to objectives. Although those responsible for selecting projects may learn to achieve a measure of consistency in the choice of 'switching values', there is no truly objective method of weighting competing types of project benefits, such as (a) augmenting, and (b) redistributing the national product.

One of the most pragmatic features of the UNIDO method is its choice of unit of account, or numeraire. Whereas a rival method of comprehensive appraisal (the Little-Mirrlees method) effectively values all project costs and benefits at 'world prices', the UNIDO choice of numeraire is the more realistic one of net aggregate consumption benefits at *domestic* prices (either the market price or the equivalent shadow price, as appropriate).

Comprehensive methods of project analysis, such as the UNIDO method, have been the subject of substantial criticism on both theoretical and practical grounds. The use of distributive weighting systems has been a particularly heavily criticised aspect. Although a number of more disaggregated approaches to project appraisal have been advanced by the critics, none of these has been proved to be indisputably superior to CBA. Pearce and Nash (1981), chapters 1 and 3, contains an excellent discussion of the issues involved.

6.3.2 Pecularities of Agricultural Projects:
The Case for Simpler Methods of Appraisal

None of the main methods of project appraisal has been developed specifically for application to agricultural projects. So, are agricultural projects sufficiently peculiar or different from other projects to render these methods unsuitable for use in agriculture? Certainly there is some substance in the argument that agricultural projects 'flaunt their difficulties' (Hirschman, 1968, ch. 5).

Good agricultural planning depends on a reliable data base, but this is generally lacking in LDCs. Reliable production and price data for agricultural inputs and products are hard to come by even in the modern agricultural sector. In the traditional sector the problem of inadequate data is usually far worse. All these problems add to the difficulty of making reliable estimates of agricultural project costs and benefits in the future, as required for the purpose of constructing the most basic of project planning documents, the table of resource flows.

Critics of the use of sophisticated methods of appraisal (such as UNIDO and LM) for selecting agricultural projects in LDCs maintain that, regardless of the strength of the *a priori* case for refinements such as shadow pricing and estimating the income redistribution effects of projects, there is no point in making these fine adjustments in practice unless the unadjusted, undistributed *financial* flows are adequately reliable. As far as agricultural projects are concerned, scarce planning skills are better concentrated upon improving the quality and reliability of physical production estimates, cost estimates and market price forecasts than being diverted to less urgent and fundamental planning activities of the kind highlighted by UNIDO and similar advanced appraisal methods. There is the further argument that because of their technical complexity, the application of advanced methods of appraisal may be beyond the capacity of agricultural project planners in LDCs.

Alternative methods of appraising agricultural projects. These and other criticisms of economically sophisticated methods of appraisal have led to the advocacy of simpler methods for use with agricultural projects (Carruthers and Clayton, 1977).[1] The so-called *decision matrix* method exemplifies this approach. The advantage of this method is not that the number of criteria used in appraising the project is reduced, but that the individual criteria are less complex and they are considered together without formality. The main criteria are:

(1) An expected *economic* internal rate of return (with limited shadow pricing) but hedged with details of the forecast error distribution.
(2) The expected *financial* internal rates of return (similarly hedged)
    (i) to farmers (or other private beneficiaries) and
    (ii) to government.
(3) Additional employment created per unit of investment.
(4) Share of project income accruing to the poorest quartile (or other predetermined proportion) of the population.
(5) Is the project located in a special development area? (Yes or No).

The cost of the project in terms of foreign exchange is an additional criterion applied to nearly all projects. On shadow prices, Carruthers and Clayton recommend 'a simple pragmatic approach, with prices adjusted in the right direction to the approximate order of magnitude.' Thus the shadow wage will usually be below the level of the money wage, whereas the shadow price of foreign exchange will normally be above the level of the official exchange rate. The actual amount of the discount (shadow wage) or premium (shadow exchange rate), which can be no more than a rough approximation, can be left to judgement and experience.

As already indicated, when using the decision matrix approach, the relative weighting of the various criteria is left to the informal judgement of politicians advised by planners.

Other alternatives to the UNIDO method of project selection are more narrowly based on a single major appraisal criterion, though not to the complete exclusion of other criteria. Thus, *programming models* are used to *maximise* a single priority objective (such as national per capita income) subject to *minimum* performance levels in associated objectives (such as minimum wages in a particular group). Even simpler, is the earlier mentioned *practitioner's approach*, which is heavily biased towards selection on the basis of a single criterion – such as the economic internal rate of return – but subject to a measure of informal 'social interpretation' (Benjamin, 1981, ch. 11). Part IV of Benjamin's book contains a number of case studies of agricultural projects using this approach.

In the end, the choice of method for appraising agricultural projects is likely to be influenced by many considerations – not least, the character of government's programme of economic and social development, the state of the agricultural information base, and the amount and quality of specialised project planning expertise at government's disposal. It is clear that, in principle, economically sophisticated methods, such as the UNIDO model, are superior to simpler but cruder methods. Whether the more advanced methods are superior *in practice* depends on the ability of planners and politicians to apply and use them correctly. Their ability should improve with experience and as better resources for planning become available. In fairness to the economists who have developed the more advanced appraisal methods it may be added that, despite their apparent preoccupation with the 'fine tuning' of project benefits and costs, they are not unaware of other equally important aspects of the planning process. For example, Square and van der Tak (1976), remark that

The judicious use of shadow prices is an important means of assessing the economic merits of a project ... but it is not a substitute for careful analysis of its technical features, investment and operating costs, organisational arrangements, market prospects, financial results and many other considerations relevant to the outcome of the project, (p. 14).

6.3.3 Ex ante *and* ex post *Evaluation*

Another aspect of project planning is 'learning by experience'. A useful terminological distinction may be made between *ex ante* appraisal and *ex post* evaluation of projects. (But note that this distinctive usage of the terms 'appraisal' and 'evaluation' is not universal: they are often used synonymously.) The point is that pre-implementation appraisal and post-implementation evaluation are complementary with evaluation providing feed-back information. Carruthers and Clayton (1977) distinguish two distinctive types of feedback:

(1) Feedback to the project itself, using comparisons between the

initial objectives, predicted performance and the actual achievement of objectives and performances as a yardstick.
(2) Feedback to the planning process by comparing project achievements with the goals of current policy (which may diverge from past goals due to shifting priorities).

Whereas some mistakes in project design can be learned by experience, others might be avoided by more careful preparation. For instance, an agricultural project may fail because the planners have overlooked important institutional constraints. Witness the example of an agricultural project in West Africa which assumed that on peasant family farms the labour of husbands and wives would substitute freely. The customary division of both land and labour between the sexes, and the fact that wives are not obliged to work on their husband's land, were both overlooked (Dey, 1982). This example serves to warn against the error of assuming that models of economic behaviour, particularly western models, are freely transferable amongst different countries and cultures.

6.4 Project analysis and policy reform measures

Project analysis is open to two general classes of criticism (Stewart, 1978). So called first-order criticisms question its very *raison d'être* by arguing that instead of tinkering with problems of price distortion, unemployment and externalities, by attempting to make compensating adjustments in the appraisal of investment projects, governments should attack the problems head-on with policies to remove the causes of market imperfections. Second-order criticisms are concerned with technical details of the methods used whilst accepting the principle of attempting to correct for the distorting effect of market imperfections. Having already examined the methodology of project appraisal in some detail, we do not propose to further pursue the second-order criticisms at this stage, but the first-order criticisms deserve brief amplification. Stewart places these in four major categories.

First, why cannot governments achieve their economic and social objectives by means of direct action on prices to bring them into line with social values, instead of merely adjusting prices as part of the appraisal procedure? Appropriate price, tax and tariff adjustments by government would be more efficient than applying elaborate rules to correct for 'incorrect' prices, which merely diverts attention from the need to remove market distortions.

Government's apparent inability to enforce its will might be interpreted as casting doubt on whether its publicly stated policy objectives (forming the basis of project appraisal) are its real objectives. If the real objectives are different (and remain concealed) then project appraisal is no more than political propaganda.

Second, if government cannot achieve success by direct market intervention, perhaps because its policy fails to command an adequate public consensus, is indirect action – using the artificial device of project appraisal – any more likely to be successful?

Third, there are the linked questions of whether the rules of project selection specified by appraisal methodology are compatible with how governments reach decisions in the real world and, even if they are, whether the government's value-judgements reflect society's social values. Regarding the rules of decision-making, it is unrealistic to assume either that economic decisions are made purely in the light of economic criteria, or that the politicians who make decisions are motivated only by altruism. Short-term political advantage may loom larger than longer-term goals of economic and social welfare in deciding where and when projects are approved, and politicians with national responsibilities may still be influenced by regional loyalties. Regarding the value-judgements which are such a prominent feature of comprehensive methods of project appraisal and selection, it is generally assumed that government judgements reflect social consensus values. But to the extent that reality is represented not by national unity and social consensus but by a plural society with conflicting interest, what purport to be social decisions are really government decisions (reflecting the values of the ruling party and its supporters).

Fourth, is the criticism that, because of its static approach to resource allocation, project analysis is incapable of being adapted to dynamic adjustments necessitated by uncertainty. Comparisons of *ex post* evaluation with *ex ante* appraisal show that major divergencies frequently occur between expectations and actual fulfillment due to unforeseen (and often unforeseeable) events.[2] This, say the critics, makes nonsense of appraisal. Moreover, due to its pre-occupation with improving static resource allocation, project appraisal misses the contributions of 'learning by doing', and improved management, to long-term growth and development.

What are the counter-arguments in response to these criticisms of project analysis and appraisal? On the question of the choice between direct and indirect measures to achieve government's development objectives, no government is omnipotent: an elected government must rule by persuasion. Because of the 'isolation paradox', private citizens and firms tend to be more myopic than the government. So, for example, it is extremely difficult for government to close the 'savings gap' by direct intervention (i.e. fiscal and monetary policy). On the question of why indirect measures can succeed where direct intervention has failed, it may be argued that, in a *mixed* economy, public sector investment and employment policy can be used to steer the economy in the direction desired by government. Public sector policy affects the private sector too, though indirectly, because of competition.

In reality, the criticism that project appraisal mis-specifies the nature of government's decision-making process, and that appraisal depends on value-judgements is an argument about the justification for planning. In a *laissez-faire* economy public investment projects would be ruled out *ex hypothesi*. In the real world virtually all LDC governments have rejected *laissez-faire* in favour of national economic planning, backed by public sector investment projects. Whilst the decision-making rules used in project planning models might be politically naive, they do at least attempt to rationalise the decision-making process in pursuit of government plans for economic and social development. In the final analysis it is for government to decide whether to accept or reject the project analyst's advice on *how* it should reach its decisions. As to the legitimacy of value-judgements, because of intangible costs and benefits, no method of project selection can be entirely objective. Project analysis merely attempts to systematise and improve the consistency of value-judgements.

Finally, in response to the criticism that project appraisal neglects dynamic factors affecting economic growth and development, it may be claimed that a good project plan anticipates the possibility of change by subjecting the results of appraisal to sensitivity analysis, by avoiding unnecessary plan rigidities and by provision for regular project monitoring and *ex post* evaluation.

It appears to us that the balance of the argument is in favour of systematised project planning rather than leaving public investment plans to the *ad hoc* decisions of government. At present, few LDCs have had sufficient experience of project appraisal and evaluation – especially in agriculture – to have advanced beyond a rather low level of performance. However, with better basic data and greater skill born of experience, their proficiency should improve. Moreover, the discipline which project planning imposes on its practitioners is a valuable means of gaining a better insight into how the economy works.

References

Bandyopadhyaya, K. (1976), *Agricultural Development in China and India*, Allied Publishers, New Delhi.
Benjamin, M.P. (1981), *Investment Projects in Agriculture*, Longman, Harlow.
Carruthers, I.D. and Clayton, E.S. (1977), 'Ex-post evaluation of agricultural projects – its implications for planning', *J. Agric. Econ.*, **XXVIII** (3).
Dervis, K., de Melo, J. and Robinson, S. (1982), *General Equilibrium Models for Development Policy*, Cambridge University Press, Cambridge.
Dey, J. (1982), 'Development planning in The Gambia: the gap between planners' and farmers' perceptions, expectations and objectives', *World Development* **10** (5).
Eckstein, O. (1965), *Water Resource Development: the Economics of Project Evaluation*, Harvard, University Press, Cambridge, Mass.
Ellman, M. (1979), *Socialist Planning*, Cambridge University Press, Cambridge.
Ghatak, S. (1978), *Development Economics*, Longman, London and New York.
Heal, G. (1973), *The Theory of Economic Planning*, North-Holland, Amsterdam.
Healey, D.T., 'Development Policy: New Thinking about an Interpretation', *J. Econ. Lit.*, September, 1972.
Hirschman, A.O. (1968), *Development Projects Observed*, Brookings Institution, Washington, DC.
Hopper, W.D. (1968), 'Investment in agriculture: the essentials for pay-offs', in *Strategy for Conquest of Hunger*, Proceedings Rockefeller Foundation, New York.
Howe, C. (1978), *China's Economy*, Cambridge University Press, Cambridge.
Johnston, B. with Page, J.M. and Warr, P. (1972), 'Criteria for the design of agricultural development strategies', *Food Research Studies in Agricultural Economics, Trade and Development*, vol. XI, pp. 27–58.
Joshi, V. (1972), 'The rationale and relevance of the Little–Mirrlees Criterion', *Bul. Oxford Inst. Econ. & Stats.*, **34** (1).
Little, I.M.D. and Mirrlees, J.A. (1968), *Manual of Industrial Project Analysis in Developing Countries*, vol. II, OECD Development Centre, Paris.
Mishan, E.J. (1972), *Elements of Cost-Benefit Analysis*, Allen & Unwin, London.
Owen, W.F. (1971), *Two Rural Sectors: their Characteristics and*

Roles in the Development Process, International Development Research Center, Occasional Paper 1, Indiana University, Bloomington.

Padoul, G. (1975), 'China 1974: problems not models', *New Left Review*, 89, January–February.

Paine, S. (1976), 'Balanced development: Maoist conception and Chinese practice', *World Development*, vol. 4, no. 4.

Panne, van de C. (1976), *Linear Programming and Related Techniques*, 2nd edn, North-Holland, Amsterdam.

Pearce, D.W. and Nash C.A. (1981), *The Social Appraisal of Projects*, Macmillan, London.

Sinha, R.P. (1975), 'Chinese agriculture: a quantitative look', *J. Dev. Studies*, **11**, 3, pp. 202–23.

Squire, L.A. and van der Tak, H.G. (1975), *Economic Analysis of Projects*, Johns Hopkins University Press, Baltimore, for the World Bank.

Stewart, F. (1978), 'Social cost-benefit analysis in practice: some reflections in the light of case studies using Little–Mirrlees techniques', *World Development*, **6** (2).

Stewart, F. and Streeten, p. (1972), Little–Mirrlees methods and project appraisal', *Bull. Oxford Inst. Econ. & Stats.*, **34** (1).

Timmer, C.P. (1974), 'Choice of technique in rice milling in Java', *Research and Training Network*, The Agricultural Development Council, September, pp. 1–10.

Symposium on the Little–Mirrlees manual of industrial project analysis in developing countries (1972), *Bull. Oxford Inst. of Econ. Stat.*, February.

UNIDO (1972), *Guidelines for Project Evaluation*, United Nations, New York.

Notes

1. A substantial number of purely theoretical limitations of comprehensive methods of project appraisal has been discussed in the literature: see, for example, the 'Symposium on the Little–Mirrlees manual of industrial project analysis in developing countries', *Bull. Oxford Inst. of Econ. Stat.*, February 1972. However, these general criticisms lie outside the purview of this book since none of them relates specifically to *agricultural* project appraisal.
2. Apart from 'random shocks' such as the natural disasters (e.g. earthquakes, droughts and floods) and sudden shifts in the external terms of trade or access to overseas markets, two of the major reasons for divergencies between *ex ante* expectations and *ex post* performance are (a) externalities such as linkage effects (see, for example, Stewart and Streeten, 1972; and Joshi, 1972); and (b) large investment projects can cause actual prices and costs to vary substantially from their expected values.

12 Agricultural Development: an Overview

The dominant feature of the LDCs compared with the rest of the world is their relative poverty, the amelioration of which is a central objective of economic development. For reasons discussed at greater length in chapter 2, poverty is also characteristic of traditional agriculture, the predominant form of agriculture in most LDCs. Inadequate access to land and capital for the majority of farmers, combined with technological backwardness or even stagnation, are two of the principal reasons for agricultural poverty. It follows that in order to ameliorate agricultural poverty through raising farm output and income, some relaxation of resource constraints is needed combined with a higher rate of technological innovation. However, due to the risks and uncertainties of agriculture, particularly in physically underdeveloped economies in tropical environments, poor peasant farmers understandably tend to be risk-averse. Hence policy measures to reduce the risks and uncertainties of agricultural production and marketing are also important for accelerating agricultural development.

Development theory postulates a progression from the type of economy where primary production, including agriculture, is dominant, through a stage where the majority of GNP and gainful employment derives from secondary industry (i.e. manufacturing) to a final stage in which tertiary or service industries predominate. Moreover, it is virtually axiomatic that, to sustain economic development in the long run, most LDCs must undertake sectoral diversification to reduce their dependence on agriculture. But, as discussed at length in chapter 3, the feasibility of sectoral diversification is usually critically dependent on a growing marketable surplus of agricultural products. However, due to the influence of labour of surplus development models, the importance of the agricultural sector as a source of redundant labour waiting to be mobilised for 'costless' industrial development has been exaggerated. Furthermore, due to the combination in many LDCs of continuing rapid population growth, with severe constraints on the growth of employment *outside* agriculture, the onset of a decline in the absolute size of the agricultural sector labour force is likely to be delayed until a relatively late stage of development. In the interim, the creation of additional employment within agriculture – by encouraging farmers to adopt appropriately labour-intensive methods of increasing output, for example – is a policy priority.

Agricultural rent and agricultural production surplus are closely analagous concepts. The amount of rent (or surplus) available *either* for investment in agriculture itself, *or* to be taxed away to fund non-agricultural investment depends upon agricultural productivity, consumption within the agricultural sector, and possibly also on the land tenure system. For the reasons discussed in chapter 4, share tenancy is not necessarily inefficient. But in LDCs, the market in agricultural land is frequently flawed by unequal bargaining power between landowners and tenants, as well as by other market imperfections. Although dual economy models have been useful in helping to conceptualise the meaning of agricultural surplus, failure to integrate the theory of capital accumulation with the theory of choice of technique helps to explain their limited predictive power. A further shortcoming of such models is that they neglect analysis of the forces generating productivity gains and development within the agricultural sector, as discussed in chapter 5.

In chapter 6 we refer to the substantial body of literature which exists on the efficiency of peasant agriculture. One view is that, despite their poverty, peasant farmers are allocatively efficient in the neo-classical mode. An opposing view is that, because of poverty, risk-minimisation takes precedence over profit-maximisation. Moreover, despite the fruits of learning from experience, gaps in the spread of knowledge can result in substantial *technical* inefficiency. Those adhering to the 'poor but efficient' viewpoint conclude that technological advance is virtually the sole means of raising output and incomes in peasant agriculture. But opponents of this view believe that significant increases in output can be achieved through the more efficient use of *existing* technology. This may be feasible by providing farmers with better safeguards against extreme fluctuations in prices and yields and by improving both the quantity and the quality of agricultural extension services. We find the evidence in favour of holding the view that peasant producers are *allocatively* efficient reasonably convincing, but subject to expected economic returns being discounted in some degree to allow for risk and uncertainty. But the evidence supporting the view that peasant agriculture is also *technically* efficient is much less convincing. We consequently support the view that important though technological advance may be in hastening agricultural development, this is not the *only* means of improving agricultural output and incomes: progress can also be achieved by improving production incentives, as well as by encouraging farmers to make better use of present techniques.

Despite the foregoing caveat, more research is needed to hasten socially beneficial agricultural technical change in LDCs. Due to the unfavourable labour/land ratio in many countries, research programmes need to emphasise biological innovations to promote land-augmenting technical change. Governments have a vital role to play in establishing and funding national research institutions and in ensuring that the objectives of research programmes are not biased in favour of private vested interests. The scope for conducting research at a purely national level in poor countries is greatly limited by financial and other constraints such as skilled manpower. One means of surmounting this hurdle is international co-operation in promoting and funding agricultural research to benefit LDCs. Such research has been notably successful in recent years, as exemplified by the dramatic yield increases obtained from new varieties of wheat and rice developed by CIMMYT and IRRI. Substantial scope exists for extending international co-operation in agricultural research to other crops and livestock products, as well as to regions of the world that have not yet been significantly affected by the Green Revolution.

The role of machinery investment, particularly field mechanisation, in agricultural development is controversial. On the one hand, extra machine-power can help to raise productivity and reduce the drudgery of hand labour for workers remaining in agriculture. On the other hand, forms of mechanisation which are potentially labour-displacing threaten to aggravate unemployment. However, because of inter-sectoral income and employment linkages, the displacement of labour from agriculture does not necessarily imply an overall reduction of employment, particularly in countries with a domestic farm machinery manufacturing capacity. Serious doubt exists concerning the feasibility of 'selective mechanisation' as a practical means of avoiding labour displacement by tractors and other farm machinery. Not only would it be difficult to identify machinery types that are invariably labour-displacing (or land-augmenting), it would be equally difficult to control the use of any particular machinery type. But governments can influence the rate of farm mechanisation *unselectively*, for better or for worse, by controlling the distribution of agricultural machinery and the terms on which it is acquired by farmers. Due to the predominance of imported farm machinery in LDCs, the government's import policy is critically important. In the past, governments have probably tended to encourage rather than discourage farm machinery imports, but without necessarily recognising and evaluating the full economic and social consequences. The major conclusion on farm mechanisation policy in LDCs is that although the substitution of capital for labour in agriculture is inevitable in the long of run, the rate at which mechanisation occurs is for public decision. Governments may choose to encourage or discourage mechanisation unselectively by applying specific taxes or subsidies at appropriate rates. Numerous non-specific government policies, such as the exchange rate, also influence the rate of farm mechanisation.

As discussed in chapter 7, there is little or no convincing evidence of genuine perverse supply response in LDC agriculture, despite the plausibility of the supposition that as their income rises peasant farmers might choose to consume more leisure as well as extra material goods. But, as expected, cash crops usually exhibit a more pronounced supply response than do food crops grown primarily for subsistence consumption. Moreover, due to farmers' crop substitution opportunities, it is easier to raise the output of individual crops by raising their price than to induce an overall increase in agricultural output by improving the agricultural sector's terms of trade. Non-price variables such as resource endowments and the choice of production techniques are equally, if not more, important than prices in determining levels of aggregate agricultural output in LDCs.

Gross inequality in the distribution of land gives rise to the demand for land reform which is often regarded as a precondition of significant agricultural development. But, for reasons discussed in chapter 8, mere redistribution of landownership is insufficient to transform agriculture. To achieve such a transformation, a comprehensive 'agrarian reform' is needed which, as well as land redistribution, also includes a full array of extension, credit and marketing services. The demise of the large private landlord sometimes leaves a vacuum in the provision of some or all of these services, and new institutions, such as service co-operatives, are needed to fill the gap. Very small-scale farmers and landless labourers usually derive few direct benefits from land reform. Other measures are needed to improve their livelihood and incomes, including better non-agricultural employment opportunities. More generally, in some LDCs the population density is so high that it is quite unrealistic to see land reform as a panacea for abolishing rural poverty and unemployment, simply because insufficient land is available to provide every rural household with an adequate-sized plot. Again, it is necessary to look to the general economy, rather than to agriculture *per se*, for a resolution of the problem. Alternatives to land reform, such as co-operative farming and tenancy reform, may help to raise agricultural output and incomes, but fail to assuage 'land hunger'. More countries have enacted land reform legislation than have actually implemented successful land reforms. Thus strong government direction and leadership are needed not only to enact the legislation but also to carry it out.

The root of the 'credit problem' in LDC agriculture is farm tenants' lack of direct access to organised money and commodity markets which, in any case, tend to be rather poorly developed in the countries concerned. Credit policy reforms include land reform (to improve the credit standing of small farmers), expansion of institutional credit agencies (to compete with private moneylenders) and the *raising* of deposit interest rates (to encourage more saving). The role of credit in agricultural development is especially important in facilitating the adoption of more appropriate technology to raise productivity. This also includes the substitution of capital for labour, especially in the long of run, with the more rapid growth of non-agricultural employment opportunities. There is also much scope for linking agricultural credit and marketing programmes to mobilise a larger agricultural surplus.

In non-subsistence economies, agricultural marketing services link food consumers with agricultural producers. The development and improvement of these services therefore form an important component of overall agricultural development. Some reforms in agricultural marketing, such as infrastructural development and the provision of better market information, are pre-eminently a government responsibility. But since some marketing functions tend to be performed more efficiently in the private sector, a good case exists for collaboration between the government and private enterprise in improving the range and quality of marketing services.

A fully convincing theoretical explanation of the relationship between population growth and food supplies in LDCs has yet to be developed. The classical or Malthusian model has been found wanting in several respects, not least by the historical experience of what are now termed the developed countries. But, as discussed in chapter 9, contra-Malthusian theories, like Ester Boserup's, suffer from the same defect of assuming that the relationship between food and population is unidirectional: a better and more comprehensive theory might need to take account of feedback effects. Theoretical reasoning backed by substantial empirical evidence point to the conclusion that hunger is primarily a function of poverty rather than a consequence of inadequate aggregate food supplies. The distribution of food supplies in LDCs roughly coincides with the distribution of income. Although increasing aggregate food production and curbing excessive population growth may help to ameliorate malnutrition in the long term, income redistribution and special programmes to raise per capita food consumption within target groups appear to be the only viable *short-term* instruments for relieving hunger in the Third World.

Few economies are entirely self-sufficient and external trade can play an important role in LDC development, both through the working of the principle of comparative advantage and because of the facilities afforded by trade for entering export markets and importing capital and scarce technical skills. In many LDCs agricultural export earnings form a major proportion of the country's total export earnings. Moreover, agricultural export prices and earnings are prone to substantial fluctuation due to a combination of factors affecting both supply and demand. The problem of agricultural export instability is exacerbated by illiberal trade policies in developed countries with respect to agricultural imports from LDCs. Commodity price stabilisation and the removal of trade barriers both require international action. Experience with international commodity agreements suggests that, like producer cartels generally, ICAs are inherently unstable, so that any temporary success in raising or even stabilising prices is likely to be shortlived. Schemes of international compensatory finance to 'top up' abnormally low primary product export revenues, though not beyond criticism, may offer a more promising solution to commodity instability. These issues are discussed at greater length in chapter 10.

To plan or not to plan? Despite the association of macro-economic planning with socialism, most present-day LDCs, regardless of political ideology, make development plans, including sectoral plans for the development of agriculture. The debate about whether governments *ought* to undertake planning, rather than leaving the course of economic growth and development to the interplay of free market forces, therefore seems somewhat arid. Economists are better employed in helping governments to plan more successfully. National economic planning originated in the Soviet Union between the two world wars and, in the context of agricultural planning, it is instructive to contrast the 'Soviet tribute model', as imposed in Russian agriculture during the 1930s, with the more recent 'Chinese model'. We make this comparison in the first part of chapter 11.

Apart from setting aggregate targets which are consistent with the overall plan frame, and formulating matching policies of implementation, the core of planning for the development of agriculture in LDCs consists of project appraisal and project selection. This is the subject of the second part of chapter 11. A vast and complex literature on project appraisal in LDCs has accumulated in recent years. The objective of project appraisal is to measure total net social benefit. However, a particularly contentious issue concerns the project analyst's ability to measure divergencies between private and social benefits or costs. Even though casual observation may confirm that because of numerous market imperfections in LDCs, prices of both factors and products are commonly distorted, the planning benefits of having reliable estimates of the underlying 'shadow prices' may fail to match their costs of preparation in terms of scarce planning expertise. The debate on this issue continues. Due to the nature of biological processes, as well as for other reasons, agricultural projects are notoriously difficult to appraise with confidence. This has led some authorities to advocate the use of simpler and less complex techniques of appraisal for agricultural projects. In practice, the choice of method for agricultural projects is likely to be influenced by numerous considerations, including the amount and quality of project planning expertise at the disposal of the government. The precise method of appraisal seems less important than that the selection of competing projects should be based upon the application of some reasonably uniform, consistent and economically rational procedure. As might be expected, the record of project planning in LDCs consists of mixture of successes and failures. The *ex post* evaluation of projects has an important contribution to make to improved planning performance in the future.

For the foreseeable future, the economic and social welfare of much of the world's population inhabiting the developing countries will continue to depend upon progress in agriculture. For that reason, social equity demands the apportionment of adequate resources of finance and skilled manpower to LDC agriculture. Ensuring that adequate resources are allotted to agriculture also serves the interests of food consumers everywhere, assuming that adequate security of supply is sought. However, realism compels us to recognise that agriculture is only one of many claimants on such limited national or international funds as are available for development. Moreover, in the long run, the relative importance of agriculture is bound to diminish in virtually all countries. Thus, at some stage of development, it is equally inevitable that agriculture must yield capital and manpower to the industrial sector. But finding the optimum allocation of investment between industry and agriculture during a specific planning period is one of the most difficult policy and planning problems confronting LDC governments. Present models of economic development are regretably of little assistance in solving this problem.

The transformation of LDC agriculture requires brain-power as well as adequate capital and manpower. LDC agricultural policy should therefore include educational measures to improve agriculture's generally poor public image and social status in order to attract larger numbers of the most able and ambitious people into entering careers associated with agriculture. The 'dignity of farming' needs to be restored not only for this reason, but also to raise the self-respect of countless numbers of people in LDCs who have no effective choice of career except on the land.

Index

activity analysis, 327–9
agrarian reform
  defined, 220
agri-business sector, 37
agricultural capital
  market imperfections, 16, 20
—co-operation, 141, 242–3, 367
—credit
  sources of, 16
  transactions costs of supplying, 16
  policy, 368
—development, financial policies for, 235–7
—exports
  and foreign exchange contribution, 67–8
—goods
  trade in, 279
—import substitution, 68
—inputs
  supply bottlenecks, 16–17
—information and extension
  barriers to spread of, 12–17
  reasons for concentration in government provision of, 13–14
  supply costs, 14
—labour
  diminishing returns to, 18, 263
  displacement of, 12

landless labourers, 24, 224, 270, 368
  opportunity cost of, 8–9
  productivity, 8–10, 17–18, 33, 36, 147, 253–6
  surplus, 50, 329, 365
  transfer to industry, 50 *et seq.*
  turning point in growth, 49–50
—land
  distribution of, 18–21, 221, 367
  effect of technological change on price of, 146
  market imperfections, 20, 365
  productivity of, 17, 139, 222, 253–6
  systems of land use, 257–8
  use intensity and labour productivity, 258–60
—marketing
  government participation in, 247–9
  infrastructural deficiency in, 240
  reform of, in LDCs, 241 *et seq.*
—markets
  barriers to spread of information in, 22
  imperfections of, in LDCs, 239–41
  narrowness in LDCs, 22
—mechanisation
  policy in LDCs, 166 *et seq.*, 367

role in agricultural development, 366–7
—prices
  policy, 245
  stabilisation of, 244–7
—production surplus, 32, 45, 221, 254, 365, 368
—productivity
  and rent, 365
  total factor, 139
—project planning, 329 *et seq.*
—projects
  alternatives to UNIDO appraisal method, 357–8
  peculiarities of, 356
—research and information policy, 164–6
—sector
  declining relative importance in the economy, 30–1, 37, 43
  share of GDP growth, 27–32
  terms of trade, 239, 245, 260, 367
—self-sufficiency, 296–9
—surplus, 82–4
—technology
  appropriate/inappropriate, 11–12
  as production shift variable in contra-Malthusian model, 257
  backwardness of, in traditional agriculture, 10–11, 364
  barriers to adoption, 11–17
  choice of, 10, 327–9
  international transfer of, 11, 154, 261, 366
  new, adoption of, 14–17, 40, 145, 165
  new, generation of, 147 *et seq.*, 257, 264
  stagnation in, 10
—trade
  and, comparative advantage, 40–1, 68–9
  benefits of, 41, 66–9
—yields
  effects of weather, 21
  problems of forecasting, 21
agriculture
  biological basis of, 21
  competitive structure of, 22
  factor market distortions, 18–21
  poverty in, 13, 17, 364
  productivity of, 36, 43
  resources structure of, 31
  semi-subsistence, 6
  subsistence, 6, 42, 145
  systems of, and relative factor security, 7
  traditional, 4–18, 252
agro-industries
  relative importance in LDCs, 37
appropriate technology, 44, 165

balanced growth, 108
Bangladesh
  agricultural unemployment in, 60–2
benefit-cost rates, 330–1
Berry and Soligo model, 55–7
bi-modal strategy, 318–22
Boserup, E.,
  contra-Malthusian model, 147, 256 *et seq.*, 275–8
buffer stocks and buffer funds, 245–7

capital, 227
  scarcity of, 228
  social opportunity cost of, 332–3
captial contribution
  equity of, 43–4
  means of transfer, 44–8
cartels, 299–300
CGIAR, 154

Chakrabarti, S.K., 34–5
China
  guaranteed employment in, 273
  population control in, 263
  taxation of agriculture, 48
  use of surplus agricultural manpower, 63
Chinese model, 315–17
Choice of technique, 327–9, 365, 367
CIMMYT, 154, 366
Cobb-Douglas production function, 119
cobweb model, 174–77
  stability of the equilibrium, 177
  stable, 177
  unstable, 177
collective farming, 142, 218
commodity trade
  concentration in, 280
  cartels in, 299–300
Commonwealth, 300
comparative advantage
  static versus dynamic, 68–9
compensating financing schemes, 306–7
competition
  in agricultural markets, 240–1
contra-Malthusian model of economic development, 253, 256 et seq., 369
co-operative farming, 271, 225, 368
cost benefit analysis, 330, 335
costs
  of adopting new agricultural technology, 15–16, 261
credit
  price of, in agriculture, 16
Cummings, 197

discount rate
  choice of, 331, 332–3
  synthetic, 333
  use in project appraisal, 352–4
disguised unemployment, 98
  and shadow wage rate, 336
  defined, 52
  empirical evidence for, 58–63
  Nurkse's model, 52
  Sen's model, 52 et seq.
dual economy models, 97, 105, 119–21
  assumptions of, 97–8
  Fei–Ranis model, 105–12
  Jorgenson model, 112–14
  Kelley, Williamson, Cheetham model, 114–19
  Lewis model, 97–105
  remarks on, 119–21
dualism
  economic, 5, 12, 42, 44, 365
  labour market, 8–10, 17

ecological disequilibrium, 262–3
economic growth
  agriculture's contributions to, 26 et seq.
  and poverty, 271
economic location theory
  and classical model of economic development, 255–6
economic rent, 75–6
  origin of, 76
Edel, M., 35
efficiency, in agriculture
  and farm size, 136 et seq.
  and land reform, 222
  concepts of, defined, 123–4
  norms of, 136
Engel's Law, 6, 31
European Economic Community, 306
exchange entitlements

failure, as cause of famine, 266 et seq.
expected price, 183–5
　Cagan's model of, 183–4
　Nerlove's model of, 183–4
export instability and economic growth, 289

factor contributions
　capital, 43–8
　labour, 49–66
factor prices
　distortion of, in LDC agriculture, 138
family farm, 5
family formation
　economic motivations for, in peasant agriculture, 270
family labour
　characteristics of, 7–8
family size
　correlation with farm size, 5
famine
　cause of, 267–8
　'boom' and 'slump', 268
farm size
　and farm income, 18
　and land productivity, 138–40, 222
　and land-use intensity, 18
　policy, 140
farm unions
　and farm size, 18
　improvement of, as project planning objective, 342
　level of, compared with non-farm sector, 13, 17
　methods of raising, 12
farmers, in traditional agriculture
　access to information, 12
　as price-takers, 22
　economic goals of, 131, 259–60

literacy of, 12
weak bargaining position of, 17, 240–1
feudalism
　characteristics of, 18
　defined 25n1
financial policies, 235–7
financial surplus, 75, 82, 85
food
　income elasticity of demand, 31–2
　maldistribution of supplies, 266–9
　marketable surplus of, 218, 220, 221, 222
　per capita consumption, 266, 272, 274
　price elasticity of demand, 33–4
　redistribution of supplies, 271–5
　stamps, 274
foreign exchange contribution, 66–9
fragmentation of farms, 5

geographic concentration in trade, 280
Green revolution
　and differential efficiency of large and small farms, 140
　distribution of benefits, 146–7
growth models, 97

Helleiner, G.K.
　land surplus economy model, 65–6
Hirschman, A.O., 38–9
human capital, 43, 63–4
human fertility
　and economic growth, 269–70
　high rate in low-income agriculture, 18

rural-urban differential in, 269-70
HYVs, 43

income
 redistribution as planning objective, 341-3
income-sharing
 on family farms, 8
inflation
 and food supplies, 33-6
imports
 of food, 32, 67-8, 261, 264
India
 designed agricultural unemployment in, 58-60
 distribution of cultivation rights in, 19
 effect of agricultural technical change on employment, 160-4
 'fair-price' shops in, 274
 famine in Bengal, 268
 farming efficiency in, 125-6
 landless rural households in, 19
 taxation of agriculture, 47
input-output analysis
 use in estimating indirect employment effects of technological change, 162
 use in estimating inter-industry linkage coefficients, 39
institutional change
 theory of induced, 152-4
institutionalists, 330
interest rate, 231-4
inter-industry linkage
 defined, 37-8
 employment linkages, 39, 65
 income generation linkages, 39, 41-2
 production linkages, 37-9
inter-locking factor markets,

Bardhan's theory, 20
intermediate technology, 44, 165
Intermediate Technology Group, 12
internal rate of return (IRR), 330-1
International Coffee Agreement, 303
International Grains Agreement, 303
International Monetary Fund, 306
International Sugar Agreement, 303
International Tea Agreement, 303
International Wheat Agreement, 303
inter-sectoral terms of trade
 manipulation of, to extract agricultural surplus, 46
investment criteria
 choice of, 330-1
IRRI, 154, 366
Israel
 land settlement in, 225

Krishna, R., 161-4, 191-3
Kuznets curve hypothesis, 273
Kuznets, Simon, 26-8, 43, 64

labour-displacing technical change
 defined, 158
labour reward system
 in present agriculture, 7-8
labour surplus, 75, 82, 85
land-augmenting technical change
 defined, 158
land reform
 and 'agrarian reform', 218-20, 367
 and distribution of benefits of technological progress, 146
 benefits and costs of, 221-4, 226

effect on production and productivity, 139, 223
effect on size of agricultural surplus, 45
limitations of, 224
meaning of, 218–20
motivations for, 18–19, 220–1, 367
redistributive effects, 20
land settlement, 225
land tenure system, 45, 217–18, 365
leisure and agricultural output, 109–11
leisure
marginal utility of, in agriculture, 10
leisure preference
of peasant farmers, 56–8, 258–9, 261
loanable funds, 327
Lewis, W.A.,
model of economic development, 54

malnutrition
ameliorisation of, 269 et seq.
and distribution of food supplies, 266–7, 369
definition, 265
link with low productivity and agricultural poverty, 17, 63, 369
measurement of, 266
Malthus, theory of population and demographic features of traditional agriculture, 18
Malthusian model of economic development, 253 et seq., 369
marginal propensity to consume in agriculture, 42
market contribution, 39–43

marketable surplus, 40, 44, 46, 342, 365
marketed surplus, 185, 189–200
estimation of, 191
marketing boards
fiscal function of, 244
functions and limitations of, 242–3
marketing co-operatives
functions and limitations of, 242–3
meta-production function, 148 et seq.
model
Chinese, 315–17
Fei–Ranis, 105–12
Jorgenson, 112–14
Lewis, 97–105
moneylenders, 328, 329

national parameters
use in project appraisal, 347–9
net present value (NPV), 330–1, 352, 354
New International Economic Order, 306
Nurkse, R., 52
nutritional education, 274

oppurtunity costs
of domestic food supplies, 32
of food imports, 32
ordinary lease, 87
owner-occupier, 86

Parikh, 197
peasant farmer
objective function of, 6, 15
planning techniques, 322–9
activity analysis of, 327–9
'cattle feed' cost minimisation, 325–7

input–output analysis, 322–4
linear programming, 324–5
population density
  and land reform, 368
  and land use intensity, 257 *et seq.*
population growth
  and food supplies, 252 *et seq.*, 368–9
  control of, 262, 269–70
  cultural restraints on, 262, 264,
  high rate in traditional agriculture, 18, 252, 365
'poor and efficient' hypothesis, 125 *et seq.*, 365
Prebisch's hypothesis, 294–6
price stabilisation, 286–9
price stabilisation schemes
  buffer stocks and buffer funds, 245–7
  limitations in reducing agricultural price uncertainty, 23
primary commodity prices
  fluctuations in, 281–2
product contribution, 27–39
profit maximisation
  as farming goal, 6
project appraisal
  and policy reform, 359–61
  and project evaluation, 358
  criticisms of comprehensive methods of, 356–8
  defined, 330
  Little–Mirrlees (LM) method, 340, 356
  objectives of, 330, 370
  UNIDO method of, 345 *et seq.*
project cycle, 344
project evaluation, 358–9, 370
property rights
  in land, 7

protein–energy malnutrition (PEM), 272

quasi-rent, 77–8

Ranis–Fei model, 57
regional income multiplier
  application in project appraisal, 352
rent, 76–82
  and agricultural productivity, 365
  and inter-sectoral capital transfers, 45
  and land reform, 221–2
  and location theory, 255–6
  and security of land, 7
  distortion caused by market imperfections, 20
  influence of technological change, 146
rental market, 79–82
resource flows
  table of, used in project appraisal, 346–7
resources structure
  of peasant agriculture, 7–8
risk and uncertainty
  and adoption of new technology, 11, 14–16
  and project appraisal, 333–5
  and resource allocation efficiency, 366
  nature and causes of, in agriculture, 21–3
  respecting prices, 22
  respecting yields, 21
risk aversion, 6, 17, 22, 131–2, 264, 364
Ricardian corn rent, 78–9
rural interest rate determination, 231–5

INDEX

rural money markets, 228
  functions of 288–31
rural–urban migration
  theory of, 8
  policy, 70
Russia
  taxation of agriculture, 47–8

savings gap, 332, 336, 361
selective mechanisation
  defined, 166
  feasibility as policy instrument, 167–8, 366–7
semi-feudal agriculture, 329
  characteristics of, 329
Sen, A.K.,
  model of disguised unemployment, 52 *et seq.*
sensitivity analysis, 329, 334–5, 361
shadow pricing
  costs of, 356, 370
  of foreign exchange, 339–40, 350
  of investment, 337–8, 350
  of wages, 338–9, 350
  principles of, 337
share tenancy, 86, 226, 365
  economic case against, 89–90
  Marshallian theory of, 89, 91–4
  theory of, 88
Soviet planning, 311
  'tribute' model, 312–15
STABEX, 306–7
stabilisation
  of commodity prices, 369
  of food prices, 36
starvation
  risks of, in traditional agriculture, 21
strategy
  bi-modal, 318–22
  uni-modal, 318–22

subsidies
  on food prices, 36
supply response, 172–98
  labour market, 185
  normal, 179
  perverse, 179
surplus labour, 98–9, 105, 112
switching values (of project appraisal parameters), 354

taxation
  of farmers and landowners, 46–7
  land tax, 47
technological change in agriculture
  biological and mechanical, 151 *et seq.*
  distribution of effects, 144–7
  effects on employment, direct and indirect, 160 *et seq.*
  embodied and disembodied, 144
  factor bias in, 155 *et seq.*
  land-augmenting and labour-displacing, 145, 158 *et seq.*, 366, 367
  nature of, 142 *et seq.*
  theory of induced, 148 *et seq.*
technological stagnation
  consequences of, 17–18
  reasons for 11–17
tenancy reform, 226, 368
time preference
  social rate of, 332–3
Todaro's rural–urban migration model, 8, 64
trade
  commodity concentration in, 280
  features of, in agricultural goods, 280–2
  geographic concentration in, 281
  in agricultural goods, 279
  policies in developed countries, 282–6

traditional agriculture
  definition, 10

UNCTAD, 301–6
UNIDO method of project appraisal
  appraisal procedure, 349–52
  data requirements, 346
  presentation and interpretation of results, 352–5
urbanisation effect, 31

wages
  and abundance of labour, 7
  subsistence, 8
weather
  effects on agricultural yields, 21
working hours
  customary, in peasant agriculture, 52–3, 258–9
work sharing
  on family farms, 8